S0-BLI-388

CROWN ETHERS AND PHASE TRANSFER CATALYSIS IN POLYMER SCIENCE

POLYMER SCIENCE AND TECHNOLOGY

Editorial Board:

William J. Bailey, *University of Maryland, College Park, Maryland*
J. P. Berry, *Rubber and Plastics Research Association of Great Britain,*
            Shawbury, Shrewsbury, England
A. T. DiBenedetto, *The University of Connecticut, Storrs, Connecticut*
C. A. J. Hoeve, *Texas A & M University, College Station, Texas*
Yōichi Ishida, *Osaka University, Toyonaka, Osaka, Japan*
Frank E. Karasz, *University of Massachusetts, Amherst, Massachusetts*
Oslas Solomon, *Franklin Institute, Philadelphia, Pennsylvania*

Recent volumes in the series:

A Continuation Order Plan is available for this series. A continuation order will bring delivery of each new volume immediately upon publication. Volumes are billed only upon actual shipment. For further information please contact the publisher.

CROWN ETHERS AND PHASE TRANSFER CATALYSIS IN POLYMER SCIENCE

Edited by

Lon J. Mathias
University of Southern Mississippi
Southern Station, Mississippi

and

Charles E. Carraher, Jr.
Wright State University
Dayton, Ohio

PLENUM PRESS • NEW YORK AND LONDON

7118-4867

CHEMISTRY

Library of Congress Cataloging in Publication Data

International Symposium on Crown Ethers and Phase Transfer Catalysis in Polymer
  Science (1982: Las Vegas, Nev.)
  Crown ethers and phase transfer catalysis in polymer science.

  (Polymer science and technology; v. 24)
  "Proceedings of an International Symposium on Crown Ethers and Phase Transfer
Catalysis in Polymer Science, sponsored jointly by the Divisions of Polymer Chemistry
and Organic Coatings and Plastics Chemistry, held at the American Chemistry Society
national meeting, March 30–April 5, 1982, in Las Vegas, Nevada"—T.p. verso.
  Bibliography: p.
  Includes index.
  1. Polymers and polymerization—Congresses. 2. Ethers—Congresses. 3. Catalysis
—Congresses. I. Mathias, Lon J. II. Carraher, Charles E. III. American Chemistry Soci-
ety. Division of Organic Coatings and Plastics Chemistry. V. Title. VI. Series.
QD380.I566 1982   668.9   83-21268
ISBN 0-306-41462-7

Proceedings of an international symposium on Crown Ethers and
Phase Transfer Catalysis in Polymer Science,
sponsored jointly by the Divisions of Polymer Chemistry and
Organic Coatings and Plastics Chemistry,
held at the American Chemistry Society national meeting,
March 30–April 5, 1982, in Las Vegas, Nevada

©1984 Plenum Press, New York
A Division of Plenum Publishing Corporation
233 Spring Street, New York, N.Y. 10013

All rights reserved

No part of this book may be reproduced, stored in a retrieval system, or transmitted,
in any form or by any means, electronic, mechanical, photocopying, microfilming,
recording, or otherwise, without written permission from the Publisher

Printed in the United States of America

QD380
I566
1982
CHEM

PREFACE

Phase transfer catalysis or interfacial catalysis is a synthetic technique involving transport of an organic or inorganic
salt from a solid or aqueous phase into an organic liquid where
reaction with an organic-soluble substrate takes place. Over
the past 15 years there has been an enormous amount of effort
invested in the development of this technique in organic synthesis. Several books and numerous review articles have appeared
summarizing applications in which low molecular weight catalysts
are employed. These generally include either crown ethers or
onium salts of various kinds. While the term phase transfer
catalysis is relatively new, the concept of using a phasetrans-
fer agent (PTA) is much older. Both Schnell and Morgan employed
such catalysts in synthesis of polymeric species in the early
1950's. Present developments are really extensions of these
early applications.

It has only been within the last several years that the
use of phase transfer processes have been employed in polymer
synthesis and modification. Similarly, the use of polymer-bound
phase transfer agents is also a recent development. These and
related areas have nonetheless enjoyed explosive growth as measured by the number of publications and the variety of applications which have appeared. Several reviews dealing with these
polymer-related investigations have been published.[1-6] In fact,
one of us began a series of review articles on these new developments,[7-8] but it was felt that the rate of growth and diversity of
this field required a much more comprehensive coverage. A symposium
entirely devoted to phase transfer catalysis and crown ethers in.
polymer science was held at the 1982 National American Chemical Society meeting in Las Vegas. This symposium was jointly sponsored
by the Divisions of Polymer Chemistry and Organic Coatings and
Plastics Chemistry. This book is a collection of expanded papers
based on the talks given at that symposium.

There exists a great deal of art mixed with the science of
employing PTAs. New PTAs are continually being introduced,
many with the capacity of performing a specific role very effi-

v

ciently. Our understanding of their application to polymers and
the use of polymeric PTAs in place of low molecular weight systems is growing rapidly despite the greater complexity and unique
properties of polymeric substrates and catalysts. These unique
properties can and do offer increased opportunities for enhanced
rates, yields, selectivities and catalyst reuse which make applications in industry attractive. This book details many of these
unique opportunities with applications useful for gram to ton
syntheses, and attempts to balance art with science regarding the
behavior, activity and mechanistic aspects of PTA's.

   This is the first book which comprehensively summarizes the
various areas of polymer science involving this topic. Consists of
five sections: polymer modification reactions, polymer synthesis
employing phase transfer catalysis, ion binding and chelating
polymers, mechanistic aspects of polymer-bound catalysts, and finally the use of polyoxyethylene as a PTA. The papers in these
sections range from very specific applications through overviews to
evaluations of fundamental principles. A number of review papers
are included which summarize previous work in selected areas. Our
goal in this book was to provide reasonably comprehensive background
reviews combined with state-of-the-art descriptions of current research efforts.

   This book was only possible with the assistance of the individual contributing authors. Their efforts cannot be understated,
and the editors wish to express their heartfelt appreciation. In
addition, the Divisions of Polymer Chemistry and Organic Coatings
and Plastics Chemistry of the ACS are thanked for financial
assistance for the symposium.

Lon J. Mathias and
Charles E. Carraher Jr.

REFERENCES

1. J. Dockx, Synthesis <u>1973</u>, 441.
2. N.K. Mathur and R.E. Williams, J. Macromol. Sci. Rev.
      Macromol. Chem. <u>C15</u>, 117 (1976).
3. J. Smid, S.C. Shah, R. Sinta, A.J. Varma, and L. Wong, Pure
      and Appl. Chem. <u>51</u>, 111 (1979).
4. G. Manacke and P. Reuter, Pure and Appl. Chem. <u>51</u>, 2313, 1979.
5. E. Chiellini, R. Solaro and S. D'Antone, Makromol. Chem.
      Suppl. <u>5</u>, 82 (1981).
6. A. Akelah and D.C. Sherrington, Chem. Rev. <u>81</u>, 557 (1981).
7. L.J. Mathias and K.B. Al-Jumah, Polym. News <u>6</u>, 9 (1979).
8. L.J. Mathias, J. Macromol. Sci. - Chem. <u>A15</u>, 853 (1981).

CONTENTS

POLYMER MODIFICATION

POLYMER SYNTHESIS

MECHANISTIC ASPECTS

ION BINDING AND CHELATING POLYMERS

POLYOXYETHYLENE

CHEMICAL MODIFICATION OF POLYMERS VIA PHASE TRANSFER CATALYSIS

Jean M. J. Fréchet

Department of Chemistry
University of Ottawa
Ottawa, Ontario, K1N-9B4, Canada

INTRODUCTION

The preparation of functional polymers is an area of organic-polymer chemistry which continues to receive much attention in view of the numerous new and imaginative applications which are discovered for specialty polymers with reactive functionalities. The two main approaches which can be used in the preparation of functional polymers consist of the polymerization or copolymerization of suitably functionalized monomers or the chemical modification of pre-formed polymers. The first approach is often considered to be the most attractive due to its apparent simplicity, although it is often ill suited for the preparation of polymers with fairly complex functionalities. In some cases, even simple polymers such as poly(vinyl alcohol) are only accessible via a chemical modification route. In other cases it may be desirable to effect a simple chemical modification reaction to prepare a less common or more reactive polymer such as poly(iodomethyl styrene) from a more readily accessible but less reactive precursor such as poly(chloromethyl styrene).

The study of polymer-supported reactions in organic chemistry (Ref. 1-6) is a field which has enjoyed rapid growth in the past two decades and has required the preparation of a large number of specialty polymers carrying various functionalities; much of this preparative work has been carried out through the modification of a few reactive polymers derived from polystyrene (Ref. 7). The chemical modification route is particularly attractive in this instance as polystyrene resins contain aromatic rings which can be modified readily, often by electrophilic aromatic substitutions, while the rest of the molecule (polymer backbone) is relatively

1

inert. In a typical case, a polystyrene resin is modified through
a reaction which introduces functional groups on some of the
aromatic rings, leaving a number of other styrene units unchanged
in the final polymer. In many instances, such reactive polymers
can also be prepared easily by copolymerization of a functional
derivative of styrene with styrene and, if a crosslinked product is
desired, with a difunctional vinyl compound such as divinylbenzene.
However, while the chemical modification of a porous crosslinked
polystyrene bead will result in the incorporation of reactive
groups only in the more accessible sites of the beads, no such
control of the placement of reactive ends can be achieved in many
co- or terpolymerizations. Typically, if the various monomers have
very different reactivity ratios, one of the monomers may become
segregated in one area of the bead where it is less accessible to
outside reagents. Such is the case with the terpolymerization of
styrene, divinylbenzene and p-styryldiphenyl phosphine: as both the
divinylbenzene and the p-styryl-diphenyl phosphine are more
reactive than styrene, the phosphine monomer becomes incorporated
near the 'core' of the resin bead in a region of high crosslink
density which results in its limited acccessibilty for further
reactions. On the other hand, the chemical modification approach
may present some drawbacks as the reactions are seldom quantitative
and thus the end-product in a chemical modification process is
usually a polymer in which some, but not all, of the original
functional groups have been modified (Ref. 3). Therefore, the
modified polymer may contain 'impurities' in the form of unreacted
groups or other functionalities resulting from side-reactions.
Obviously, the amount of such impurities will depend on the nature
of the starting polymer and on the reaction or reaction sequence
which is performed. In some cases very few impurities will be
found in the final polymer, while in some other cases their
presence, even in appreciable amounts, may not affect the end-use
or the usefulness of the polymer. One of the greatest advantages
of the chemical modification route remains that it can be used to
prepare polymers or copolymers which would be difficult or even
impossible to prepare by any other method.
    This review will focus on the use of phase transfer catalysis
for the chemical modification of polymers which already contain
reactive groups. As numerous excellent reviews (Ref. 8) of phase
transfer catalysis exist, the basic principles of the method will
not be included here. In addition no attempt will be made to
ensure encyclopedic coverage of the field but selective coverage
will be given to areas with which this author is most familiar.

APPLICATION OF PHASE TRANSFER CATALYSIS TO THE MODIFICATION OF
POLYMERS: GENERAL CONSIDERATIONS AND EARLY WORK

    Phase transfer catalysis has been used extensively in organic
synthesis and many of the findings which are reported in the

abundant literature on the use of this technique can be applied to
the chemical modification of polymers once it has been recognized
that polymers possess a number of special features which may affect
their reactivity or the course of a given reaction. For example,
in a number of cases the reactions will involve three distinct
phases instead of the more usual two in most classical organic
reactions. A similar situation prevails when polymers are used as
phase transfer catalysts in triphase catalysis (Ref. 9). As was
mentioned above, the usual methods of purification - distillation,
crystallization, chromatography, etc. - cannot be used to purify
polymers after chemical modification; thus, the reactions must be
designed while taking into account the possible inclusion of
impurities in the final polymer, or the fact that it may still
contain some unreacted functionalities. A third problem can be
particularly troublesome in some reactions such as those involving
polyfunctional reagents as these may react through more than one of
their reactive ends, which may result in a drastic alteration or
crosslinking of the starting polymer. Similar problems also arise
in classical organic syntheses involving polyfunctional molecules,
but purification is usually possible. Finally, a potentially very
critical problem which is common to all chemical modification
reactions is the analysis of the final product. This may be
relatively simple with soluble polymers but it may become an
exceedingly difficult task with crosslinked polymers, and much
effort has to be devoted to the evaluation and analysis of the
final product after chemical modification of insoluble resins.
A survey of the literature shows that in a number of instances,
functional polymers prepared by chemical modification have been
ill-characterized and may in fact have been quite different from
their assigned structures. The finding that a polymer does not
react as claimed in the literature, or as expected from its assumed
structure can sometimes be attributed to one of two main culprits:
the lack of thorough characterization of the polymer which may
contain impurities, etc, or the inability to perceive that organic
reactions with polymeric substrates require some special attention
to accomodate the special features of polymeric materials (Ref. 3,
6).

Although most of the work on the use of phase transfer
catalysis for the chemical modification of polymers was carried out
in the last five years, the first report of the application of this
technique to the modification of reactive polymers originated ten
years ago from the laboratory of Okawara, one of the great
innovators of polymer chemistry, and was concerned with the
chemical modification of poly(vinyl chloride) (Ref. 10, 11). This
was followed by simultaneous brief reports by Roovers (Ref. 12) and
by Roeske et al. (Ref. 13) describing the use of 18-crown-6 in the
reaction of carboxylates with poly(chloromethyl styrene). Our own
first report (Ref.14) in this field described the use of phase
transfer catalysis to control site-site interactions in the

reaction of the same polymer with 1,4-butanedithiol in basic medium, and was followed shortly by several other reports (Ref. 15-18) showing the general applicability of phase transfer catalysis in polymer modification. These included the generation of polymer supported ylides (Ref. 15), nucleophilic displacements on polymers (Ref. 16-18), displacements with polymeric nucleophiles (Ref. 16-18), addition reactions (Ref. 16), etc. Since then, we have used phase transfer catalysis routinely in the preparation of a large number of new functional polymers for a variety of applications such as reactive polymers for organic syntheses, polymeric separation media, disinfectant polymers, resist materials, etc. Although much of this work remains unpublished, some of it will be included in this review to demonstrate the versatility of chemical modification of polymers via phase transfer catalysis.

CHOICE OF REACTION CONDITIONS

As was mentioned earlier, most of the results reported in the literature on phase transfer catalysis in organic synthesis can be adapted for use in the chemical modification of polymers after taking into account the special features of the reactive polymers.

Reactions with linear polymers.

These reactions may be attempted using a suspension of the finely powdered polymer in the reaction medium. Although this technique has been used successfully, for example in the surface modification of poly(vinyl chloride) powder (Ref. 10), it only leads to partial conversion as many of the polymer's reactive groups are inaccessible to the reagent and since reaction rates are generally quite low. The modification of such linear polymers in the solid state or perhaps even in a melt is nevertheless a tempting goal as it may provide an avenue for the removal or destruction of small amounts of reactive impurities in bulk polymers. We have been looking at this problem and have obtained some encouraging preliminary results. Higher conversions and faster reaction rates can be obtained with solutions of linear polymers with liquid-liquid (Ref. 12, 17, 19, 20) or liquid-solid (Ref. 19, 20) phase transfer catalysis. In liquid-solid phase transfer reactions, the addition of a trace amount of water may result in a noticeable increase in the extent and rate of reaction, although it may also result in increased side-reactions such as hydrolysis (Ref. 20)

A common problem with the modification of polymers is that they may croslink due to interactions between their reactive sites during reaction. For example, we have observed (Ref. 17) that since soluble poly(chloromethyl styrene) (P)-CH_2Cl reacts much faster with acetate ion than with hydroxide ion under phase transfer conditions, a convenient route to prepare the hydroxymethyl polymer (P)-CH_2OH, is to treat a solution of (P)-CH_2Cl

in an organic solvent such as o-dichlorobenzene with both acetate
and hydroxide ions. Acetate ion rapidly displaces the chloride
ions to yield Ⓟ-CH$_2$OAc which is then saponified to the
hydroxymethyl polymer. The reaction is however accompanied by a
troublesome side-reaction which results in crosslinking of the
final product. This crosslinking reaction, which is presented
schematically in Figure 1, is the result of the attack of a
Ⓟ -CH$_2$-O$^-$ site, resulting from the saponification of Ⓟ-CH$_2$OAc,
on a neighboring unreacted chloromethylated site. Attack of the
same reactive intermediate on an acetylated site would cause no net
change in composition and would not result in crosslinking.

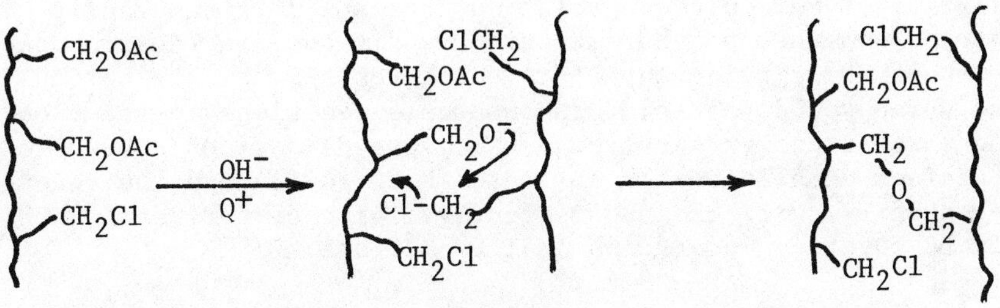

Figure 1.

A similar observation was made by Nishikubo and coworkers
(Ref. 20) for the reaction of Ⓟ-CH$_2$Cl with potassium hydroxide.
No reaction occurred under solid-liquid phase transfer conditions
with powdered KOH and a solution of the polymer in anhydrous
toluene, but when a 50% aqueous solution of KOH was used, the
reaction did proceed, albeit not to completion, to yield an
insoluble gel as crosslinking was observed due to interactions
between neighboring sites. It should be mentioned however, that
soluble poly(hydroxymethyl styrene) can be prepared from its
chloromethylated precursor without any crosslinking in a one-pot
reaction (Ref. 17) by treating a solution of the chloromethylated
polymer first with acetate ion under phase transfer conditions,
then by adding hydroxide ion to the reaction mixture once the rapid
displacement of chloride is complete, or by carrying out the
reaction in DMF-aqueous sodium hydroxide (Ref. 21, 50).

Another instance in which a crosslinking side-reaction can be
expected to occur, at least to some extent, is in the reaction of a
polymer with a substrate which contains two or more identical
reactive groups. For example, Nishikubo and coworkers (Ref. 20)
have shown that some crosslinking occurs in the reaction of
malononitrile, which has two active hydrogens, with
poly(chloromethyl styrene) in basic medium under phase transfer
conditions.

Reactions with Crosslinked Polymers.
These reactions usually involve three distinct phases as the

polymer itself is totally insoluble. In general, chemical
modification reactions are most successful with polymers which
swell extensively in the reaction medium to expose the reactive
sites located inside the polymer particles, or with polymers which
have their reactive sites located in the pores of a high surface
area macroporous resin. Therefore, reactions involving such
polymers will usually require that a good swelling solvent be used
where applicable. In addition, it can be expected that as
diffusion phenomena become very important for reactions within the
pores of the polymer particles, the reactions will be slower than
comparable reactions in solution. We have frequently noted that
although crosslinked polymers often have lower reactivities than
similar small molecules in solution, the polymers can easily
withstand somewhat harsher reaction conditions and thus reactions
can usually be carried out at temperatures slightly higher than
those which would be used for comparable reactions in solution.
Using phase transfer catalysis, excellent degrees of conversion can
be obtained with insoluble polymers (Ref. 14-18) and the reactions
are generally quite clean, frequently giving better results than
would be obtained under classical conditions.

    Although side reactions which cause additional crosslinking
are difficult to detect in polymers that are already crosslinked,
they undoubtedly occur when polyfunctional reagents are used since
the polymer chains can move fairly freely within the crosslinked
network, and thus, reactive sites in close proximity to one another
can interact easily. In our study (Ref. 16, 18) of the reaction of
excess malononitrile with a 20% chloromethylated crosslinked
polystyrene resin in the presence of base and tetrabutyl ammonium
hydroxide, we routinely obtained conversions of more than 95%
(calculated from N and Cl analyses for a monosubstitution)
indicating that although a small amount of additional crosslinking
might have occurred, and indeed probably did occur, the reaction
was nevertheless of practical value.

Choice of a Phase transfer Catalyst and Other Factors Influencing the Reaction.

While a number of different phase transfer catalysts such as
tetrabutyl ammonium halides (Ref. 10, 11), tetrabutyl ammonium
hydroxide (Ref. 16-18), tetrabutyl ammonium hydrogen sulfate (Ref.
21-22), Adogen-464 (Ref. 16, 18), tetrabutyl phosphonium bromide
(Ref. 19, 20), 18-crown-6 (Ref. 12, 13, 19, 20), cryptand [222]
(Ref. 21, 22), etc., have been used in the chemical modification of
polymers, few systematic studies of the influence of the catalyst
on the reactions have been done. It is presumed that the same
considerations which govern the choice of a phase transfer catalyst
for classical organic synthesis also apply in the case of reactions
with polymers.

Nevertheless, the excellent studies of Nishikubo and coworkers

(Ref, 19, 20) are well worth describing here in some detail, as they focus on the choice of a phase transfer catalyst and on several other factors which are of importance in the chemical modification of soluble poly(chloromethyl styrene). In a first series of experiments involving solid–liquid reactions with potassium acetate as a nucleophile, Nishikubo and coworkers observe that while reactions carried out without any added phase transfer catalyst do not work in apolar solvents, satisfactory results can be obtained in DMF (Table 1); this confirms previous observations (Ref. 7, 14, 23, 24) which suggest that DMF and DMSO are excellent solvents for nucleophilic displacements on poly(chloromethyl styrene).

Table 1. Reaction of Potassium Acetate with (P)-CH$_2$Cl under Solid–Liquid Phase Transfer Conditions

SOLVENT	T ^{o}C	CATALYST*	DEGREE OF SUBST.**
Toluene	30	none	0%
DMF	30	none	77%
Toluene	30	18-crown-6	24%
o-dichlorobenzene	30	18-crown-6	37%
DMF	30	18-crown-6	99%
o-dichlorobenzene	50	18-crown-6	57%
o-dichlorobenzene	80	18-crown-6	88%
Toluene	30	TMAC	0%
Toluene	30	TBAB	63%
Toluene	30	TBPB	73%

* TMAC = tetramethylammonium chloride, TBAB = tetrabutyl-ammonium bromide, TBPB = tetrabutylphosphonium bromide.
** Degree of Substitution calculated from Cl analysis and expressed in mole%

Table 2: Reaction of Potassium Acetate with (P)-CH$_2$Cl under liquid–liquid phase transfer conditions

CONDITIONS	CATALYST*	DEGREE OF SUBST.
Toluene, 0.4M CH$_3$COOK/H$_2$O	18-crown-6	Trace
Toluene, 0.4M CH$_3$COOK/H$_2$O	TBAC	Trace
Toluene, Conc.CH$_3$COOK/H$_2$O	TBAB	50%
Toluene, Conc.CH$_3$COOK/H$_2$O	TBPB	82%
Toluene, Conc.CH$_3$COOK/H$_2$O	18-crown-6	Trace
Toluene, Conc.CH$_3$COOK/H$_2$O	DCHC	6%

* TBAC = Tetrabutylammonium chloride, TBAB = Tetrabutyl-ammonium bromide, TBPB = Tetrabutylphosphonium bromide, DCHC = Dicyclohexyl-18-crown-6.

In the presence of 18-crown-6 the degree of conversion
increases with increased solvent polarity, best results being
obtained in DMF (Table 1); as expected, the influence of
temperature is also quite noticeable. Table 1 shows that the
nature of the catalyst and the type of phase transfer reaction,
solid-liquid or liquid-liquid, are very important factors.
Short-chain tetraalkyl ammonium salts (methyl, ethyl or propyl)
have no catalytic activity, while tetrabutyl ammonium or
phosphonium salts have good activities; several other phase
transfer catalysts were also included in this study but will not be
reviewed here. Reactions with aqueous solutions of potassium
acetate (Table 2) confirm that best results are obtained when a
concentrated solution of the salt is used. The scale of catalytic
activity for these liquid-liquid reactions is the following:

tetrabutylphosphonium bromide > tetrabutyl ammonium bromide >>
dicyclohexyl-18-crown-6 > 18-crown-6.

Table 3. Reaction of P-CH Cl with malononitrile: conversion
as a function of the catalyst.

CONDITIONS (Base)	18-crown-6	TBAB	TBPB
Solid-liquid (KOH)	22%	45%	39%
Solid-liquid (KOH) (trace of water)	76%	76%	87%
Liquid-liquid (KOH) (dilute aqueous)	2%	66%	–
Liquid-liquid (Potassium t-butoxide)	96%	95%	–

With a more powerful nucleophile such as potassium
thioacetate, the reaction is much easier proceeding even with a
dilute aqueous solution of the nucleophile under phase transfer
conditions. Best results are obtained under solid-liquid phase
transfer catalysis with essentially complete conversion obtained
with both tetrabutyl ammonium chloride and 18-crown-6. A critical
evaluation of this and further data reported by Nishikubo and
coworkers (Ref. 19, 20) leads to the following general conclusions:

* Tetrabutyl phosphonium and, to a lesser extent, tetrabutyl
ammonium salts are generally the best catalysts.

    * Crown ethers seem to be effective only in solid-liquid reactions.

    * As expected, strong nucleophiles react better than weak nucleophiles and can be used even in fairly dilute solutions; higher reaction rates are obtained at higher temperatures.

    It should be noted that other factors such as the presence of traces of water may have a great influence on the reaction, both in terms of percent conversion and occurence of side-reactions such as hydrolysis of the product. Table 3 supports this point and shows that a trace amount of water accelerates the reaction of malononitrile with soluble poly(chloromethyl styrene) in the presence of base and a phase transfer catalyst. The conclusion of Nishikubo and coworkers following this study should be quoted here: "the hydrophobic interactions of the reagent, catalyst, and base have a strong influence on the degree of the alkylation reaction of [poly(chloromethyl styrene)] with active methylene compounds". It is perhaps unfortunate that the degrees of substitution reported by Nishikubo et al. in these two studies (Ref. 19, 20) are based only on halogen analyses which show the removal of chloride from the chloromethyl groups but do not confirm the introduction of new functionalities. Thus, it is difficult to quantify side-reactions such as the crosslinking reaction which is observed with malononitrile. In contrast, our yield and conversion data for similar reactions (Ref. 16, 18) on a crosslinked resin are based on the results of both chlorine and nitrogen analyses and confirm that relatively few side-reactions occur.

    Finally, it is appropriate to say a few words on the choice of solvent for the chemical modification of polymers under phase transfer catalysis. As was mentioned earlier, numerous reactions which do not proceed in non-polar solvents such as toluene or dichloromethane in the absence of a phase transfer catalyst do proceed satisfactorily in DMF. Thus, many research groups, including ours, have used DMF extensively in polymer modifications with or without added catalyst, with increases in reaction rates and conversions being observed in the former case. As DMF is often a solvent for both the polymer and, in many instances, at least some of the reagent, it is debatable whether or not the term "phase transfer catalysis" applies (Ref. 50). More important perhaps is the fact that considerable amounts of dimethylamine can be produced through decomposition of DMF when the solvent is treated with concentrated aqueous base in the presence of a phase transfer catalyst. Obviously this may lead to undesirable side-reactions with incorporation of dimethylamine moieties into the modified polymers (Ref. 50).

REACTIONS ON CHLOROMETHYL SUBSTITUTED POLYSTYRENE (I)

Due to its ready availability, high reactivity and other interesting properties, chloromethyl substituted polystyrene (I) has been the most extensively studied substrate in chemical modification via phase transfer catalysis. Reactions have been carried out both on soluble poly(chloromethyl styrene) (Ref. 12, 17, 19-21), which can be obtained either by direct polymerization of the monomer or by partial chloromethylation of linear polystyrene, and on crosslinked partly chloromethylated styrene-divinylbenzene resins (Ref. 13-18, 22, 25-30). In most cases, the latter resins have been 1% or 2% crosslinked gels which are highly swellable and possess a high reactivity almost comparable to that of the soluble polymer. However, as much of our current research is concerned with the preparation of functional polymers for use as separation media or for water or waste treatment, we have also used macroporous resins such as Amberlite XE-305 and comparable polymers prepared in our laboratory (Ref. 16, 18, 28) which are better suited for larger scale filtrations and do not require swelling of their pores prior to reaction. It should be noted that, in our hands, these macroporous resins always show a lower reactivity than their 1% or 2% crosslinked swellable counterparts, and often have a tendancy to undergo significant mechanical breakdown on repeated handling.

As a very large number of modification reactions have been carried out on polymer I, the readers are referred to Table 4 which lists most of the types of structures which have been prepared from I under phase transfer conditions. It many cases the polymers have not been fully characterized but, in most instances, reasonable evidence supporting the proposed structures is given.

In general, reactions with monofunctional molecules such as carboxylate (Ref. 12, 13, 16, 17, 19, 28), nitrile (Ref. 16), phenolates (Ref. 16, 19, 21, 28), thiocyanate (Ref. 21, 26), phtalimide (Ref. 21, 27), carbazole (Ref. 22), thioacetate (Ref. 19), N,N-diethyl dithiocarbamate (Ref. 19), etc. are easily performed and a variety of reaction conditions and catalysts can be used successfully. For example, Scheme 1 shows the reaction of polymer I with methoxy triethylene glycol.

$$\text{(P)}-CH_2Cl \ + \ HO-(CH_2CH_2O)_3CH_3 \xrightarrow[Q^+]{OH^-} \text{(P)}-CH_2O(CH_2CH_2O)_3CH_3$$

SCHEME 1.

The reaction can be done either in a solid-liquid-solid system, with a suspension of finely powdered base in an organic solvent containing tetrabutyl ammonium chloride or 18-crown-6, or in a liquid-solid-liquid system with a concentrated aqueous solution of

Table 4. Chemical Modification of Chloromethyl Polystyrene

(P)-CH$_2$Cl under Phase Transfer Catalysis: Structure

of the Polymers after Modification. (Reference)

(P)-CH$_2$-S-(CH$_2$)$_4$-SH (14)

(P)-CH$_2$-S-(CH$_2$)$_4$-S-CH$_2$-(P) (14)

(P)-CH$_2$-S-(CH$_2$)$_3$-CH$_3$ (16)

(P)-CH$_2$CH(CN)$_2$ (16, 18, 20)

(P)-CH$_2$CH(COOR)$_2$ (16, 20)

(P)-CH$_2$CH-COOR (16)
　　　　|
　　　　CN

(P)-CH$_2$-C(COOR)$_2$ (20)
　　　　|
　　　　CH$_3$

(P)-CH-CN (18)
　　|
　　CH$_2$CN

(P)-CH$_2$-O-⟨◯⟩-X (16, 19)

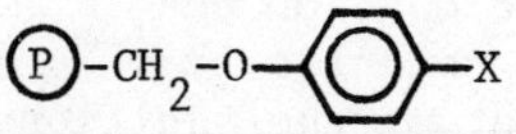

(P)-CH$_2$-O-CO-CH$_3$ (16, 17, 19)

(P)-CH$_2$-OH (16, 17, 20)

(P)-CH$_2$-O-CH$_2$-(P) (17, 20)

(P)-CH$_2$-O-CO-(CH$_2$)$_{16}$CH$_3$ (12)

(P)-CH$_2$-O-CO-CH=CH-Ph (12)

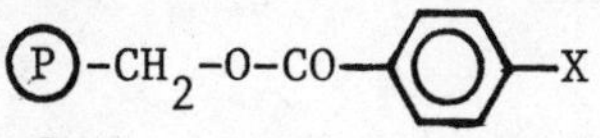

(P)-CH$_2$-O-CO-⟨◯⟩-X (12)

(P)-CH$_2$-O-CH$_2$-Ph-OCH$_3$ (21)

(P)-CH$_2$-PHTALIMIDE (21, 27)

(P)-CH$_2$-SCN (21, 26)

(P)-CH$_2$-SH (17, 26)

(P)-CH$_2$-CN (16)

(P)-CH$_2$-S-S-CH$_2$-(P) (17)

(P)-CH$_2$-CROWN (57)

(P)-CH$_2$-S-CO-CH$_3$ (19)

(P)-CH$_2$-S-CS-N(C$_2$H$_5$)$_2$ (19)

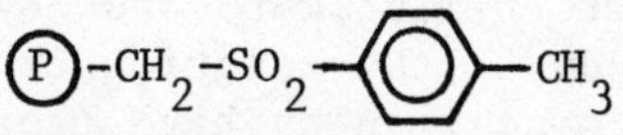

(P)-CH$_2$-SO$_2$-⟨◯⟩-CH$_3$ (19)

(P)-CH$_2$-CH-Ph (21)
　　　　|
　　　　CN

(P)-CH$_2$-CARBAZOLE (22)

(P)-CH$_2$-INDENE (21)

(P)-CH$_2$-I (25)

(P)-CH$_2$-O-CO-⌇⌇-NH-BOC- (13)

(P)-CH$_2$-S-CH$_2$CH$_2$-OH (28)

(P)-CH$_2$-S-CH$_2$-CH(OH)-CH$_2$OH (28)
　　　　　　　　　　　　　　　　　(29)

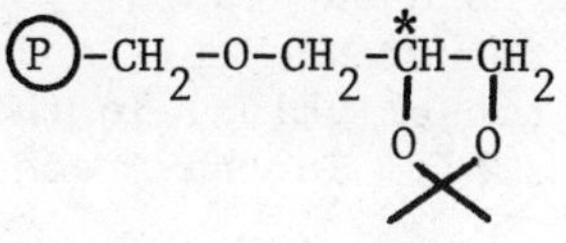

(P)-CH$_2$-O-CH$_2$-$\overset{*}{C}$H-CH$_2$ (28)
　　　　　　　　　O　O
　　　　　　　　　　╳

(P)-CH$_2$-O-CO-$\overset{*}{\Box}$N (28)
　　　　　　　　　　 |
　　　　　　　　　　 H

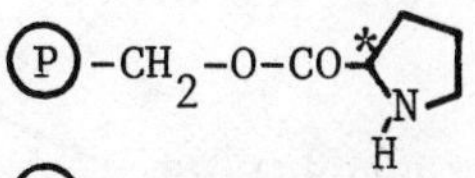

(P)-CH$_2$-O-CH$_2$-CH-Ph (28)
　　　　　　　　　　*|
　　　　　　　　　　N=R

(P)-CH$_2$-O-TYROSINE (28)

base in the presence of tetrabutyl ammonium hydroxide or tetrabutyl phosphonium bromide (Ref. 28).

As side-reactions such as the hydrolysis of the chloromethyl groups often occur, it is generally wise to follow the modification reaction not only through a measurement of the disappearance of chlorine from the starting polymer, but also through measurements of the incorporation of the new functionalities on the polymer. In this respect, gravimetric data are often invaluable although it must be remembered that, with crosslinked resins, extensive washings are required to remove all soluble impurities from the polymer.

Reactions with polyfunctional molecules are much more difficult to control and, in the case of linear I, they often lead to crosslinked materials as it is frequently difficult to prevent reactions by two or more of the reactive groups of the reacting molecules. An excellent example of this behavior can be found in the report of Nishikubo and coworkers (Ref. 20) which describes their study of the reaction of linear I with malononitrile in basic medium; these results have been described earlier in this review. With crosslinked polymer I, the problem is less severe as a small amount of additional crosslinking does not affect the properties of the polymer as drastically as it does with a linear polymer and thus, the usefulness of the final polymer may not overly be affected. For example, the reaction of 1-2% crosslinked I with malononitrile gives a product which contains at least 95% of the theoretical amount of nitrogen, indicating that most of the reaction consists of the simple replacement of chloride by the anion on malononitrile. Although no measurement of the additional crosslinking resulting from reaction of the polymer bound malononitrile moieties could be made, it is expected that some did occur, but this had little apparent effect on the reactivity of the malononitrile polymer as shown by its easy transformation into the corresponding 1,3-diamine (Ref. 18).

In fact, we have successfully used phase transfer catalysis to reduce the occurrence of site-site interactions (which cause additional crosslinking) in the reaction of polymer I with 1,4-butanedithiol in the presence of base according to Scheme 2 (Ref. 14).

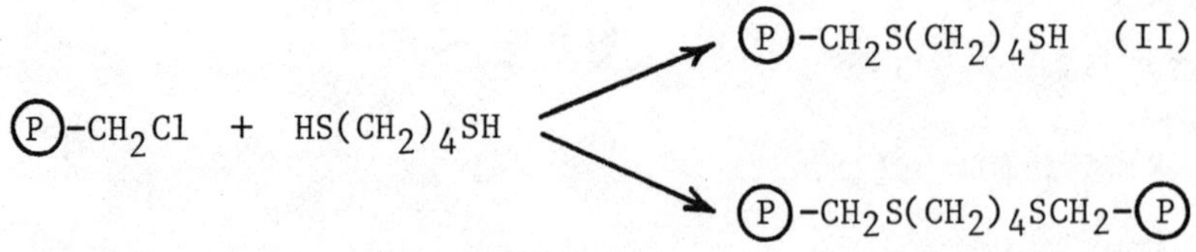

$$\text{\textcircled{P}}-CH_2Cl \ + \ HS(CH_2)_4SH \ \Bigg\langle \begin{array}{l} \nearrow \ \text{\textcircled{P}}-CH_2S(CH_2)_4SH \quad (II) \\ \\ \searrow \ \text{\textcircled{P}}-CH_2S(CH_2)_4SCH_2-\text{\textcircled{P}} \\ \qquad\qquad\qquad\qquad\qquad (III) \end{array}$$

SCHEME 2.

While under classical conditions extensive site interactions result
in a minimum of 26% double binding of the dithiol with formation of
III rather than the desired II, the extent of double coupling can
be reduced to ca. 5% under phase transfer conditions. The success
of this reaction is attributed partly to the fact dianions are not
transported as easily as monoanions through the organic phase (Ref.
8) and to a reduction in the occurrence of proton transfer
reactions which transform polymer II into its reactive anion.

With a number of polyfunctional molecules such as
2-mercaptoethanol which do not have identical reactive groups, it
is often possible to obtain reaction at one end of the molecule
only through a proper control of the reaction conditions (Ref. 28).
For example, in the case of the reaction of 2-mercaptoethanol with
I, the sulfur nucleophile is much more reactive than the oxygen
nucleophile and reaction with an excess of 2-mercaptoethanol, with
respect to base, leads to binding through sulfur only. A similar
reaction has also been used effectively by Hodge and coworkers and
by ourselves to prepare a polymer supported diol IV from I and
3-mercapto-1,2-propanediol:

$$\text{P}-CH_2Cl \;+\; HS-CH_2CH(OH)CH_2OH \;\longrightarrow\; \text{P}-CH_2S-CH_2CH(OH)CH_2OH \quad (IV)$$

While our reaction was a solid-liquid-solid reaction, Hodge's
involved the use of concentrated aqueous sodium hydroxide and
tetrabutyl ammonium hydroxide. It is noteworthy that the final
polymer (IV) is much more effective in its use as a
polymer-supported protecting group for carbonyl compounds (Ref. 29)
than a somewhat similar polymer-bound diol prepared under standard
conditions (Ref. 30, 31).

With other substrates such as L-proline, it is much more
difficult to control the reaction as both the amino group and the
carboxylate are able to displace chloride ions from I in basic
medium; we have however been able to prepare both products under
controlled reaction conditions (Ref. 28). Similarly, an amino-acid
such as tyrosine has three potential reactive sites, but conditions
can be found to attach it to polymer I through either the phenolic
or the amino group (Ref. 28). With 2-chloro-3-hydroxy pyridine,
binding through oxygen can be obtained readily under phase transfer
conditions to yield a polymer which is of interest as a regenerable
polymeric protecting group (Ref. 28b).

Reactions such as those used for the preparation of II, III,
or IV should always be carried out under an inert atmosphere to
prevent the formation of disulfides which might adversely affect
the reactions. This is particularly important in the preparation
of a polymer such as mercaptomethyl polystyrene using our
procedure (Ref. 17) or that of Gozdz (Ref. 26) as extensive
disulfide formation is observed when oxygen is present while the

phase transfer reaction is being carried out.

A final application of phase transfer catalysis to the chemical modification of I is the key step in the formation of a polymer bound thiothiazolone through reaction of I with carbon disulfide and glycine in basic medium with added tetrabutyl ammonium hydroxide. The open chain intermediate obtained in 90% functional yield can then be closed under mild conditions to afford the thiothiazolone which is currently being tested as a regenerable polymeric protecting group.

Although phase transfer catalysis is certainly an important and extremely versatile tool for the chemical modification of chloromethyl polystyrene, it is not necessarily always the best method as excellent results can also be obtained for some nucleophilic displacements when DMF or even DMSO (at low temperature to avoid oxidation to the carboxaldehyde polymer) are used as solvent for the nucleophile. For example, we prefer to use a solution of sodium cyanide in DMF to prepare cyanomethyl polystyrene from I rather than using a different solvent and phase transfer conditions, and we routinely prepare iodomethyl polystyrene from I by reaction with sodium or potassium iodide in acetone rather than under the conditions of Gozdz (Ref. 25). Recent work by Bied-Charreton et al. (Ref. 32) has also shown that excellent results could be obtained even under classical conditions in the transformation of I into its malononitrile derivative if the chloromethylated polymer is first transformed into the more reactive iodomethyl derivative; this is in sharp contrast with earlier data from the same laboratory (Ref. 33).

REACTIONS ON POLYMERIC SULFONIUM OR PHOSPHONIUM SALTS

The use of Wittig reactions for the chemical modification of polymers has been known for some time. In 1971, Fréchet and Schuerch (Ref. 34) reported on the preparation of polymer bound cinnamyl chloride (VI) and cinnamyl alcohol (VIII) from the corresponding polymer-bound phosphonium salt (V) or aldehyde (VII) (Scheme 3). The starting polymers V and VII were themselves obtained by chemical modification of 2% crosslinked polystyrene (Ref. 34). The Wittig reaction has also been used extensively for the preparation of alkenes from polymer-supported phosphonium salts (Ref. 35-39). The purpose of these reactions done under classical conditions, was not the chemical modification of the polymer but its use in organic synthesis as polymer supported reagents have some advantages over conventional reagents (Ref. 1-4). Similarly, a polymer containing sulfonium ylid moieties was prepared by Tanimoto et al. in 1968 (Ref. 40-41) via the chemical modification of a crosslinked polymer with pendant benzyl methyl sulfide functionalities; the resulting polymer, which had a fairly low reactivity, was tested in epoxidation reactions but was not very useful as its use led to extensive side reactions which prevented its recycling.

Phase transfer catalysis, which proved extremely useful in classical ylid reactions with both phosphonium and sulfonium salts (Ref. 8, 42-45), was first used with a polymer by Farrall, Durst and Fréchet in 1978 (Ref. 15) according to scheme 4. In this reaction, the polymeric sulfonium salt (IX), which is suspended in a dichloromethane solution of the carbonyl compound, is treated with aqueous sodium hydroxide in the presence of tetrabutyl ammonium hydroxide to give over 95% yield of the desired epoxide together with a polymeric by-product (X) which can be recycled and reused repeatedly without any loss of activity. In contrast, the same polymeric reagent (IX) used under classical conditions affords lower yields of epoxides and loses its activity rapidly on repeated recycling. This last observation shows clearly that phase transfer catalysis may contribute significantly to the prevention of side reactions in some modifications of polymers.

The use of polymers in phase transfer catalyzed Wittig reactions was first reported by Hodge and coworkers (Ref. 46) in 1980. Soluble as well as crosslinked polymeric phosphonium salts (XII) were used in reactions with aldehydes to afford various olefins according to Scheme 5. In these reactions cetyl trimethyl ammonium bromide or tetrabutyl ammonium iodide were used as phase transfer catalysts. The author noted that the reactions also proceeded, albeit more slowly in the case of crosslinked polymers, in the absence of any added catalyst, as the polymer itself acted as a phase transfer catalyst. Two other reports from the laboratory of Hodge (Ref. 29, 47) are especially interesting in the framework of this review as they utilize phase transfer catalysis and Wittig reactions to prepare several interesting polymers, some of which are similar to VI and VIII which we had prepared ten years earlier under classical Wittig conditions. These reactions, shown in Scheme 6, were usually performed in dichloromethane with cetyl trimethyl ammonium bromide as the phase transfer catalyst. High yields were obtained in the preparation of polymers containing phosphines (XIVa), crown ethers (XIVb) or ferrocene (XIVc) pendant groups.

Finally, very recent work from our laboratory (Ref. 48) has focused on the use of phase transfer catalysis for the preparation of reactive polystyrene resins with vinyl or oxirane pendant groups (Scheme 7) by reaction of V with aqueous formaldehyde in the presence of sodium hydroxide and tetrabutyl ammonium hydroxide, or by reaction of aldehyde resin VII with trimethylsulfonium chloride under similar conditions. Success of the latter reaction hinges on the use of trimethylsulfonium chloride rather than the corresponding iodide. When the iodide is used the reaction is

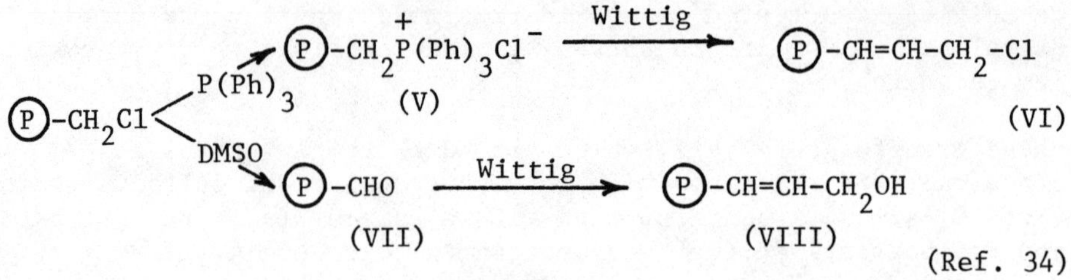

$$\text{(P)}-CH_2Cl \underset{DMSO}{\overset{P(Ph)_3}{\big<}}$$

(P)$-CH_2\overset{+}{P}(Ph)_3Cl^-$ $\xrightarrow{\text{Wittig}}$ (P)$-CH=CH-CH_2-Cl$
(V)   (VI)

(P)$-CHO$ $\xrightarrow{\text{Wittig}}$ (P)$-CH=CH-CH_2OH$
(VII)   (VIII)

(Ref. 34)

SCHEME 3

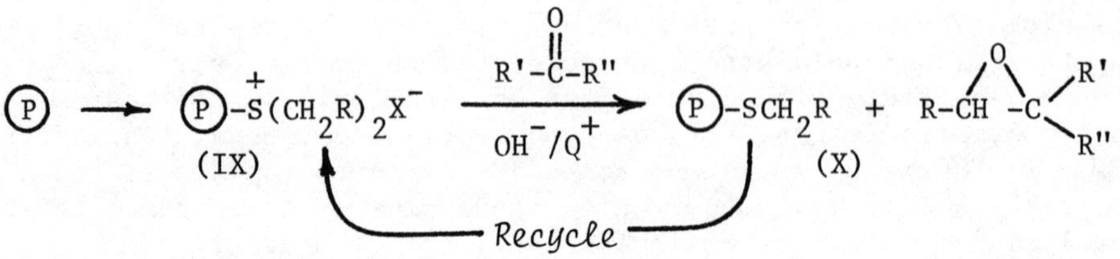

(P) $\longrightarrow$ (P)$-\overset{+}{S}(CH_2R)_2X^-$ $\xrightarrow[OH^-/Q^+]{\overset{O}{\overset{\|}{R'-C-R''}}}$ (P)$-SCH_2R$ $+$ $R-\overset{O}{CH}-\overset{R'}{\underset{R''}{\overset{|}{C}}}$
          (IX)   (X)

$\longleftarrow$ $\mathit{Recycle}$ $\longrightarrow$

SCHEME 4

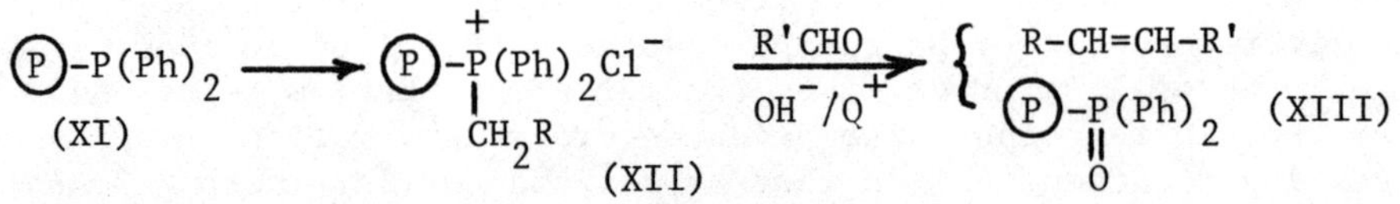

(P)$-P(Ph)_2$ $\longrightarrow$ (P)$-\overset{+}{\underset{CH_2R}{P}}(Ph)_2Cl^-$ $\xrightarrow[OH^-/Q^+]{R'CHO}$ $\left\{ \begin{array}{l} R-CH=CH-R' \\ \text{(P)}-\overset{\|}{\underset{O}{P}}(Ph)_2 \quad \text{(XIII)} \end{array}\right.$
(XI)   (XII)

SCHEME 5

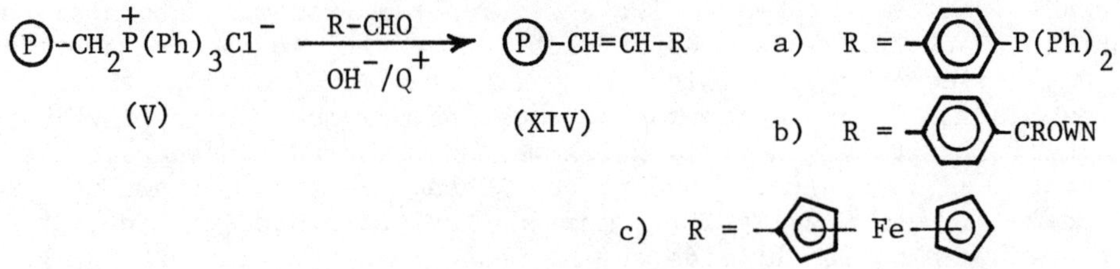

(P)$-CH_2\overset{+}{P}(Ph)_3Cl^-$ $\xrightarrow[OH^-/Q^+]{R-CHO}$ (P)$-CH=CH-R$ a) $R = $ —⬡—$P(Ph)_2$
(V)   (XIV)   b) $R = $ —⬡—CROWN

c) $R = $ —⬠— Fe —⬠—

SCHEME 6

(P)$-CH_2\overset{+}{P}(Ph)_3Cl^-$ $+$ HCHO (aq.) $\xrightarrow[OH^-]{Bu_4NOH}$ (P)$-CH=CH_2$
(V)   (XV)

(P)$-CHO$ $+$ $(CH_3)_3\overset{+}{S}Cl^-$ $\xrightarrow[OH^-]{Bu_4NOH}$ (P)$-CH\overset{\diagdown}{\underset{O}{\diagup}}CH_2$
(VII)   (XVI)

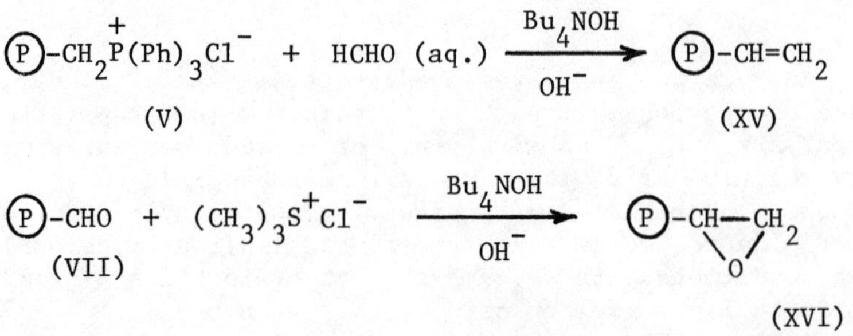

SCHEME 7

extremely slow and opening of the epoxide is observed. This
behavior of iodides is not uncommon and it is generally best to
avoid iodides in phase transfer reactions.

CHEMICAL MODIFICATION OF POLY(VINYL HALIDES)

    Although very little data concerning the chemical modification
of poly(vinyl halides) can be found in the literature, it is an
area of particular importance as the technique has great potential
for the removal of the small amounts of reactive impurities which
are found in the polymers and contribute to their relatively poor
thermal and photochemical stabilities. The presence of loosely
bound chlorine in poly(vinyl chloride) has long been known (Ref.
49). The loosely bound atoms result from defects which occur
during the polymerization reaction; for example, elimination of
hydrogen chloride creates allylic sites which are particularly well
suited for further eliminations. These render the polymer more and
more unstable as additional unsaturation is introduced in the
polymer chains. It would be of considerable value to develop a
mild method to effect the replacement of such labile allylic
chlorine atoms by other groups which could contribute to an
increased stability of the polymer. Several groups including ours
are actively seeking a solution to this challenge. Since at the
present time little data on the use of phase transfer catalysis to
remove only the loosely bound halogen atoms of poly(vinyl halides)
has been released, the following section will be devoted mainly to
a review of reactions in which attempts have been made to
drastically modify the polymers through the introduction of large
numbers of new functionalities.

    Okawara and coworkers (Ref. 10) first attempted to use phase
transfer catalysis to modify a finely dispersed poly(vinyl
chloride) powder in aqueous medium using nucleophiles such as
azide, dithiocarbamate, or thiophenoxide ions in the presence of
tetrabutyl ammonium salts. While a maximum conversion of 10% was
obtained with the first two nucleophiles, thiophenoxide afforded a
30% conversion. Although the authors indicate that the reaction
took place only at the surface of the polymer particles, the fairly
high conversion obtained with thiophenoxide might suggest
otherwise. A second report from the same laboratory (Ref. 11)
focuses on reactions of poly(vinyl chloride) solutions with sodium
azide in the presence of various catalysts. As expected, the
reaction is strongly catalyzed by cationic surfactants such as
dimethyl distearyl ammonium chloride or tetrabutyl ammonium
chloride which both afford essentially complete conversion to the
azido polymer. In contrast, tetrabutyl ammonium iodide is totally
ineffective.

We have recently completed a study of the chemical
modification of poly(vinyl bromide) (Ref. 50, 51) which was aimed

at testing the chemical modification route for the preparation of
polymers containing sulfone side-chains on a polyvinylic backbone,
for potential application as resist materials. The replacement of
bromine by various nucleophiles such as methyl sulfinate, methyl
thiolate or the anion of 2-mercaptoethanol occurred very slowly in
halogenated hydrocarbons but much more rapidly in DMF (Scheme 8).
The reactions were generally accompanied by some polymer
decomposition with elimination of hydrogen bromide resulting in the
formation of double bonds on ca. 10-20% of the original repeating
units. Conversion to the methyl sulfone polymer did not exceed 70%
while the water-soluble 2-hydroxythioethyl polymer XIX was obtained
in 90% conversion. A similar reaction with the non-nucleophilic
trifluoromethyl sulfinate ion gave extremely low conversions (ca.
20%) with little decomposition of the polymer.

Some very recent studies by Lewis at al. (Ref. 52, 53) have
provided an initial report on the use of crown ethers for the
replacement of the labile chlorine atoms of poly(vinyl chloride) by
acetate ions. In this ongoing study, the authors are studying the
kinetics of the modification process and the relative stabilities
of the modified polymers under pyrolytic conditions, in an attempt
to determine ideal conditions for the substitution of the labile
chlorine atoms only, as more extensive substitutions result in some
polymer degradation.

CHEMICAL MODIFICATION BY NUCLEOPHILIC DISPLACEMENTS ON
MISCELLANEOUS POLYMERS

Several reports on the chemical modification of polyepichlorohydrin under phase transfer catalysis have appeared in the last
five years. An earlier account of the preparation of poly-
(glycidyl cinnamate) by reaction of polyepichlorohydrin with
potassium cinnamate probably does not qualify as a phase transfer
catalyzed reaction, as Nishikubo and coworkers (Ref. 54) attempted
the reaction only in the presence of methyltriethyl ammonium iodide
which is known (Ref. 8) to have no catalytic activity in phase
transfer reactions. Interesting results were later reported by
Boileau and coworkers for the chemical modification of
polyepichlorohydrin with carbazole in the presence of various phase
transfer catalysts. With 5% tetrabutyl ammonium hydrogen sulfate
as catalyst, best results were obtained in DMF with conversions as
high as 41% vs. less than 5% in several other polar solvents.
Increases in catalyst concentration resulted in modest increases in
conversion with 56% removal of chlorine obtained using an equimolar
amount of the tetrabutyl ammonium salt. Better results were
obtained using dicyclohexyl-18-crown-6 or cryptand [222] which were
effective (52 and 66% conversions respectively) at 5 mole %
concentrations. Reactions with other amines such as dimethylamine
or isobutylamine gave low conversions (<10%) while 2,3-dimethyl

indole was found to be more reactive than carbazole. A significant problem in all of these reactions is that the polyepichlorohydrin undergoes extensive chain degradation during modification. We have observed (Ref. 50) a similar behavior in our study of the reaction of polyepichlorohydrin with acetate and hydroxide ions under phase transfer conditions. Low conversions were obtained in halogenated hydrocarbons while high conversions, accompanied by extensive chain cleavage, were observed when DMF was used as solvent for the reaction. Attempts to catalyze the Finkelstein reaction of poly-epichlorohydrin with sodium iodide in butanone using 15-crown-5 or methyltributyl ammonium iodide were unsuccessful (Ref. 56), and the reaction was also accompanied by extensive chain degradation. It seems likely that the problem of chain degradation will remain critical in all attempts to effect the chemical modification of polyepichlorohydrin.

Ricard et al. (Ref. 57) have reported on the preparation of a polystyrene resin containing pendant 1,3-diamine groups by a phase transfer catalyzed reaction of malononitrile with a chloro-acetylated polystyrene, followed by reduction of the dinitrile. Other reactions of diamines with chloromethyl polystyrene were also reported by the same authors but the published data seem to suggest that fairly low yields and significant double-binding were obtained. Nevertheless the method is valuable as two reactive polymer-bound crown compounds were prepared.

Our attempts to prepare useful materials by chemical modification of poly(vinylidene bromide) (Ref. 50) were not very successful, although, in reactions with the anions of methyl mercaptan or 2-mercaptoethanol under phase transfer conditions, degrees of substitution as high as 55% were obtained, while a reaction with the less nucleophilic sodium methyl sulfinate gave only 46% conversion. The reactions were accompanied by extensive elimination and the final modified polymers were always highly colored.

CHEMICAL MODIFICATION OF NUCLEOPHILIC POLYMERS

This is an area which has received much less attention as few nucleophilic polymers, other than those containing amine groups, are available. However, it is likely that in the coming years reactions in which polymers react with electrophiles will gain much importance as a large number of new polymer structures will become accessible through the use of one or two reactive nucleophilic polymers. Typically, amine containing polymers have been used mainly under classical conditions in reactions with electrophilic reagents and, although we have experimented with phase transfer catalysis in some applications, reactions under classical conditions were generally quite satisfactory (Ref. 57, 58).

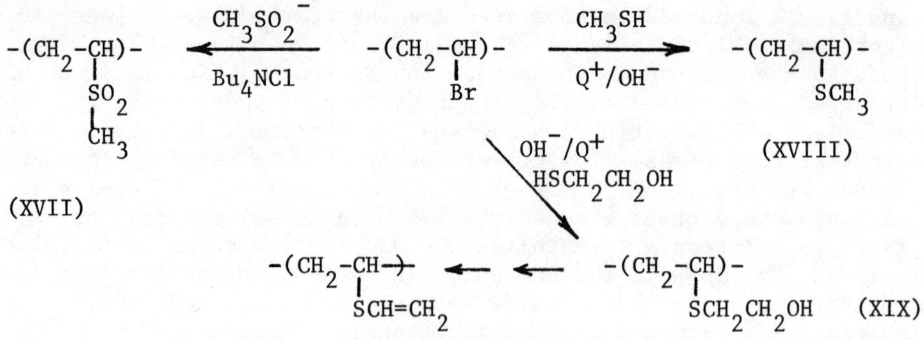

$$-(CH_2-CH)- \xleftarrow[Bu_4NCl]{CH_3SO_2^-} -(CH_2-CH)- \xrightarrow[Q^+/OH^-]{CH_3SH} -(CH_2-CH)-$$
$$\quad\;\; SO_2 \qquad\qquad\qquad\qquad Br \qquad\qquad\qquad\qquad\qquad SCH_3$$
$$\quad\;\; CH_3$$

$$\searrow \; OH^-/Q^+ \qquad\qquad (XVIII)$$
$$\qquad\;\; HSCH_2CH_2OH$$

(XVII)

$$-(CH_2-CH)- \;\leftarrow\; \leftarrow\; -(CH_2-CH)-$$
$$\quad\;\; SCH=CH_2 \qquad\qquad\qquad SCH_2CH_2OH \quad (XIX)$$

SCHEME 8

Table 5. Polymers prepared by chemical modification of (P)-OH
and (P)-SH under phase transfer conditions (Reference).

(P)-OCH$_2$CH=CH$_2$	(16, 59)	(P)-S(CH$_2$)$_4$Br	(16)
(P)-OCH$_2$CH$_2$⟨⟩-NO$_2$	(16, 59)	(P)-S(CH$_2$)$_4$S-(P)	(16)
(P)-OCH$_2$-⟨⟩$_X$	(62)	(P)-SCH$_2$CH$_2$OH	(16)
(P)-O(CH$_2$CH$_2$O)$_3$CH$_3$	(62)	(P)-S(CH$_2$CH$_2$O)$_n$CH$_2$CH$_2$OH	(16)
(P)-O-CO-R	(62)	(P)-S(CH$_2$CH$_2$O)$_3$CH$_3$	(62)

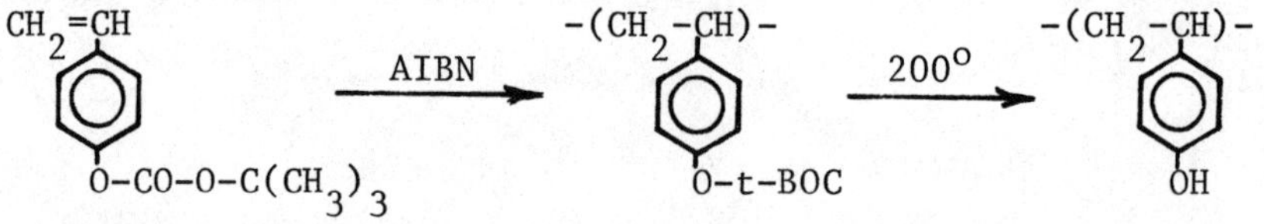

$$CH_2=CH \qquad\qquad\qquad -(CH_2-CH)- \qquad\qquad -(CH_2-CH)-$$
$$\bigcirc \xrightarrow{AIBN} \bigcirc \xrightarrow{200^\circ} \bigcirc$$
$$O-CO-O-C(CH_3)_3 \qquad\qquad O-t-BOC \qquad\qquad\qquad OH$$

SCHEME 9

$$(P)-CH_2SH \;+\; \bigcirc=O \xrightarrow[Q^+]{OH^-} (P)-CH_2-S-\bigcirc=O$$

$$(P)-SH \;+\; CH_2=CH-COOCH_3 \xrightarrow[]{OH^-/Q^+} (P)-S-CH_2CH_2-COOCH_3$$

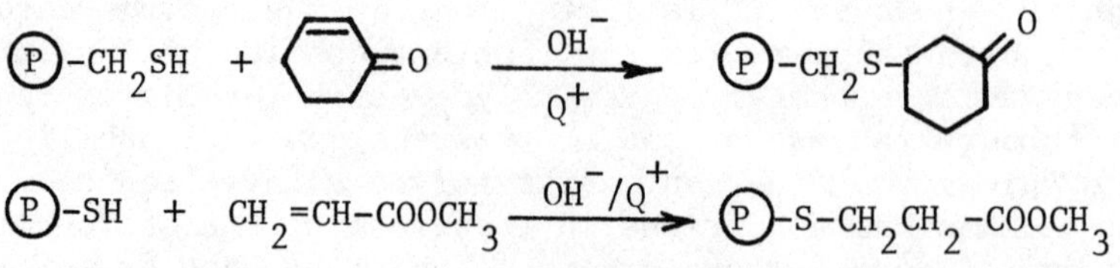

SCHEME 10

Over the past few years we have developed polymers containing vinyl phenol, (P)—OH, and vinyl thiophenol, (P)—SH, functionalities (Ref. 16, 17, 59, 60). These polymers are extremely efficient nucleophiles and can be used in nucleophilic displacements (Ref. 16, 17, 59) or Michael additions (Ref. 61, 62). Most of the reactions with these polymers were carried out using crosslinked (P)—OH or (P)—SH resins obtained by chemical modification of polystyrene-divinylbenzene resins (Ref. 17, 63), although we have just started using a new much improved (P)—OH resin obtained by suspension copolymerization (Ref. 59). The resin is prepared from styrene, divinylbenzene and p-t-butoxycarbonyl styrene; the t-BOC protected polymer is then transformed into (P)—OH (Scheme 9) by thermolysis of the t-BOC groups (Ref. 59, 60), a very clean reaction which is accompanied by evolution of carbon dioxide and isobutylene. The resulting porous polymer has a high reactivity and is free from the additional ether crosslinks which are present in the resin obtained by chemical modification. It is best to use the t-BOC protected monomer rather than p-hydroxystyrene itself for the suspension copolymerization, as it leads to a final product which is free of the oxidized impurities which are always present in resins prepared by polymerization of p-hydroxystyrene (Ref. 59).

Reactions with (P)—OH can be done using solid-liquid-solid or liquid-solid-liquid reaction conditions, the former being advantageous in a number of cases. We have tested successfully several phase transfer catalysts including tetrabutyl ammonium hydroxide, chloride, bromide, or hydrogen sulfate, tetrabutyl phosphonium bromide, Adogen 464, and 18-crown-6. In many instances the reactions are quantitative as evidenced by the complete disappearance of the hydroxyl bands in the infrared spectrum. A survey of the types of reactions we have carried out to date is shown in Table 5. We are currently modifiying (P)—OH through the introduction of chiral groups for application of the modified polymer as a chiral support. Similar reactions were also performed with (P)—SH and generally gave extremely high yields due to the strength of the thiolate nucleophile. In these reactions, care was taken to exclude oxygen from the reaction medium to avoid the formation of additional crosslinks through the formation of disulfide links between sites on the polymer. A number of reactions with (P)—SH are shown in Table 5.

Similarly, polystyrene resins with hydroxymethyl or mercaptomethyl functionalities have been prepared and tested in phase transfer catalyzed reactions with electrophiles such as p-nitrophenethyl bromide (Ref. 16, 17). As expected, the mercaptomethyl resin was the most reactive, giving high yields of the modified polymer while generally more modest yields were obtained with the hydroxymethyl polymer.

The thiol resins ⑫—SH and ⑫—CH₂SH have also been used in
Michael additions in the presence of tetrabutyl ammonium hydroxide
and a source of protons. A large number of Michael acceptors were
used successfully and gave the adducts in essentially quantitative
yields; Scheme 10 shows two reactions involving methyl acrylate and
cyclohexen-2-one. Practical applications of this reaction to the
extraction of unsaturated lactones from solutions have been
investigated (Ref. 61, 62).

We have also recently described the use of a polystyrene resin
with pendant sulfinate groups (Ref. 64) in phase transfer catalyzed
reactions with various electrophiles. The reactions afford
sulfones in excellent yields, often nearly quantitative, even
though sulfinates are not reputed to be very good nucleophiles. In
contrast, the same reactions without any added phase transfer
catalyst give only very low conversions. The sulfinate resins have
also been used extensively in a number of Michael additions and
have proved to be excellent for use as regenerable separation media
in the removal of allergenic substances from some plant extracts
used in the perfume and cosmetics industry (Ref. 61, 62).

Finally, we have used a cyanomethylated polystyrene resin,
⑫—CH₂CN, as a nucleophile in the preparation of a polymer-bound
dinitrile by reaction with chloroacetonitrile and base in the
presence of a phase transfer catalyst (Ref. 16, 18); the reaction
gave a very high yield of the desired product.

MISCELLANEOUS REACTIONS

A number of other chemical modifications of polymers have been
performed under phase transfer conditions including the cleavage of
peptides from a solid support in a Merrifield solid phase synthesis
(Ref. 65), and the hydrolysis of methyl methacrylate in the
presence of catalysts such as polyethylene glycol or 18-crown-6
(Ref. 66). Hradil and Svec (Ref. 67) have very recently completed
a study of the reaction of hydrolyzed copoly(glycidyl methacrylate
ethylene dimethacrylate) with propane sultone in the presence of
tetrabutyl ammonium hydroxide. While the reaction gave only 25%
yield in the absence of catalyst, a drastic improvement to 68%
conversion was observed when a phase transfer catalyst was added.
An electron microprobe study showed that, in addition to providing
higher overall yields for the reaction, the use of phase transfer
catalysis resulted in a polymer with a better distribution of the
reactive groups throughout the polymer beads rather than just at
the surface as was the case with the non-catalyzed reaction.
Although the authors did not report on the possibility of
interaction of the reactive sites leading to additional
crosslinking, or on partial hydrolysis of the polymer in the
strongly basic medium, some physical changes in the polymer

particles were observed with a small 20-33% decrease in bead
specific surface areas as well as a reduction in pore volumes.

REFERENCES

1.   "Polymer-supported Reactions in Organic Synthesis". P. Hodge
and D.C. Sherrington Eds., Wiley, London (1980).

2.   N.K. Mathur, C.K. Narang and R.E. Williams, "Polymers as Aids
in Organic Synthesis", Academic Press, New York (1980).

3.   J.M.J. Fréchet, Tetrahedron, 37, 663 (1981).

4.   A. Akelah and D.C. Sherrington, Chem. Rev., 81, 557 (1981)

5.   C.C. Leznoff, Acc. Chem. Res., 11, 327 (1978).

6.   J.I. Crowley and H. Rapoport, Acc. Chem. Res., 9, 135 (1976).

7.   J.M.J. Fréchet and M.J. Farrall, "Chemistry and Properties of
Crosslinked Polymers" (S.S. Labana, Ed.), Academic Press, New York,
59 (1977).

8.   E.V. Dehmlow, Angew. Chem. Int. Ed., 13, 170 (1974). W.P.
Weber and G.W. Gokel, "Phase Transfer Catalysis in Organic
Synthesis", Springer-Verlag, Berlin (1977).   C.M. Starks and C.L.
Liotta, "Phase Transfer Catalysis", Academic Press, New York
(1978).   L.J. Mathias, J. Macromol. Sci.-Chem., A15, 853 (1981).

9.   S.L. Regen, J. Am. Chem. Soc., 97, 5956 (1975).   D.C.
Sherrington "Polymer-supported Reactions in Organic Synthesis" (P.
Hodge and D.C. Sherrington, Eds.), Wiley, London, 157 (1980).

10. M. Takeishi, Y. Naito and M. Okawara, Angew. Makromol. Chem.,
28, 111 (1973).

11. M. Takeishi, R. Kawashima and M. Okawara, Makromol. Chem.,
167, 261 (1973).

12. J.E.L. Roovers, Polymer, 17, 1107 (1976).

13. R.W. Roeske and P.D. Gesellchen, Tetrahedron Lett., 3369
(1976).

14. M.J. Farrall and J.M.J. Fréchet, J. Am. Chem. Soc., 100, 7998
(1978).

15. M.J. Farrall, T. Durst and J.M.J. Fréchet, Tetrahedron Lett.,
203 (1979).

16. J.M.J. Fréchet, M.D. de Smet and M.J. Farrall, J. Org. Chem.,
44, 1774 (1979).

17. J.M.J. Fréchet, M.D. de Smet and M.J. Farrall, Polymer, 20,
675 (1979).

18. J.M.J. Fréchet, M.D. de Smet and M.J. Farrall, Tetrahedron
Lett., 137 (1979).

19. T. Nishikubo, T. Iizawa, K. Kobayashi, and M. Okawara,
Makromol. Chem. Rapid Commun., 1, 765 (1980).

20. T. Nishikubo, T. Iizawa, K. Kobayashi, and M. Okawara,
Makromol. Chem. Rapid Commun., 2, 387 (1981).

21. T.D. N'Guyen and S. Boileau, Tetrahedron Lett., 2651 (1979).

22. T.D. N'Guyen, A. Deffieux, and S. Boileau, Polymer, 19, 423
(1978).

23. M.B. Shambhu and G.A. Digenis, J. Chem. Soc. Chem. Commun.,
619 (1974).

24. C.H. Bamford and M. Lindsay, Polymer, 14, 330 (1973).

25. A.S. Gozdz and A. Rapak, Makromol. Chem. Rapid Commun., 2, 359
(1981).

26. A.S. Gozdz, Makromol. Chem. Rapid Commun, 2, 595 (1981).

27. A.S. Gozdz, Polymer Bulletin, 5, 591 (1981).

28. J.M.J. Fréchet, E. Bald, J. Halgas, and P. Lecavalier,
unpublished data. (b) J.M.J. Fréchet and G. Darling, unpublished
data. (c) J.M.J. Fréchet and W. Amaratunga, unpublished data.

29. P. Hodge and J. Waterhouse, Polymer Prep., 23, 142 (1982).

30. C.C. Leznoff and J.Y. Wong, Can. J. Chem., 51, 3756 (1973).

31. C.C. Leznoff and S. Greenberg, Can J. Chem., 54, 3824 (1976).

32. C. Bied-Charreton, M. Frostin-Rio, D. Pujol, A. Gaudemer, R.
Audebert and J.P. Idoux, J. Molec. Catal., in press (1982).

33. C. Bied-Charreton, J.P. Idoux and A. Gaudemer, Nouv. J. Chim.,
78, 303 (1978).

34. J.M.J. Fréchet and C. Schuerch, J. Am. Chem. Soc., 93, 492
(1971).

35. F. Camps, J. Castells, J. Font and F. Vela, Tetrahedron Lett., 1715 (1971).

36. J. Castells, J. Font, and A. Virgili, J. Chem. Soc. Perkin I, 1, (1979).

37. S.V. McKinley and J.W. Rakshys, J. Chem. Soc. Chem Commun., 134 (1972).

38. W. Heitz and R. Michels, Angew. Chem. Int. Ed., 11, 298 (1972).

39. W. Heitz and R. Michels, Liebigs Ann. Chem., 227 (1973).

40. S. Tanimoto, J. Horikawa and R. Oda, Kogyo Kagaku Zasshi, 70, 1269 (1967).

41. S. Tanimoto, J. Horikawa and R. Oda, Yuki Gosei Kagaku Kyodai Shi, 27, 989 (1969).

42. G. Markl and A. Merz, Synthesis, 295 (1973).

43. W. Takagi, I. Inoue, Y. Yano and T. Okonogi, Tetrahedron Lett., 2587 (1974).

44. A. Merz and G. Markl, Angew. Chem. Int. Ed., 12, 845 (1973).

45. Y. Yano, T. Okonogi, M. Sunaga and W. Takagi, J. Chem. Soc. Chem. Commun., 527 (1973).

46. S.D. Clarke, C.R. Harrison and P. Hodge, Tetrahedron Lett., 21, 1375 (1980).

47. P. Hodge and J. Waterhouse, Polymer, 22, 1153 (1981).

48. J.M.J. Fréchet and E. Eichler, Polymer Bulletin, in press (1982).

49. A.H. Frye, R.W. Horst, J. Polym. Sci., 40, 419 (1959).

50. J.M.J. Fréchet, J. Macromol. Sci.-Chem., A15, 879 (1981).

51. J.M.J. Fréchet, Org. Coat. Plast. Chem., 42, 268 (1980).

52. J. Lewis, M. Naqvi and G.S. Park, Makromol. Chem. Rapid Commun., 1, 119 (1981).

53. J. Lewis, M. Naqvi and G.S. Park, Polym. Prep., 23, 140 (1982).

54. T. Nishikubo, T. Ichijyo and T. Takaoka, Nippon Kagaku Kaichi, 35 (1973).

55. T.D. N'Guyen, Thesis (3eme cycle), Universite Pierre et Marie Curie, Paris 6 (1978).

56. E. Schacht, D. Bailey, and O. Vogl, J. Polym. Sci. Polym Chem Ed., 16, 2343 (1981).

57. A. Cheminat, C. Benezra, M.J. Farrall, and J.M.J. Frechet, Can J. Chem., 59, 1405 (1981).

58. J.M.J. Fréchet, M.J. Farrall, C. Benezra, and A. Cheminat, Polym Prep., 21, 101 (1980).

59. J.M.J. Fréchet, E. Eichler, C.G. Willson and H. Ito, manuscript submitted for publication.

60. H. Ito, C.G. Willson, J.M.J. Fréchet, M.J. Farrall and E. Eichler, manuscript submitted for publication.

61. J.M.J. Fréchet, A. Hagen, C. Benezra, and A. Cheminat, Pure and Applied Chem., in press.

62. J.M.J. Fréchet, A. Hagen and F. Bouchard, unpublished data.

63. M.J. Farrall and J.M.J. Fréchet, J. Org. Chem., 41, 3877 (1976).

64. A.J. Hagen, M.J. Farrall and J.M.J. Fréchet, Polymer Bulletin, 5, 111 (1981).

65. J.P. Tam, W.F. Cunningham, B.W. Erickson, and R.B. Merrifield, Tetrahedron Lett., 4001 (1977).

66. A.S. Gozdz, Makromol. Chem. Rapid Commun. 2, 443 (1981).

67. J. Hradil and F. Svec, Polymer Bulletin, 6, 565 (1982).

CHEMICAL MODIFICATION OF CHLOROMETHYLATED POLYSTYRENE WITH

PHOSPHINE OXIDES USING PHASE-TRANSFER CATALYSIS

Thanh Dung N'Guyen [a], Jean-Claude Gautier [b] and
Sylvie Boileau

[a]Laboratoire de Chimie Macromoléculaire associé au CNRS
Collège de France, 11 place Marcelin Berthelot, 75231
Paris Cédex 05, France ; [b]Centre de recherche du
Bouchet, SNPE, BP n°2, 91710 Vert le Petit, France

INTRODUCTION

Applications of soluble as well as cross-linked polymeric reagents are numerous, namely in the field of organic synthesis, metal chelation or pharmacology [1]. Chemical modification is an important procedure for the preparation of polymeric reagents. Phase
transfer catalysis which has been widely used in organic chemistry,
has found an increasing number of applications in the chemical modification of polymers having reactive functional groups, over the
past few years [2].

Tertiary phosphine oxides of the type $R_2R'PO$ are generally prepared by alkylation of the Grignard reagent $R_2P(O)Mg\ X$ formed by
reaction of diethyl phosphonate with RMgX [3-8]. Another convenient
procedure involves the reaction of alkyl halides with alkali salts
of dialkyl phosphine oxides [3,9-11]. These methods of preparation
take time and require dry reagents and dry solvents. We wish to
describe the preparation of tertiary phosphine oxides from a secondary phosphine oxide and an alkyl halide using phase transfer catalysis conditions [12-15]. After our first paper on this topic [12],
the synthesis of novel tertiary (carbamoylmethyl) phosphine oxides
using liquid-liquid phase transfer catalysis has been described [16].
We have studied in detail the reaction of dioctyl phosphine oxide
(DOPO) with benzyl chloride. Moreover, we have prepared polymers
containing pendant phosphine oxide groups by reaction of DOPO with
soluble and cross-linked chloromethylated polystyrenes as a continuation of our work on chemical modification of polymers under
PTC conditions [17,18].

EXPERIMENTAL

DOPO was prepared by reaction of diethyl phosphite with octyl magnesium chloride in refluxing THF [19] and was recrystallized in hexane. Alkyl halides, phase transfer catalysts (SO_4HNBu_4 : TBAH, dicyclohexyl-18 crown-6 (DCHE) and kryptofix [222] and [222 BB]) and solvents were pure grade commercial products. They were used without further purification. Soluble and cross-linked chloromethylated polystyrenes (Dow Chemical Co. and Fluka) were purified by dissolution or swelling in chloroform and precipitation in methanol twice before drying under high vacuum.

Typically a heterogeneous mixture containing DOPO (5.5 mmol) in toluene (15 ml), benzyl chloride (8.2 mmol), 50% aq. NaOH (5 ml) and TBAH (0.16 mmol) was stirred at 65°C for 3 h under nitrogen. After cooling, the organic layer was diluted with methylene chloride (25 ml), separated, washed with a slightly acidic aqueous solution (30 ml of HCl 0.1 N) and with water, and dried over $MgSO_4$. The solvents were removed under reduced pressure to give a 100 % yield of benzyl dioctyl phosphine (based on DOPO).

The procedure for the chemical modification of polymers was previously described [17,18]. Composition of the polymers was determined by elemental analysis and the degree of substitution of chloromethylated groups was calculated from the percentages of P and of remaining Cl. It is necessary to carry out the elemental analysis of the modified polymers under carefully controlled conditions in order to avoid moisture because these polymers are hygroscopic. Molecular weights of the soluble samples were measured by osmometry in toluene at 37°C.

RESULTS AND DISCUSSION

Results of the reaction of DOPO with some alkyl halides under PTC conditions are shown in Table 1. In the case of $\emptyset CH_2Cl$, no reaction occurs in the absence of catalyst even after 6 h at 80°C whereas a quantitative yield of benzyl dioctyl phosphine oxide is obtained with a small amount of catalyst.

The rate of chlorine substitution in $\emptyset CH_2Cl$ by DOPO, in the toluene phase, was followed by VPC :

$$\emptyset CH_2Cl \; + \; [\,(Octyl)_2PO^-,Q^+\,] \; \xrightarrow{\;k\;} \; (Octyl)_2P(O)CH_2\emptyset \; + \; Cl^-,Q^+$$

The rate of disappearance of $\emptyset CH_2Cl$ is equal to :

$$-d[\emptyset CH_2Cl]\,/dt = \; k\,[\emptyset CH_2Cl]^x\,[(Octyl)_2PO^-,Q^+]^y \tag{1}$$

Table 1.   Reaction of DOPO with alkyl halides under
           PTC conditions at 65°C
(DOPO/toluene/50% aq. NaOH = 5.5 mmol/15 ml/5 ml ;
[DOPO]/[RX] = 1/1.5 ; [DOPO]/[XR'X]= 2.5/1 ;
[TBAH]/[DOPO] = 0.05)

Run	RX	Time h	Yield %
1 [a)	ϕCH_2Cl	6	0
2 [b)	ϕCH_2Cl	3	100
3 [c)	$ClCH_2-C_6H_4-CH_2Cl$	3	98
4	$CH_2=CH-CH_2Br$	4	61
5	$ClCH_2-CH_2-CN$	5	74
6	$Cl(CH_2)_7CH_3$	6	39

[a) without catalyst, 80°C ; [b) [TBAH]/[DOPO] = 0.03 ; [c) [TBAH]/
[-CH_2Cl] = 0.05

If we assume that x = 1 and $[(Octyl)_2PO^-,Q^+]^y$ = Cte, equation
(1) becomes after integration :

$$L_n \frac{[\phi CH_2Cl]_o}{[\phi CH_2Cl]} = k't \qquad (2)$$

Plots of $L_n \dfrac{[\phi CH_2Cl]_o}{[\phi CH_2Cl]}$ versus time are shown in Fig. 1 for experi-

ments made with different catalysts in the following conditions :
toluene/50% aq. NaOH : 15ml/5ml; [DOPO] = 5.47 mmol ; [ϕCH_2Cl] =
8.20 mmol ; [catalyst] = 0.16 mmol ; θ = 65°C. Straight lines are
observed, the slope of which gives k'. This result indicates a
simple kinetic behaviour for this type of reaction. A first order
with respect to ϕCH_2Cl concentration is observed and the concentration of active species derived from DOPO remains constant. This
last point is not true after 1 h reaction with quaternary ammonium
salts as catalysts, presumably because a decomposition of this
type of compounds occurs. The efficiency of catalysts follows the
order :
               TBAH > [222] > [222 BB] > DCHE

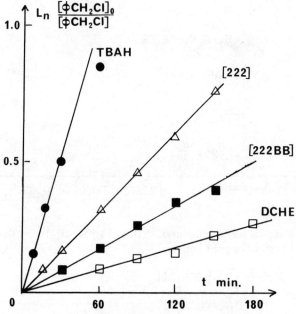

Fig. 1.   Plot of L_n $(\emptyset CH_2Cl)_o/(\emptyset CH_2Cl)$ versus time in the organic phase.

The results concerning the reaction of DOPO with chloromethylated polystyrenes under PTC conditions are shown in Table 2. Degrees of substitution are quite high and depend on the experimental conditions : catalyst, reaction time, stirring speed. The main absorption bands observed on the IR spectra of the modified polymers are located at 1165 cm^{-1}, $\nu(P=O)$, and between 2860 and 2960 cm^{-1} which correspond to the octyl groups.

The degree of functionalization D.F. which is the fraction of aromatic rings carrying the substituents in the polymer remains nearly constant after the chemical modification with DOPO under the PTC conditions used here.

Moreover, the molecular weights $\overline{M}_n$ measured by osmometry and the composition of the polymers treated under PTC conditions have been examined. The results are shown in Table 3.

In the absence of substituent, there is no significant change of $\overline{M}_n$ when the polymer is stirred in a mixture of toluene and 50% aq. NaOH at 60°C for 3 h, in the absence as well as in the presence of TBAH (runs 0,11 and 12). Composition of the polymer remains constant when no catalyst is added (run 11) whereas some chloromethyl groups are transformed into CH_2OH groups in the presence of catalyst (run 12). This is due to the reaction of CH_2Cl groups with tetrabutylammonium hydroxide. It is very important to note that this reaction does not occur in the presence of the substituent (run 7). Moreover, in the case of run 7, there is a good agreement between $\overline{M}_p$ measured by osmometry and $\overline{M}_p$ calculated from the value of unmodified polymer and the degree of substitution.

These results indicate that no appreciable degradation and side reactions occur under the PTC conditions we have employed.

In conclusion, we have presented a new method for the preparation of tertiary phosphine oxides of the type $R_2R'PO$ from a secondary phosphine oxide and an alkyl halide with excellent yields, using phase transfer catalysis. Furthermore, this method can be successfully applied to the chemical modification of soluble and cross-linked polystyrenes. Examination of the molecular weights, $\overline{M}_n$, and of the composition of the modified polymers indicates that no side reactions or degradation occur in the conditions we have used. Further work in this field is in progress.

Table 2

Reaction of DOPO with chloromethylated polystyrenes under PTC conditions at 65°C (DOPO = 1.76 mmol ; toluene = 50 ml ; polymer = 1.4 mmol Cl; 50% aq. NaOH = 4 ml ; magnetic stirring).

Runs	PSt-CH$_2$Cl D.F. a)	Catalyst (mol %)[b]	Time h	D.S. [c]	Functional yield % [d]
7	0,34 (sol.)	TBAH (5)	3	57	100
8 [e]	0.34 (sol.)	[222] (5)	16	77	100
9 [f]	0.33 (sol.)	TBAH (5)	3	86	100
10	0.67$_5$ (resin)	TBAH (20)	6	95	98

a) Degree of functionalization ; b) fraction of aromatic rings carrying the substituents ; c) from the Cl content of polymer ; d) based on P and Cl contents of the polymer ; d) (Final D.F./ Initial D.F.) x 100 ; e) 60°C ; f) 400 r.p.m.

Table 3

Molecular weights, $\overline{M}_n$, and composition of polymers treated under PTC conditions.

Run	Treatment	$\overline{M}_n$ osm.	Composition of polymers a)			
			a	b	c	d
0	no	174 000	1.90	1.00	0.0	—
11	Toluene/50% aq. NaOH, 3 h, 60°	190 000	1.90	1.00	0.00	—
12	Toluene/50% aq. NaOH, 5 mol % TBAH, 3h, 60°C	180 000	1.90	0.80	0.18	—
7	Toluene/50% aq. NaOH, [DOPO]/[-CH$_2$Cl] = 1.26, 5 mol % TBAH, 3 h, 65°C	235 000 b)	1.90	0.43	0.00	0,57

$$(CH_2-CH)_a(CH_2-CH)_b(CH_2-CH)_c(CH_2-CH)_d$$

a) determined from elemental analysis :

b) theoretical value calculated from $\overline{M}_n$ osm. of unmodified polymer and the degree of substitution : 242 000

REFERENCES

1.   For reviews see : K. Geckeler, V. N. R. Pillai and
      M. Mutter, Adv. Polym. Sci., 39, 65 (1981) ; P. Hodge
      and D.C. Sherrington, "Polymer-supported Reactions in
      Organic Synthesis", John Wiley and Sons, New York, 1980;
      N. K. Mathur, C. K. Narang and R.E. Williams, "Polymers
      as Aids in Organic Chemistry", Academic Press, New-York,
      1980.
2.   See for instance : J. M. Fréchet, present book.
3.   G.M. Kosolapoff and L. Maier, "Organic Phosphorous
      Compounds", Wiley Interscience, New York, vol. 3,
      1972, p.368.
4.   J. J. Richard and C.V. Banks, J. Org. Chem., 28
      123(1963).
5.   T. H. Sidall III and M. A. Davis, J. Chem. and Eng.
      Data, 10,303 (1965).
6.   I. M. Downie and G. Morris, J. Chem. Soc., 5771(1965).
7.   H. M. Priestley, J. Chem. and Eng. Data, 12, 618(1967).
8.   H.R. Hays,J. Org. Chem., 33,3690(1968).
9.   A. K. Hoffmann and A. G. Tesch, J. Am. Chem. Soc.,
      81, 5519(1959).
10.   L. Horner, P. Beck and V. G. Toscano,Chem.Ber.,94,1317(1961).
11.   R. B. Wetzel and G. L. Kenyon, J. Am. Chem. Soc.,
      94, 1774(1972) and J. Org. Chem., 39, 1531(1974).
12.   S. Boileau, J. C. Gautier and T. D. N'Guyen, Eur. Pat.
      31 761 (1980).
13.   T. D. N'Guyen, Thèse Doctorat d'Etat, Paris (1981).
14.   T. D. N'Guyen, J.C. Gautier and S. Boileau, Polym.
      Prep., 23(1), 143(1982).
15.   T. D. N'Guyen, S. Boileau and J. C. Gautier, Bull. Soc.
      Chim. France, in press.
16.   K. M. Kenn, N. V. Nguyen and D. J. Cross, J. Org. Chem.,
      46, 5188(1981).
17.   T. D. N'Guyen, A. Deffieux and S. Boileau, Polymer,
      19, 423 (1978).
18.   T. D. N'Guyen and S. Boileau, Tetrahedron Letters,
      2651 (1979).
19.   R. H. Williams and L. A. Hamilton, J. Am. Chem. Soc.,
      74,5418(1952) and 77, 3411 (1955).

PHASE TRANSFER CATALYSED POLYMER-SUPPORTED WITTIG REACTIONS

Charles R. Harrison, Philip Hodge,* Barry J. Hunt,
Ezzatollah Khoshdel, and Janette Waterhouse
Chemistry Department
University of Lancaster
Lancaster, Great Britain

In recent years organic chemists have shown considerable interest in polymer-supported reactions.[1,2] Such reactions have several attractive features one of which is that isolation of the products is simplified because the supported species can easily be filtered off from the non-supported species. This article is concerned with olefin synthesis by phase transfer catalysed polymer-supported Wittig reactions. Two main types are considered, those in which the phosphine is the supported species (see Scheme 1) and those in which the alkyl halide is the supported species (see Scheme 2). An example of the third type in which the carbonyl compound is the supported species is also given.

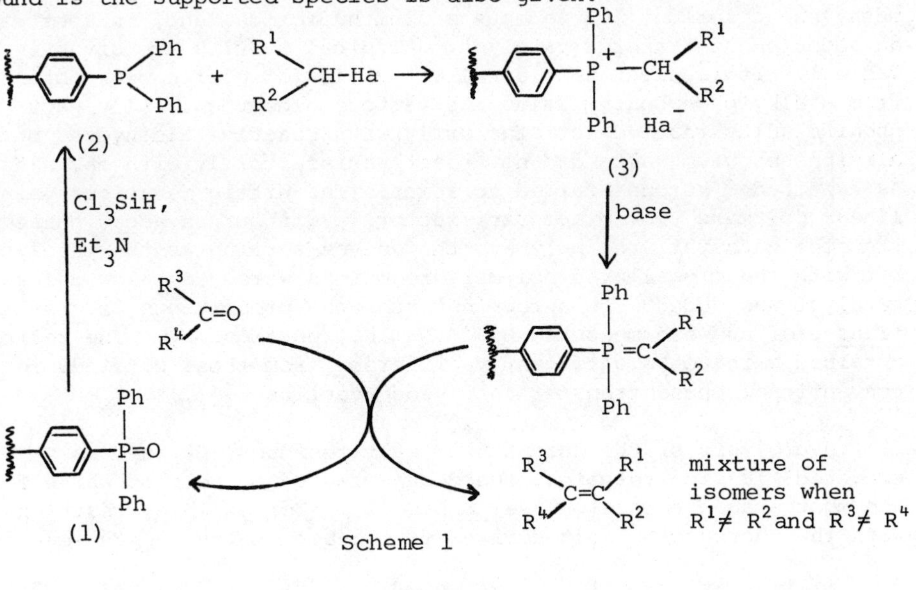

Scheme 1

35

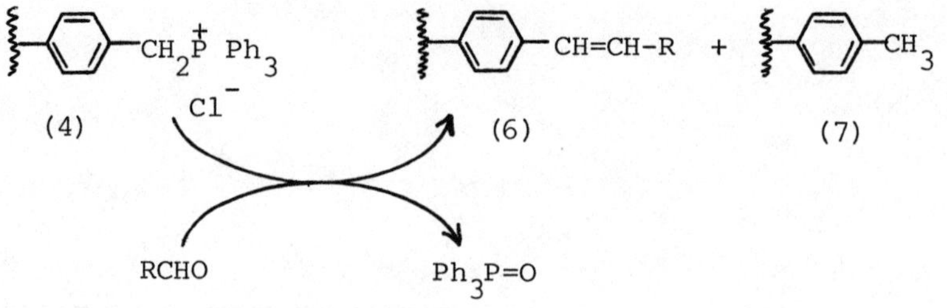

Scheme 2

Polymer-supported Wittig reactions of the type outlined in Scheme 1 have been studied previously by several research groups,[3,4,5] but the reactions were not phase transfer catalysed. One advantage[3] of such Wittig reactions is that the olefin produced is easily separated from the phosphine oxide by-product (1). Linear or cross-linked polymers containing phosphine residues (2), prepared by the chemical modification of preformed polystyrenes[3,4,6] or by copolymerisation,[4] reacted smoothly with alkyl halides to give the phosphonium salts (3). Treatment of the latter with various bases, for example, n-butyl lithium in dioxan[3] or sodium dimsyl in dimethyl-sulphoxide/tetrahydrofuran,[4] gave the ylids and these reacted with the carbonyl compounds to give olefins. Although no systematic attempts were made to optimise the reaction conditions, the yields of olefin were generally good.[7] The spent polymer was successfully regenerated by reaction with trichlorosilane and triethylamine.[3]

Phase transfer catalysed reactions of the type outlined in Scheme 1 were carried out by vigorously stirring together a mixture of the supported phosphonium salt (3), the carbonyl compound in methylene chloride, 50% aqueous sodium hydroxide, and, in most cases, an added phase transfer catalyst.[8] Typical results are given in Table 1. It is clear that excellent yields of olefin were obtained from aralkylphosphonium salts and various aldehydes. Allylphosphonium salts reacted satisfactorily with reactive aldehydes, but alkylphosphonium salts did not react satisfactorily with any aldehyde. As expected,[9] ketones failed to react. The Wittig reactions using linear polymers took place satisfactorily without an added phase transfer catalyst, the polymers themselves serving as the catalyst but with the crosslinked polymers reactions were very slow unless a catalyst was added. In a few instances the proportions of cis- and trans-olefins were measured by [1]H N.M.R. spectroscopy. The ratios obtained were not significantly different from those obtained in conventional phase transfer catalysed reactions.[9]

Hydrolysis of the phosphonium salt residues (3) was, as expected,[9] a side reaction, the group cleaved being that which formed the most stable carbanion (see Scheme 3). For example, reactions with the phosphonium salt derived from 2-bromomethylnaphthalene gave

Table 1.　Reactions of Polymer-supported Phosphonium Salts with Carbonyl Compounds under Phase-transfer Conditions (See Scheme 1)[a]

Halide used to prepare Phosphonium Salt	Polymer[b]	Carbonyl Compound	Catalyst[c]	Reaction Time	Yield of Olefin[d]
Benzyl chloride	C	9-formyl-anthracene	C	2h.	98%[e]
Benzyl chloride	C	9-formyl-anthracene	none	2h.	35%[f]
Benzyl chloride	L	9-formyl-anthracene	none	2h.	92%[e]
Benzyl chloride	C	$p\text{-}CH_3\cdot C_6H_4\cdot CHO$	C	16h.	100%[g]
Benzyl chloride	C	$p\text{-}Cl.C_6H_4\cdot CHO$	T	4h.	97%[h]
Benzyl chloride	C	ferrocene carboxaldehyde	C	3h.	42%
Benzyl chloride	L	$C_6H_5.CH{=}CH.CHO$	none	2h.	75%[f]
Benzyl chloride	C	$n\text{-}C_7H_{15}.CHO$	T	3h.	93%[f]
Benzyl chloride	C	cyclohexanone	T	3h.	0
4-t-Butylbenzyl chloride	C	formaldehyde	C	3h.	95%[i]
2-Bromomethyl-naphthalene	C	β-naphthaldehyde	C	3h.	65%[j]
2-Bromomethyl-naphthalene	L	9-formyl-anthracene	none	2h.	100%
9-Chloromethyl-anthracene	C	benzaldehyde	C	3h.	20%[k]
Allyl bromide	C	$p\text{-}Cl.C_6H_4\cdot CHO$	none	17h.	78%
Allyl bromide	C	9-formyl-anthracene	T	4h.	trace
n-Hexyl bromide	X	$p\text{-}Cl.C_6H_4\cdot CHO$	T	4h.	trace
Methyl iodide	C	$p\text{-}Cl.C_6H_4\cdot CHO$	T	19h.	0

Notes for Table:

[a] A mixture of the phosphonium salt (1.5 mmol.), the carbonyl compound (1.0 mmol.), methylene chloride (10ml.), 50% aqueous sodium hydroxide (3ml.), and, if necessary, a catalyst (0.03 mmol.) was vigorously stirred under N_2 at 20°.　　　　　(continued)

b
  C = Polystyrenes crosslinked with 1% DVB with 3.0-3.5 mmol/g of phosphine.
  L = Linear polystyrene, MW = 150,000, with 2.7 mmol/g of phosphine
  X = Amberlite XE-305, a macroporous polystyrene, with 2.3 mmol/g of phosphine.

c
  C = Cetyltrimethylammonium bromide. T = Tetrabutylammonium iodide.

d
  Unless indicated otherwise the yields are based on the carbonyl compound and are of isolated material with satisfactory ^{1}H nmr spectrum and/or physical properties.

e
  Only trans-olefin formed.

f
  Determined by ^{1}H nmr analysis.

g
  Cis: trans-ratio, 43:57.

h
  Cis: trans-ratio, 56:44.

i
  Plus 16% 4-t-butyltoluene   )

j
  Plus 4% β-methylnaphthalene) Yields based on phosphonium

k
  Plus 80% 9-methylanthracene) salt

some 2-methylnaphthalene. Hydrolysis was relatively unimportant when benzyl phosphonium salts were used because high yields of olefin were obtained when only 1.5 molar equivalents of salt was used and any toluene formed was readily removed by evaporation.

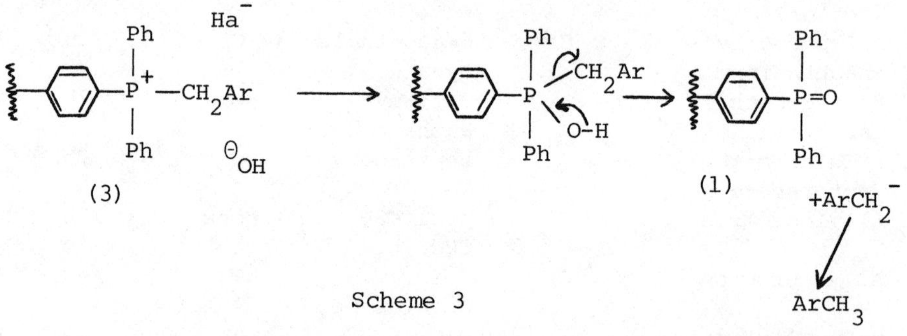

Scheme 3

Since the above phase transfer catalysed reactions worked particularly well with phosphonium salts prepared from benzyl halides, it was considered of interest to investigate reactions using phosphonium salts (4) prepared from chloromethylated polystyrenes, that is reactions where the supported species was the alkyl halide (see Scheme 2). Both linear and crosslinked chloromethylated polystyrenes reacted smoothly with triphenylphosphine to give polymers with residues (4).[10] An alternative way of preparing the linear polymer is by copolymerisation of styrene and the salt (5).[11] When the salts (4) were treated with various aldehydes in methylene chloride and

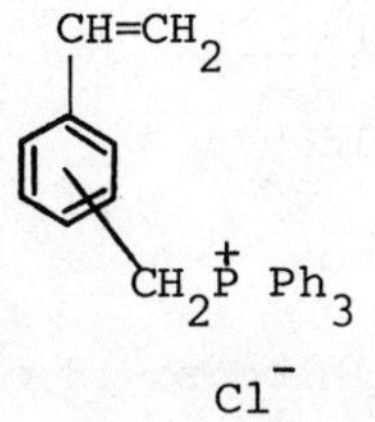

$$CH=CH_2$$

$$CH_2\overset{+}{P} Ph_3$$

$$Cl^-$$

(5)

50% aqueous sodium hydroxide at 20°, Wittig reactions occurred to give polymers with olefinic residues (6).[10] Some typical results are summarised in Table 2. As in the earlier reactions no added phase transfer catalyst was required when linear polymers were used,

Table 2.   Phase Transfer Catalysed Wittig Reactions (see Scheme 2)

Polymer[a]	Carbonyl Compound[b]	Reaction Time (h)	Yield[c] (%)
Using Aqueous Sodium Hydroxide[d]			
2	Formaldehyde	60	13
L	Benzaldehyde	3	55
2	Benzaldehyde	52	76
L	p-Chlorobenzaldehyde	3	40
2	p-Chlorobenzaldehyde	17	97
2	p-Nitrobenzaldehyde	23	49
1	Thiophen-2-aldehyde	70	83
2	Ferrocene-2-carboxaldehyde	36	62
2	p-Diphenylphosphinylbenzaldehyde	48	80
1	4-Formylbenzo-15-crown-5	48	ca.50
1	4,4'-Diformyldibenzo-18-crown-6	60	ca.25
1	4-t-Butylcyclohexanone	72	0
1	Thiophen-2-aldehyde and p-Nitrobenzaldehyde	48	65 27

(continued)

Using Solid Potassium Carbonate[e]

1	p-Bromobenzaldehyde	30	79
2	Ferrocenecarboxaldehyde	21	54
1	4,4'-Diformyldibenzo-18-crown-6	48	ca. 40
1	p-Bromoacetophenone	52	17

Notes for Table:

[a] L = Linear polystyrene (Mn = 111,000) containing 1.5 mmol. of residues (4) per g.
1 = 1% Crosslinked polystyrene containing 1.3-2.0 mmol. of residues (4) per g.
2 = 2% Crosslinked polystyrene containing 1.5-2.0 mmol. of residues (4) per g.

[b] Mole ratio of formaldehyde to residues (4) was ∿17:1. With other carbonyl compounds the mole ratio was 1.0-1.5:1.

[c] Determined by elemental analyses and/or weight changes. The yields are those for the Wittig reaction. The rest of the residues (4) were hydrolysed to residues (6).

[d] The carbonyl compound in methylene chloride, the polymer, and 50% aqueous sodium hydroxide solution were vigorously stirred at 20°. With linear polymers no catalyst was added but with crosslinked polymers cetyltrimethylammonium bromide was added.

[e] The carbonyl compound in tetrahydrofuran, the polymer, solid potassium carbonate and 18-crown-6 or dicyclohexyl-18-crown-6 were heated under reflux.

but with crosslinked polymers the reactions were very slow in the absence of an added catalyst. Hydrolysis of the phosphonium salt was again a competing reaction and in this system those phosphonium salt residues that did not take part in the Wittig reaction were converted into hydrocarbon residues (7).

These Wittig reactions can be used to introduce various interesting residues into polystyrenes. By using formaldehyde, polymers with styryl residues were obtained which could be useful for preparing graft polymers.[12] By using p-diphenylphosphinylbenzaldehyde, a polymer with phosphine residues (8) was obtained which could be useful for supporting transition metal complex catalysts.[13] By formylating benzo-15-crown-5 and dibenzo-18-crown-6 and using the products in the Wittig reaction, polymers with pendant crown ether residues were obtained, for example (9), which can be used as polymer-

(8)   (9)

supported phase transfer catalysts. By using an equimolar mixture
of thiophen-2-aldehyde and p-nitrobenzaldehyde, a polymer containing
both thienyl and 4-nitrophenyl residues (mole ratio, 70:30) was obtained demonstrating that two residues can be introduced in the one
reaction.

The advantages of using the phase transfer catalysed Wittig
reactions to modify chloromethylated polystyrenes are as follows.

i) The reaction solvent need not be dry, a cheap base is used, and
the reactions proceed at 20°.

ii) The groups introduced are bound to the polymer by a linkage that
is stable to both acid and base.

iii) Groups (4) that do not react to give olefinic residues (6), give
methyl groups (7) and the latter will not interfere in subsequent
applications.

iv) The groups are attached via a short 'spacer arm'.

v) More than one group can be introduced in the same reaction.

Reactions of the type outlined in Scheme 2 were also carried
out by stirring crosslinked polymer (4) with solid potassium carbonate and the carbonyl compound in tetrahydrofuran in the presence
of 18-crown-6 or dicyclohexyl-18-crown-6. These reactions were generally slower than those discussed above but in some cases better
yields were obtained (see Table 2). Recently we have found that
these Wittig reactions, or at least those with p-bromobenzaldehyde
and ferrocenecarboxaldehyde, can also be carried out without any
added catalyst. This surprising result is similar to some others
recently reported.[14] Although efforts were made to thoroughly dry
the tetrahydrofuran and potassium carbonate before use, it is likely
that traces of water were present. The result suggests that potassium carbonate is sufficiently soluble in tetrahydrofuran under
these conditions to reach the polymer beads and then to diffuse into
them. It is interesting to note that if the base initially reacts
with the more accessible phosphonium residues a "layer" free of
ionic groupings will be formed, but that even so in the later stages
of the reaction the potassium carbonate can still diffuse through
this "layer" without the assistance of a crown ether. Also, that
although crown ethers may help the potassium carbonate to pass into
solution, the complexes formed will be bulky and may have some

difficulty diffusing into the polymers. These points may have
implications for the mechanism of triphase catalysis.[15]

A current problem in studies of the chemical modification of
crosslinked polymers is in fully characterising the products. A
novel method for the preparation of a functionalised crosslinked
polymer would be to prepare a suitable functionalised linear polymer,
characterise it, and then, as a final step, crosslink it. The linear
polymer could be prepared by either free radical, anionic, or cationic polymerisation and could be characterised by all the usual
spectroscopic and other methods. Phase transfer catalysed reactions
are particularly attractive for the crosslinking because in a two-
phase liquid system there is, as in suspension polymerisation, the
possibility of preparing beads. As a preliminary experiment a
linear polymer containing phosphonium salt residues (4) was reacted
with terephthaldehyde using the usual methylene chloride - 50% aqueous sodium hydroxide system in the presence of an added phase
transfer catalyst. The product, obtained in good yield, was insoluble in all solvents tried though it swelled considerably in
several. The infrared spectrum was consistent with the product containing residues (6), (10), and (11). A polymer with similar properties was prepared by carrying out a Wittig reaction between a
linear polymer with aldehyde residues (12) and the bisphosphonium
salt (13). Note that this last reaction is an example of the third
type of polymer-supported Wittig reaction, in which the carbonyl
compound is the supported species. Other examples of phase transfer
catalysed crosslinking reactions we have studied are the cross-
linking of chloromethylated linear polystyrene by reaction with
quinol and aqueous sodium hydroxide or by reaction with bisphenol A
and aqueous sodium hydroxide. These reactions afford products containing crosslinks (14) and (15) respectively.

(10)   (11)

(12)   (13)

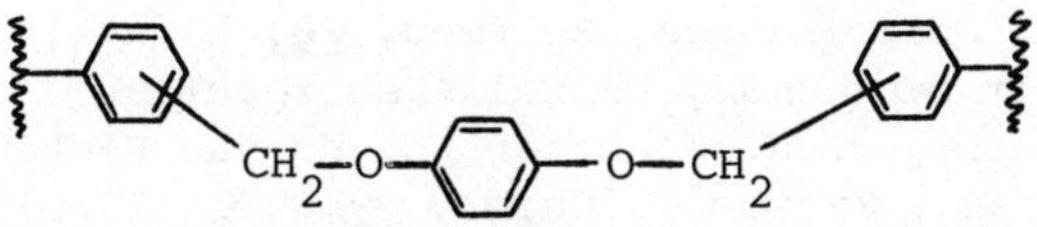

(14)

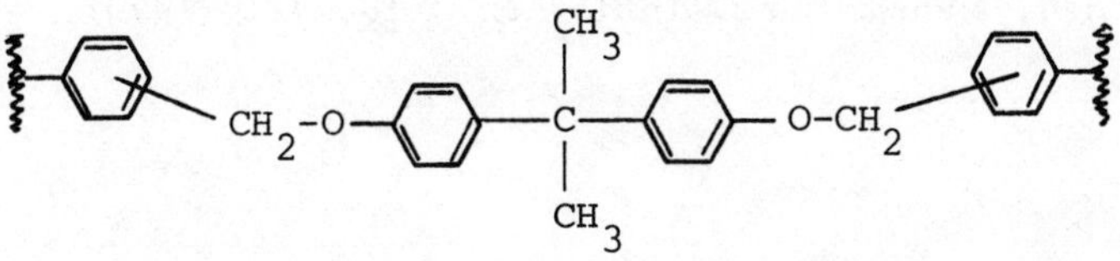

(15)

It is possible that similar crosslinking reactions could be carried out using solid potassium carbonate as the base in tetrahydrofuran or another suitable solvent, with or without an added catalyst. If only a small amount of solvent was needed, there is the possibility of using the reactions for a type of reaction moulding. Studies of such reactions are currently underway.

ACKNOWLEDGEMENT

We thank the Science and Engineering Research Council for generous financial support.

REFERENCES

1. P. Hodge and D. C. Sherrington, Eds. "Polymer-supported Reactions in Organic Synthesis", Wiley, London, 1980.
2. N. K. Mathur, C. K. Narang, and R. E. Williams, "Polymers as Aids in Organic Chemistry", Academic Press, New York, 1980.
3. W. Heitz and R. Michels, Liebigs Ann. Chem., 227 (1973).
4. S. V. McKinley and J. W. Rakshys, J. Chem. Soc. Chem. Comm., 134 (1972).
5. J. Castells, J. Font, and A. Virgili, J. Chem. Soc. Perkin I, 1 (1979) and references therein; W. Heitz and R. Michels, Angew. Chem. Int. Edn., 11, 298 (1972); and S. D. Clarke and P. Hodge, unpublished results.
6. H. M. Relles and R. W. Schluenz, J. Amer. Chem. Soc., 96, 6469 (1974).
7. P. Hodge in ref. 1, p.139-146.
8. S. D. Clarke, C. R. Harrison, and P. Hodge, Tet. Lett., 21, 1375 (1980).
9. W. Takagi, I. Inoue, Y. Yano, and T. Okonogi, Tet. Lett., 15, 2587 (1974) and G. Märkle and A. Merz, Synthesis, 295 (1973).

10. P. Hodge and J. Waterhouse, Polymer, 22, 1153 (1981).

11. P. Hodge and A. Wightman, unpublished results.

12. See, for example, J. H. Schutter, C. H. van Hastenberg, P. Piet, and A. L. German, Makromol. Chem., 89, 201 (1980).

13. C. U. Pittman, Chapter 5 in ref. 1.

14. S. Yanagida, K. Takahashi, and M. Okahara, J. Org. Chem, 44, 1099 (1979); W. M. MacKenzie and D. C. Sherrington, Polymer, 21, 791 (1980); M. Schneider, J.-V. Weber, and P. Faller, J. Org. Chem., 47, 364 (1982).

15. S. L. Regen, Angew. Chem. Intl. Edn., 18, 421 (1979).

INFLUENCE OF QUATERNARY AMMONIUM SALTS ON CELLULOSE MODIFICATION

William H. Daly, John D. Caldwell, Kien Van Phung
and Robert Tang

Department of Chemistry
Louisiana State University
Baton Rouge, Louisiana   70803

INTRODUCTION

Current interest in cellulose chemistry is focused on two
areas; a) development of cellulose solvents to provide direct
methods for solution spinning and b) development of techniques to
control distribution of substituents along the cellulose backbone.
Homogeneous substitution,i.e., random distribution of substituents
among all the anhydroglucose units in the chain,can be accomplished
by initial dissolution of the cellulose substrate followed by
reaction in a homogenious media.[1] However,solutions of native
cellulose are extremely viscous, so it is unlikely that substrate
concentration levels high enough to allow commercially feasible
production of derivatives could be employed.

Direct solid phase modification is the most practical
approach to cellulose derivation, but the products tend to exhibit
a heterogeneous distribution of substituents. For example,
cellulose ethers are produced in a multistage process. Initially,
native cellulose is treated with caustic solution($>$40%) to disrupt
the crystallinity, to swell the substrate, and to generate active
sites (sodium cellulosate). Efficient dispersion of the caustic
solution is essential if a product with a homogeneous distribution
of substituents at low degrees of substitution is desired. Pretreated cellulose is a swollen, extremely hydrophilic mass through
which hydrophobic reagents must penetrate to complete the ether
synthesis. Following the initial displacement step, the anhydroglucose unit becomes more hydrophobic and additional reagent is

45

attracted to the same site; further reaction is favored at neighboring repeat units and non-random substitution occurs. A phase
transfer catalyst could influence each of these steps by: a)
improving the caustic dispersion b) transfering the reagent into
the cellulose phase, and c) converting the cellulosate counterion
to a more hydrophobic unit that would attract hydrophobic
reagents.

RESULTS AND DISCUSSION

Phase transfer catalysts have been employed extensively in
organic synthesis,[2] but only limited applications in cellulose
modification have been reported.[3] Initially we elected to study
cellulose benzylation primarily to minimize the experimental
difficulties, i.e.,high pressure reactions, associated with
production of the more common derivatives, methyl and ethyl
cellulose. Benzyl cellulose has been the subject of many
intermittant investigations.[4] Lorand and Georgi[5] demonstrated
that benzylation is diffusion rate limited, i.e., the swollen
surface inhibited further reaction within a given cellulose
particle. The diffusion problems occurred at temperatures
exceeding 60° the benzyl chloride diffusion rate did not increase
as fast as the substitution rate, and heterogeneous mixtures
containing unreacted cellulose particles were obtained at higher
temperatures. Addition of quat salt decreased the diffusional
barrier and promoted high reaction rates at 100°. A more
homogeneous derivative is produced; clear toluene solutions of
derivatives with D. S. > 2 were obtained. A similar demonstration
of reagent phase transfer occurs in cyanoethylation of cellulose.

Catalyst Selection

Dockx has suggested that phase transfer catalysts enhance the
concentration of hydroxide ion in the non-aqueous phase.[6] In the
case of cellulose modification, this would accelerate undesirable
side reactions of the reagents without promoting any interaction
with the hydrophilic cellulose phase. Our data is consistent with
this interpretation when long chain tetraalkylammonium salts are
employed. If the alkyl chain length exceeds four carbon atoms,
lower degrees of substitution are achieved in the presence of
catalyst than are obtained in control experiments. A corresponding increase in the concentration of hydrolysis products
confirms that the main activity is in the non-aqueous phase.

Application of phase transfer catalysis to cellulose
modification requires selection of a catalysis which promotes
reactions in the hydrophilic phase. Short chain quaternary salts

exhibit minimal solubility in organic solvents, and are not considered to be effective catalysts for organic reactions. However, we have observed that tetramethylammonium chloride (TMAC) doubles the rate of benzylation and increases the extent of substitution by four-fold.[3] A more modest 50% increase in the D.S. is observed when tetraethylammonium chloride is employed as a catalyst. These results suggest that the catalyst is replacing the cellulosate counterions and is decreasing the hydrophilicity of the active sites. The tetraammonium counterion exerted a negligible influence on the catalyst activity.

Quat salts promote the transfer of caustic into the cellulose phase. Studies of both benzylation and cyanoethylation provide evidence for caustic binding. Addition of more than 1.0 ml of 50% sodium hydroxide per gram cellulose to control benzylation experiments decreases the extent of substitution and promotes hydrolysis of the reagent. In the presence of TMAC a significant increase in the initial benzylation rate is observed at a ratio of 2.0 ml/gm, indicating that a higher concentration of active cellulosate sites form. Higher caustic ratios reduce the initial rate but higher D. S. products are produced eventually. Apparently these high caustic loadings increase the hydrophilicity of the activated cellulose and reduce the rate of benzyl chloride diffusion into the matrix. This effect is accompanied by rapid hydrolysis of the benzyl chloride. Thus, the optimum caustic/cellulose ratio must be determined experimentally for each different reagent.

The addition sequence: quat salt, caustic solution, pretreatment, then reagent is critical. Addition of quat salts to caustic pretreated cellulose does not produce a significant enhancement of the extent of substitution. Similarly, addition of tertiary amines assuming that the corresponding quaternary salt of the reagent will be produced in situ is ineffective. Treatment of native cellulose with tetramethylammonium hydroxide alone did not produce an activated substrate.

The most significant variables governing the extent of cellulose alkylation are: the nature of the quat salt, the cellulose/catalyst molar ratio, reagent concentration, concentration and molar ratio of caustic, and reaction temperature. The extent of substitution in either benzylation or ethylation under a given set of reaction conditions is directly proportional to the cellulose/catalyst ratio. A 10:1 ratio was used in this work, but significant effects are observed at ratios as low as 100:1. The highly swollen appearance of cellulose pretreated in the presence of high catalyst concentrations suggests that the quat salts are penetrating the fiber matrix and disrupting the crystallinity. However, the catalytic effect of TMAC is less pronounced when

highly crystalline cotton linters are used as a cellulose
substrate. More intense treatment conditions would be required to
activate crystalline substrates.

Ethylation

We have surveyed a series of different catalyst types under
relatively standard ethylation conditions.[7]   Ethylation of
cellulose can be conducted in a two stage process, which involves
a recharge of caustic and ethyl chloride and a temperature
increase when the D.S. reaches 1.4. Wood cellulose was treated
with a fine spray of concentrated catalyst solution followed by
hot 50% caustic. The mixture was either heated in toluene or dry-
blended to produce the activated cellulose, which was charged into
a bomb reactor along with additional sodium hydroxide and ethyl
chloride. The reactor was immersed in a fluidized bed preheated
to the desired temperature, and the bath temperature was held
within a five degree range during the course of the reaction.
The extent of substitution obtained after 2-2.5 hours reaction
time is reported in Table I. The pretreatment conditions
influence the relative effectiveness of the quat salt, but
comparisons within a given pretreatment are valid.

We observed that the addition of TMAC shortened the time
required for the first stage and, depending upon the temperature,
it was possible to produce a fiber-free intermediate product with
a D.S. up to 1.95. The intermediate product isolated from control
runs exhibited extensive fiber structure and the D.S. did not
exceed 1.45. After recharging the reactor, the second stage was
completed in 2 hours in contrast to the 5 hours normally required
to reach a minimal D.S. of 2.6.

A detailed study of the first stage of ethylation was
conducted to ascertain the influence of temperature,and reaction
time on the efficiency of conversion. The temperature effect on
the extent of substution attained in two hours is illustrated in
figure 1. No substitution was observed at temperatures below 80°.
The products produced at $80-90^{\circ}$ were heterogeneous, partially
water soluble, mixtures. Raising the temperature to $100-110^{\circ}$
produced substitutions comparable to those observed at 145°
without catalyst. In fact, the final product produced after two
stages at 110° exhibited a D.S. of 2.7 and a higher viscosity
($[\eta]= 85$ cp vs $[\eta]= 22$ cp) than ethyl cellulose produced at higher
temperature and longer reaction times. Thus, reactions conducted
in the presence of TMAC may be effected at lower temperatures, and
the concommitant cellulose degradation is reduced.

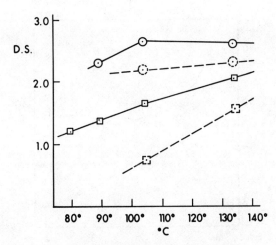

Fig. 1.   Influence of temperature on extent of ethylation;
           ▢ first stage,☉ , second stage,
           ————— catalysed, —————— control.

The influence of reaction time at 110° on ethyl chloride
consumption, D.S. and ethylation efficiency is depicted in figure 2.
The composition of the organic by-product mixture was monitored by
gas chromatography; the accuracy of this technique was not high
but general trends could be ascertained. The ratio of ethyl ether
to ethanol was approximately 10:1 indicating that hydrolysis of
ethyl chloride was slow relative to the rate of ether formation.
Therefore, two equivalents of base were consumed per mole of by-
product formed. Titration of residual sodium hydroxide provided a
technique for evaluating efficiency, which is defined as follows:

$$\text{Efficiency} = \frac{\text{mole of ethylation}}{\text{mole of EtCl consumed}} = \frac{\text{D.S. by hydrolysis}}{\text{total consumption of NaOH}}$$

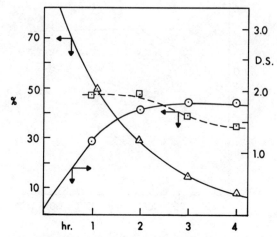

Fig. 2.   Variation of Component Composition in First Stage of
Ethylation: △ ,mole % ethyl chloride remaining;
⊡ ,ethylation efficiency; ⊙ ,degree of substitution.

Based upon D.S and ethylation efficiency the optimum reaction
time for the first stage appears to be two hours. Longer reaction
times lead to extensive hydrolysis and only small increases in the
extent of substitution. In general by-product formation occurs
subsequent to cellulose substitution so minimum reaction times
favor high efficiency.

Influence of Catalyst Structure

Selection of appropriate quaternary ammonium catalysts is
limited by their stability in the reaction medium. The cellulose
pretreatment requires heating in strong caustic solutions up to
120°; most tetraalkyl ammonium salts undergo Hoffman degradation
under these conditions. We have found two relatively stable quat
salts, tetramethylammonium chloride and N,N-dimethylmorpholinium

bromide. Longer chain tetraalkyls, N,N,-dimethylpiperidinium bromide and bis-quats are subject to extensive degradation during the pretreatment and substitution stages. However, several of these salts enhance the extent of cellulose substitution before degradation occurs.

If pretreatment is effected in toluene, the most pronounced catalytic effect is produced by the cyclic amines, N,N-dimethylmorpholinium and N,N-dimethylpiperidium chloride. The morpholinium derivative is more stable than the corresponding piperidinium salt and thus is the preferred catalyst. We were seeking to demonstrate an enhanced catalytic effect due to additional complexation of an anhydroglucose unit with the

Table I. Ethylation of Wood Cellulose in the
Presence of Quaternary Ammonium Salts

Quatenary Salt	Toluene Pretreatment*		Dry blend Pretreatment**	
	D. S.	% Ethoxyl	D. S.	% Ethoxyl
None	0.93	22.70	1.16	26.9
$(Me)_4N^+$ Cl^-	1.19	27.91	1.56	34.1
$MeP^+\phi_3$ Br^- 1.15	27.15			
Aliquat 336	0.80	19.97		
N,N,N,N',N',N'-Hexa-methylethylene ammonium chloride			0.96	23.18
Piperdine			1.05	24.7
N-Methyl morpholine	1.17	27.05	1.07	25.1
N,N-Dimethyl-morpholinium bromide	1.74	37.08	1.13	26.3
N,N-Dimethyl-piperidinium bromide	1.76	37.55	1.16	26.8
N-Methyl-morpholine-N-oxide	0.85	20.67		

* Pretreatment in toluene at 110° for 1 hr.; reaction for 2.5 hr.
**Dry-blended at $80-120^o$ for 0.5 hr.; reaction for 2 hr.

morpholino oxygen, but the similarity of the catalytic effect
exhibited by piperidinium salts eliminates this possibility.
Bis-quaternary salts were examined, but were found to be subject
to extensive degradation under the reaction conditions, and no
unusual catalytic properties were observed. N-Methylmorpholino-N-
oxide, a known cellulose solvent, failed to exert any catalytic
influence. Methyl triphenylphosphonium bromide was stable under
the reaction conditions but exhibited activity equivalent to
tetramethylammonium chloride, neither of these quats was effective
when the pretreatment was conducted in toluene. The toluene
slurry enhances the efficacy of more hydrophobic catalysts that
may ,in turn exert more influence of the latter stages of
substitution when the substrate is more hydrophobic.

      An entirely different order of reactivity was observed when
dry-blending was used in pretreatment. These results may be
partially due to problems in dispersing the catalysts in dry
cellulose powder. Tetramethylammonium chloride (TMAC) is very
soluble in water and can be dispersed using a minimum of water.
We have shown that the efficiency of the catalyst decreases when
the water content of the cellulose increases, so dilution of the
catalyst solution is counterproductive. Addition of other
diluents such as butanol to the system reduced catalyst activity
and tended to promote more by-product formation. Thus, it was not
feasible to disperse the more hydrophobic quat salts efficiently
in organic solvents. Experimental difficulties aside, however,
TMAC is uniquely effective in actvating cellulose in a dry
blend.

Cyanoethylation

      Michael addition of acrylonitrile to cellulose represents a
second class of typical derivation procedures.[8] Lower caustic
concentrations are used to activate the cellulose and the reaction
occurs a lower temperatures. The effect of various caustic
concentrations on the rate and extent of substitution was studied
in pure acrylonitrile and in acetone. Addition of TMAC to the
pretreatment step has no discernable effect upon the initial rate
of cyanoethylation. Upon addition of a twenty-fold excess of
acrylonitrile to activated cellulose, a small endothermic heat of
mixing is observed followed by a gradual increase in temperature
produced by the exothermic Michael addition. The temperature was
held at 27° to prevent a comcommitant anionic polymerization of
acrylonitrile. The influence of the TMAC becomes evident as the
reaction proceeds. Cyanoethylated cellulose imbibes the remaining
reagent when the D.S. approachs 2 and an extremely viscous mass
forms. Uncatalysed reactions stop at this point; catalysed
systems appear to autoaccelerate to degrees of substitution

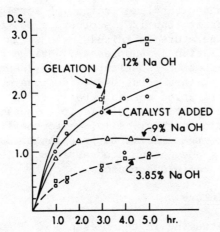

Fig.3.   Cyanoethylation of Wood Cellulose: $\square$, 5.0 g cellulose, 0.3 g TMAC 100 ml acrylonitrile stirred 3.5 hr at 27°; $\bigcirc$,control (same conditions, TMAC omitted), $\triangle$ No catalyst at 35°.

approaching the theoretical limit. The addition of TMAC can be delayed until the mixture gels and it will still promote further substitution up to the theoretical limit (Fig. 3).

Addition of a 1:1 mixture of acetone-acrylonitrile to cellulose pretreated with 10% sodium hydroxide followed by heating the mixture to 40° for 3.5 hr produced a very low D.S. (0.23) derivative. Incorporation of TMAC into the pretreatment step raised the D.S. attained under these conditions to 2.1. N,N-dimethylmorpholinium chloride was much less effective, a D.S. of 0.5 was produced. Since the pretreatment was conducted by dry-blending, these results are consistant with the unique catalytic

activity of TMAC observed in cellulose ethylation. The initial
dispersion of sodium hydroxide controls the extent of substitution. Reagent dilution decreases the effectiveness of the quat
salt, but a significant catalytic effect is still observed.

A major problem in cellulose cyanoethylation is control of
the concommitant anionic polymerization of acrylonitrile. At base
concentrations greater than 12%, exothermic polymerization occurs
rapidly if the temperature is raised to 40°. In the presence of
TMAC, the polymerization is suppressed, until a significant
increase in the extent of substitution as evidenced by gelation
occurs. However, at this point hydroxide ions must be released
and a rapid polymerization ensues. Derivatives with apparent M.S.
of 6.5 are isolated, but it is difficult to separate homopoly-
acrylonitrile from cyanoethyl cellulose to confirm if grafting has
occurred. However, the initial suppression of base catalysed
polymerization is a convincing demonstration of hydroxide transfer
and binding in the cellulose phase.

Carboxymethyl Cellulose

Cellulose pretreated in the presense of TMAC failed to
exhibit enhanced activity toward sodium trichloroacetate in an
isopropanol slurry. Products with essentially identical D.S.'s
were isolated from "catalysed" and control experiments. However,
solutions of "catalysed" CMC, (D.S.=0.7), appeared more fiber free
than those prepared from controls with similar D.S.. The
catalysed solutions (1%) exhibited slightly higher viscosities
(500 cp vs 325 cp) and were less thixotropic than standard CMC.
Efforts to ascertain if this difference in rheological behavior
can be attributed to a more homogeneous distribution of
substituents are in progress.

EXPERIMENTAL

Reagents. Bleached wood powder (Buckeye Cellulose) was used as
the cellulose source; an equilibrium water content of 8% was
determined gravimetrically. Commercially available quaternary
ammonium salts, benzyl chloride, ethyl chloride, chloroacetic acid
and acrylonitrile were used without further purification. N,N-
Dimethylmorpholinium bromide was prepared by treating N-methyl-
morpholine (50 g, 0.49 mol) with 40 ml (0.73 mol) of methyl
bromide in 100 ml methanol at 100° and 60 psi for four hr in a
bomb reactor. Venting the bomb, evaporating the excess methyl
bromiode and the methanol, washing with cold ethanol and drying
yielded 72.5 g of quat salt, m.p. 345°(decomp). Using the same
technique, N,N,-dimethylpiperidinium chloride and N,N,N,N',N',N'-

hexamethylethylenediammonium dichloride were prepared from the corresponding tertiary amines and methyl chloride.

Cellulose Pretreatment. A. Dry-Blend Technique. Powdered cellulose (36.0 g; 0.20 eq) was blended while 2.0 g (0.018 mol) of tetramethylammonium chloride in 2 ml of water was sprayed into the powder using a fine-needled syringe. A solution of 30 g sodium hydroxide in 30 ml H_2O was heated to boiling; the hot solution was added slowly to the TMAC treated cellulose and blending was continued for 0.5 hr. A total of 100 g of pretreated cellulose powder is obtained.
B. Slurry technique. A hot solution of 50% NaOH, 60 g, was dispersed in 300 ml of toluene and the slurry temperature was raised to 80°. After adding 36 g of cellulose and 0.018 mol of quat salt, the slurry was stirred at 110° for 1 hr. A quantitative transfer to a reactor provided an activated cellulose substrate for subsequent modification.

Benzyl Cellulose. Benzyl chloride (26 g 0.20 mol) was added to 5 g cellulose pretreated with 0.15 g TMAC and 10 g of 50% caustic and the pot temperature was raised to 100° during a 30 minute interval. Within 1-1 $\frac{1}{2}$ hours the swollen cellulose powder developed a doughy texture and then hardened into a solid mass which could not be dispersed with the mechanical stirrer. The reaction was terminated after 5 hours by grinding the yellow lump under 200 mL of ether. Benzyl cellulose precipitates as a white flake; the by-products which had plasticized the derivative were extracted quantitatively by the ether. The benzyl cellulose was isolated by filtration, washed with 200 mL ether and the benzyl chloride, benzyl alcohol and benzyl ether content of the combined ether washes determined by gas chromatography. The inorganic salts were removed from the product by titration with water and extensive water washing. After drying at 40° in vacuo overnight, 9.72 g of white benzyl cellulose was recovered. The degree of substitution, 2.0, calculated from the weight gain was confirmed by elemental analysis.

Ethyl Cellulose. Pretreated cellulose (100 g) was charged into a two liter stirrer pipe bomb, combined with NaOH (80 g, 2.0 M), ethyl chloride (250 g, 4.0 M) and 300 mL toluene and the mixture was heated to 130° for two hours. After cooling to 5°, the bomb was opened, excess ethyl chloride was allowed to evaporate and the residue poured into two liters of water. The toluene was steam distilled from the suspension and the ethyl cellulose was recovered by filtration. Approximately 46 g of ethyl cellulose, D.S. = 1.56, was obtained. A sample of ethyl cellulose prepared under identical conditions except omitting the catalyst was obtained in reduced yield (21 g) due to the higher water

solubility of the lower D.S. product (D.S. = 0.85). The ethoxyl
content was determined by weighing 65–75 mg of derivative into a
5 ml reaction vial. A xylene solution (2 ml) containing a
toluene internal standard and 2 ml of 50% HI solution were added
and the vial was sealed. After heating to 150° for 1 hr, the
concentration of ethyl iodide in the organic phase was determined
by gas chromatography using a 6′ OV-11 column at 90°.

Carboxymethyl Cellulose. Cellulose (5 g), pretreated as above,
was slurried in 120 mL of isopropanol and stirred at room
temperature while 2.7 g of chloroacetic acid dissolved in 10 mL of
isopropanol was added over a 5 minute interval. The mixture was
heated at 50° for three hours and 70° for one hour before
isolating the product by filtering the slurry and washing the
residue with 400 mL of 70–80% ethanol. The dried carboxymethyl
cellulose, 6.4 g, exhibited a D.S. of 0.8. (titration according
to ASTM procedure).[9]

Cyanoethyl Cellulose. A. Neat Reaction. Wood cellulose, 5.0 g
was treated with 0.15 g of tetramethylammonium chloride dissolved
in 2 mL of water. The cellulose was slurried in 120 mL of
acrylonitrile and 5 mL of 12% sodium hydroxide was added. The
slurry was stirred at 25° for 4 hours during which time the
mixture became very doughy. The reaction mixture was washed with
isopropanol until a free flowing slurry formed. The product was
recovered by filtration, 9.4 g, D.S. = 3.0, based upon a nitrogen
content of 13.02%.
B. Solvent Diluted Process. An activated substrate was prepared
by dry-blending 5 g of cellulose, 0.3 g TMAC, and 5 ml of 10%
sodium hydroxide at 25° for 15 min. A solution of acrylonitrile
(100 ml) in 100 ml acetone was added and the reactor was immersed
in a 40° oil bath. After 3.5 hr the mixture was a viscous yellow
solution containing some unreated fibers. The product was
precipitated in water, the acrylonitrile and acetone were removed
by steam distillation, and the aqueous suspension of polymer
neutralized with 5 % HCl. After isolating and drying in vacuo,
8.46 g of cyanoethyl cellulose, D.S.=2.14, was obtained.

ACKNOWLEDGEMENT

     The generous support of this research by Dow Chemical, USA is
greatly appreciated.

REFERENCES

1. A. F. Turbak, R. B. Hammer, R. E. Davies, H. L. Hergert, Chemtech., 10, 51 (1980); B. Philipp, H. Schleicher, and W. Wagenknecht, Chemtech, 7, 702 (1977); C. L. McCormick and T. S. Shen in "Macromol. Solutions-Prop. Relat. Polym.,(Pap. Symp.)" ed. by R. B. Seymour and G. A. Stahl, Pergamon, New York, 1982, p.101.

2. C. M. Starks and C. Liotta, "Phase Transfer Catalysis, Principles and Techniques", Academic Press, New York,1978; W. P. Weber and G. W. Gokel, "Phase Transfer Catalysis in Organic Synthesis" Springer Verlag, Berlin,1977.

3. W. H. Daly and J. Caldwell, J. Polymer Sci. Polymer Letters, 17, 55 (1979).

4. M. Gomberg and C. Buchler, J. Am. Chem. Soc., 43, 1904 (1921); H. Okada, Cellulose Chemie, 12, 11 (1931); D. Traill, J. Soc. Chem. Ind. Trans. and Comm. 337T (1934).

5. E. J. Lorand and E. A. Georgi, J. Am. Chem. Soc., 59, 1166 (1937).

6. J. Dockx, Synthesis, 441 (1973).

7. A. B. Savage, "Ethyl Cellulose", in Encyclopedia of Polymer Science, 3, 475 (1965).

8. N. M. Bikales "Ethers from , -Unsaturated Compounds", in "Cellulose and Cellulose Derivatives, Vol V, Part V" (N. M. Bikales and L. Segal,Eds.), Wiley-Interscience, New York, 1971, p 811.

9. ASTM D 1439-63T (1963).

SYNTHESIS OF POLYETHERS BY PHASE-TRANSFER CATALYZED

POLYCONDENSATION

Thanh Dung N'Guyen and Sylvie Boileau

Laboratoire de Chimie macromoléculaire associé au CNRS
Collège de France, 11 place Marcelin Berthelot, 75231
Paris Cédex 05, France

INTRODUCTION

Phase-transfer catalysis (PTC) has been recently applied to the preparation of polyethers by polycondensation [1-13]. This method offers several advantages over solution conducted nucleophilic displacement step-growth polymerizations. These include the substitution of inexpensive solvents for anhydrous aprotic solvents such as DMSO, DMAC etc ... , lower reaction temperatures and shorter reaction times. We wish to report our results concerning the preparation of polyethers from bisphenol A (BPA) and 1,4-dichloro-2-butene (DCB) under PTC conditions. The kinetics and the mechanism of this new type of polycondensation are described.

EXPERIMENTAL

Bisphenol A (Merck) was purified by recrystallization from toluene and trans 1,4-dichloro-2 butene (Aldrich) was redistilled before use. Phase-transfer catalysts (SO_4HNBu_4 : TBAH and kryptofix [222])as well as solvents were pure grade commercial products and were used without further purification.

Typically, a heterogeneous mixture containing BPA (3.07 mmol) dissolved in NaOH 3N (10 ml), toluene (10 ml), TBAH (0.61 mmol) and DCB (3.07 mmol) was stirred at 65°C under nitrogen for 5 h. After cooling, the organic layer was diluted with toluene (20 ml), separated from the aqueous phase, washed with slightly acidic aqueous solution (30 ml of HCl 0.1 N) and with water. The polymer was recovered by precipitation in methanol and purified by dissolution in $CHCl_3$ followed by precipitation in methanol. It was then dried under high vacuum. A 98 % yield of a white powder was obtained (based on starting materials).

Composition of the polymers was determined by elemental analysis and their molecular weights, $\overline{M}_n$, were calculated from the chlorine content. Analysis of the molecular weight distribution was examined by GPC in THF at 30°C.

RESULTS AND DISCUSSION

Some typical results are shown in Table 1. Polymers are soluble in organic solvents such as benzene, toluene, THF and chlorinated hydrocarbons. The IR spectrum of these polymers displays characteristic absorption bands at 1640 cm^{-1} (C=C stretching), 1600 and 1500 cm^{-1} (phenyl ring vibrations), 1235 cm^{-1} (phenyl ether stretching) and 1015 cm^{-1} (aliphatic ether stretching). The ^{1}H NMR spectrum of these polyethers recorded in $CDCl_3$ at room temperature exhibits multiplets between 7.20 and 6.67 ppm (aromatic protons) and peaks at 6.02 ppm (-CH=CH-), at 4.50 ppm (-CH$_2$-O) and 1.60 ppm (-CH$_3$). A small peak is observed at 4.05 ppm which can be attributed to the protons of the chloromethyl end groups. This peak is absent in the spectrum of sample 6 (Table 1).

No trace of phenolic end groups is detected (except for run 6) whereas chloromethyl end groups can be seen in the ^{1}H NMR spectrum as well as in the ^{13}C NMR spectrum. Absence of phenolic end groups in polymers is confirmed by UV spectrophotometric analysis after conversion of the phenolic groups to phenolate via quaternary ammonium hydroxide, according to Shchori and Mc Grath [14]

From these results, the following structure can be proposed for the polyethers prepared by PTC polycondensation :

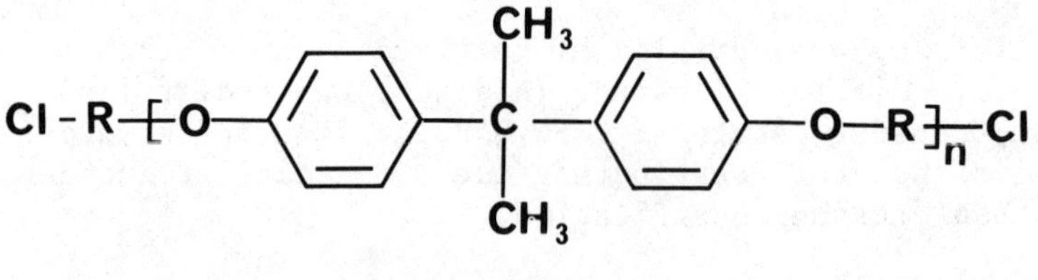

$$Cl-R \left[O-\underset{\text{(phenyl)}}{} \overset{CH_3}{\underset{CH_3}{\overset{|}{\underset{|}{C}}}} \underset{\text{(phenyl)}}{} O-R \right]_n Cl \qquad (I)$$

with   R = -CH$_2$-CH=CH-CH$_2$-

Integration of the different peaks of the ^{1}H NMR spectrum of sample 2 (Table 1) is in agreement with formula I for n = 7 thus leading to a value of $\overline{M}_n$ equal to 2100 which agrees quite well with the $\overline{M}_n$ determined from the chlorine content of polymer ($\overline{M}_n$ = 2000).

Table 1.

Preparation of polyethers from BPA and DCB in toluene-aqueous NaOH system, at 65°C, with PTC catalysts (magnetic stirring).

Run	Catalyst (mol %) a)	NaOH	Time h	Yield %	Elemental Analysis				$\overline{M}n$ b)
					C,	H,	O,	Cl	
1	-	3N	18	0					-
2	TBAH (10)	3N	5	98	79.1,	6.9,	10.8,	3.5	2000
3	[222] (5)	3N	17	85	79.6,	7.2,	10.9,	2.6	2700
4	TBAH (10)	6N	5	92	79.9,	7.0,	11.2,	2.3	3100
5	TBAH (5)	6N	5	98	79.2,	7.1,	11.6,	2.2	3200
6	TBAH (30)	25N	5	92	80.1,	7.1,	12.1,	0.0	3000 c)
7	TBAH (100)	3N	5	100	80.9,	7.1,	10.8,	1.5	4600
8	TBAH (10)	6N	24	100	80.0,	7.1,	10.9,	0.9	8200
9	TBAH (30)	6N	5	93	80.3,	7.1,	11.2,	0.8	8400

a) mol % of the chlorine content of DCB ; b) calculated from the chlorine content of polymer according to formula I ; c) determined by GPC.

Molecular weights of these polyethers were examined by GPC in THF at 30°C (using the calibration curve of polyethers formed from BPA and diglycidyl ether). The results are in good agreement with the $\overline{M}_n$ values determined from the chlorine content of polymers as shown in Table 2.

Table 2.

Molecular weights of polymers prepared from BPA

and DCB under PTC conditions (toluene/aq. NaOH,

65°C , TBAH = 10 mol %)

Run	NaOH	Time h	Yield %	$\overline{M}_n$ [a)	$\overline{Mn}$ GPC	$\overline{Mw}/\overline{Mn}$ GPC
10	1N	5	58	1400	1400	1.2
2	3N	5	98	2000	2000	1.7
11	6N	3	-	3700	3800	1.3
12	20N	5	91	5800	5200	1.3
8	6N	24	100	8200	7300	1.3

[a) calculated from the chlorine content of polymer.

These polyethers are prepared in a similar way as that described for the synthesis of phenol ethers [15] and polyhydroxyethers [6] using PTC conditions. Polycondensation takes place via alkylation of phenoxide ions with chloromethyl groups in the organic phase. A phase transfer catalyst is necessary for the transportation of the bisphenolate ion to the toluene phase in a highly reactive form. No polymer can be obtained in the absence of TBAH even after 18 h of reaction at 65°C (run 1, Table 1). Concentration of $\sim O^-, Q^+$ in the organic phase is very low in the absence of DCB (4% of TBAH in the conditions of run 2, Table 1). Thus, this type of polycondensation might be considered as a special case of an interfacial polycondensation in which the locus of the polymerization is presumably more in the organic phase than at the interface and the concentration of $\sim O^-, Q^+$ is much lower than that of chloromethyl groups. This last point can explain that the polyethers contain chloromethyl groups at each end.

The rate of the reaction of DCB with $\sim O^-, Q^+$ was followed by VPC in the organic phase, using undecane as internal standard (BPA :

3.07 mmol , DCB : 3.07 mmol, TBAH : 0.61 mmol, NaOH 6 N : 30.7 mmol
in 5 ml H_2O, toluene : 10 ml, undecane : 2.35 mmol). The rate of
disappearance of DCB is equal to :

$$-d[DCB]/dt = k \, [\sim O^-,Q^+]^x \, [DCB]^y \qquad (1)$$

If we assume that $y = 1$ and $[\sim O^-,Q^+]^x = $ Cte, equation (1)
becomes after integration :

$$L_n \, \frac{[DCB]_o}{[DCB]} = k't \qquad (2)$$

with $k' = k \, [\sim O^-,Q^+]^x \qquad (3)$

Plots of $L_n \, \dfrac{[DCB]_o}{[DCB]}$ versus time are shown in Fig. 1 for experiments
made at different concentrations of TBAH.

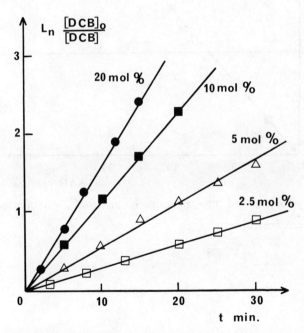

Fig. 1 Plot of $L_n([DCB]_o/[DCB])$ versus time in the organic phase.

Straight lines are observed the slope of which gives k'. The rate
constant, k', is proportional to the catalyst concentration up to
10 mol % as shown in Fig. 2

Dependence of $\overline{M}_n$ on several factors was examined. $\overline{M}_n$ increases
on increasing the catalyst concentration as shown by the results
of Table 3.

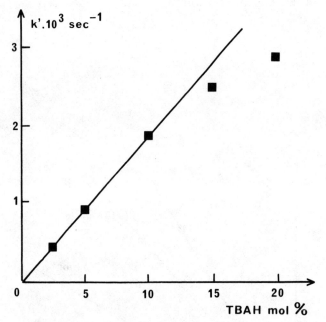

Fig. 2   Dependence of the rate constant, k', on the catalyst
concentration.

Table 3.

Influence of the catalyst concentration on the molecular weight $\overline{M}_n$ of polyethers under PTC conditions (toluene/aq. NaOH, 65°C, reaction time : 5 h, magnetic stirring).

Run	TBAH mol % [a)	NaOH	Yield %	$\overline{M}_n$ [b)
5	5	6N	98	3200
13	10	6N	98	4100
9	30	6N	93	8400
2	10	3N	98	2000
7	100	3N	100	4600

[a) mol % of the chlorine content of DCB; [b) calculated from the chlorine content of polymer according to formula I.

An increase of $\overline{M}_n$ is observed on increasing the NaOH concentration as shown by the results of Table 2 (runs 10,2, 11 and 12). However if the NaOH concentration is too high (25N, run 6, Table 1) the molecular weight $\overline{M}_n$ is lower than that observed with the same catalyst concentration but with NaOH 6N (run 9, Table 1). Moreover, some phenolic end groups have been detected by IR, NMR and UV spectrophotometry and the GPC diagram shows a bimodal distribution instead of the unimodal and narrow molecular weights distribution observed for other polyethers.

The results indicate that a large amount of bisphenolate ions are formed in the locus of polymerization under the experimental conditions used in this run (NaOH, 25N, TBAH : 30% mol) leading to some side reactions.

An increase of $\overline{M}_n$ is also observed on increasing the ratio [BPA]/[DCB] as shown in Fig.3

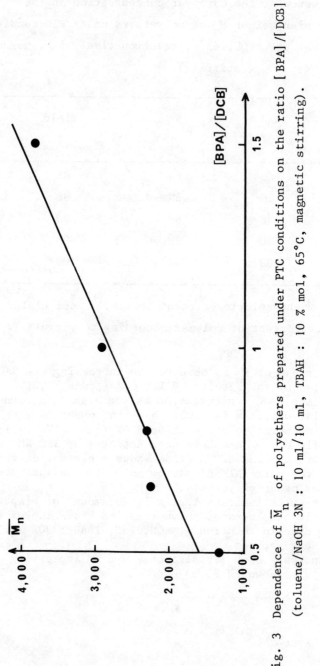

Fig. 3 Dependence of $\overline{M}_n$ of polyethers prepared under PTC conditions on the ratio [BPA]/[DCB] (toluene/NaOH 3N : 10 ml/10 ml, TBAH : 10 % mol, 65°C, magnetic stirring).

In a general manner, $\overline{M}_n$ increases on increasing the catalyst concentration , NaOH concentration, the [BPA]/[DCB] ratio as well as the stirring speed: that is on increasing the concentration of tetrabutylammonium phenolate species in the organic phase. However if the concentration of these active species is too high, a decrease of $\overline{M}_n$ is observed together with the presence of phenolic end groups in the polymers.

In conclusion, polycondensation of BPA with DCB in aqueous alkali-toluene system, at 65°C, using a quaternary ammonium salt as phase transfer catalyst leads to polyethers having chloromethyl groups at each end, in excellent yields. Molecular weights distribution of these polyethers is surprisingly quite narrow (see Table 2). Additional studies with this and other polymer systems are in progress.

REFERENCES

1. Y. Imai, M. Ueda and M. Li, J. Polym. Sci. Polym. Lett. Ed., 17, 85 (1979).
2. T. D. N'Guyen and S. Boileau, Polym. Bull., 1,817(1979).
3. A.K. Banthia, D.C. Webster and J.E. Mc Grath, Org. Coatings and Plastics Chem. Prep., 42, 127 (1980).
4. R. Kellman, D. Gerbi, R.F. Williams and J.L. Morgan, Polym. Prep., 21(2), 164(1980).
5. Y. Imai, J. Macromol. Sci. Chem., A 15, 833(1981).
6. A. K. Banthia, D. Lundsford, D.C. Webster and J.E. Mc Grath, J. Macromol. Sci. Chem., A 15, 943(1981).
7. R. Kellman , J.C. Mc Pheeters, D.J. Gerbi, R.F. Williams and R.B. Bates, Polym. Prep., 22(2),383(1981).
8. D.J. Gerbi, R.F. Williams, R. Kellman and J.L. Morgan, Polym. Prep.,22(2),385(1981).
9. T.D. N'Guyen, Thèse de Doctorat d'Etat, Paris (1981).
10. T.D. N'Guyen and S. Boileau, Polym. Prep., 23(1),154(1982).
11. J.-I. Jin and J.-H. Chang, Polym. Prep., 23(1),156(1982).
12. R. Kellman , D.J. Gerbi, J.C. Williams and R.F. Williams, Polym. Prep., 23(1), 174(1982).
13. G.G. Cameron and K.S. Law, Makromol. Chem., Rapid Comm., 3,99(1982).
14. E. Shchori and J.E. Mc Grath, Appl. Polym. Symp., 34,103(1978).
15. A. Mc Killop, J.C. Fiaud and R.P. Hug, Tetrahedron, 30,1379(1974).

EFFECT OF TRIETHYLAMINE ON THE SYNTHESIS OF GROUP IV B POLYETHERS

Charles E. Carraher, Jr., Samuel T. Bajah and
Leonard M. Jambaya

Department of Chemistry
Wright State University
Dayton, Ohio 45435

INTRODUCTION

The use of two-phase systems employing reactive reactants
where reaction occurs near the interface was popularized by Schnell
on polycarbonates,[1-5] Conix on polyesters,[6-9] Morgan on polyure-
thanes and polyamides[10-12] and Carraher on metal-containing polymers
and phosphorus-containing products.[13-15] The method has many
variations and is a highly effective procedure for the rapid
synthesis of many products. Until recently the technique emphasized
the synthesis of polymeric materials based on the Schotten-Baumann
concept on the laboratory scale. More recently similar synthetic
systems employed the addition of agents, now called phase transfer
agents, with an enlarged emphasis including many (potential) industrial scale syntheses of both small and polymeric materials.

Schnell and coworkers employed tertiary amines and quaternary
ammonium compounds to improve the synthesis of polycarbonates
and the monomer, diphenyl carbonate.[1-5] Schnell and Bottenbruch
proposed that the tertiary amines formed adducts with the acid
chloride in the organic phase forming a more reactive intermediate
(1). This is possible, but it is more likely that they are phase
transfer agents for the phenoxide.

$$R-\overset{\overset{\text{O}}{\|}}{C}-Cl:NR_3 \quad \underline{1}$$

Morgan[11] called phase transfer agents accelerators and explained their function in accord with current thought.

Carraher and coworkers employed organic bases as triethylamine
for several reasons including a) permitting base action to occur
in both the organic and aqueous phases and b) construction of
alternative interfacial systems such as the inverse (Lewis acid
contained in the aqueous phase and Lewis base in the organic
phase) and nonaqueous (Lewis acid contained in a nonpolar organic
liquid as carbon tetrachloride or long-chain hydrocarbons, Lewis
base contained in a polar organic liquid as DMSO, HMPA, TEP and
DMF) interfacial systems.[13,14]

Synthesis of Group IV B polyethers (2) was accomplished
utilizing aqueous and nonaqueous solution polycondensations and
classical and inverse (or reverse) interfacial polycondensations.[16-19]

$$Cp_2MCl_2 + HO-R-OH \rightarrow \{M-O-R-O\} \qquad \underline{2}$$

One group of studies involved identification of the reactive
species and locating, for the interfacial systems, the site of
reaction. For the synthesis of polyesters, reaction occurs within
the aqueous phase while for the synthesis of polyethers and polythioethers, synthesis occurs near the interface, but within the
organic layer. Thus an effective phase transfer agent, PTA,
will not cause polyether synthesis to change reaction phase (since
PTA agents act to bring agents into the organic phase) but would
assist diol transport across the interface with reaction occurring
further within the organic phase.

Following is a brief study illustrating the variety of results
that may occur employing similar reactions utilizing triethylamine
as the potential PTA.

EXPERIMENTAL

Polymerization procedures are similar to those described
in detail elsewhere.[16-19] Briefly, solutions of Cp_2MCl_2 are
added to rapidly (23,000 rpm no load) stirred aqueous solutions
containing diol and any added base. For interfacial systems,
Cp_2MCl_2 is contained in a water immiscible solvent whereas for
aqueous solution systems Cp_2MCl_2 is contained in water. The
reaction apparatus is similar to that described elsewhere.[20]
Solutions are added through a large-mouthed funnel placed in
a hole in the jar lid. Addition is rapid so that 100 ml of solution can be added in less than 3 seconds. Timing for each reaction
is begun after the second phase has been introduced into the
stirring jar. Syntheses employing hydroquinone or substituted

hydroquinones were accomplished under a nitrogen atmosphere utilizing equipment described elsewhere.[20]

Polymer precipitates rapidly from the reaction jar as a
tacky to powdery solid. It is recovered using suction filtration,
washed repeatedly with water and transferred to a Petri dish,
allowed to dry in the air and then added to a sample container
for further study.

Characterization procedures and results are given elsewhere.[16-19]

RESULTS AND DISCUSSION

An added base probably acts in several ways. For the synthesis of polyethers where reaction occurs very near the interface,
it is possible that the major role of sodium hydroxide is to
prime the diol for subsequent nucleophilic attack as pictured
below.

$$R-OH + OH^- \rightarrow R-O_{\overset{\displaystyle \cdot}{H} \cdots OH}{}^-$$

$$\underline{3}$$

The phase transfer activity of tertiary amines is believed
to occur because of the formation of a quaternary ammonium ion
pair through reaction with hydrogen chloride eliminated during
reaction of the alcohol and acid chloride.

$$R-OH + M-Cl \rightarrow R-O-M + HCl$$

$$R_3N + HCl \rightarrow R_3NH^+ Cl^- \qquad \underline{4}$$

The most utilized criteria for phase transfer activity is
an enhanced rate (yield at constant time) and chain length.
The Group IV B polyethers considered here are insoluble, leaving
only yield as a parameter to measure the affect of various bases
on the polymerizing process. Further, triethylamine can act
in its capacity to neutralize in both the organic and aqueous
phase.

Tables 1 and 2 contain results as a function of polymer
system and base for the titanium and zirconium polyethers. This
data is reduced to illustrate yield trends in Tables 3 and 4.

Table 1. Results as function of employed base, system and diol
          for the synthesis of titanium polyethers.[c]

	YIELD %			
	Sol-NaOH[a]	Sol-Et$_3$N[a]	IF-NaOH[b]	IF-Et$_3$N[b]
HYDROQUINONE	1	20	46	14
2,3-DICYANOHYDROQUINONE	70	70	72	63
2,5-DITERT-BUTYLHYDROQUINONE	74	71	41	4
CHLOROHYDROQUINONE	41	21	61	15
4,4'-ISOPROPYLIDENEDIPHENOL	30	57	36	5
ETHYLENE GLYCOL	0	0	24	0

Reaction conditions:

[a] Cp_2TiCl_2 (1.00 mmole) contained in 50 ml of water is added to
a stirred (2,3000 rpm) aqueous solution containing the diol
(1.00 mmole) in 50 ml water with added base (2.00 mmole) at
25°C for 1 min. stirring time.

[b] The same as a, except the Cp_2TiCl_2 is contained in 50 ml of
$CHCl_3$.

[c] Portions taken from references 17 and 18.

It is difficult to experimentally demonstrate that Et_3NHCl
is acting as a PTA in these situations. Even so several observations can be made. First, the yield trends as a function diol
are different for the two types of condensation systems, solution
and interfacial, but are similar for each system regardless of
whether the base is sodium hydroxide or triethylamine for M=Ti.
This is consistent with reaction with the reactants being Cp_2M^{+2}
and RO^- for the aqueous solution systems but Cp_2MCl_2 and ROH
for the interfacial systems and for reaction occurring within
the organic phase.[17,18,21] For M=Zr the yield trends are similar
for the aqueous solution systems, but dissimilar for the interfacial systems showing that the nature of the added base is important in determining the overall yield trends for the interfacial
systems.[16,19,22] Whether this difference is effected by the triethylamine acting as a PTA is not known. While several yields
are greater relative to sodium hydroxide systems, the majority
of yields are less, as noted below.

Table 2. Results as a function of employed base, system and
          diol for the synthesis of zirconium polyethers.[c]

	YIELD %			
	Sol-NaOH[a]	Sol-Et$_3$N[a]	IF-NaOH[b]	IF-Et$_3$N[b]
HYDROQUINONE	40		61	
2,7-DICYANOHYDROQUINONE	51	81	45	30
2,5-DITERT-BITYLHYDROQUINONE	40	66	21	26
CHLOROHYDROQUINONE	5	15	50	18
METHYLHYDROQUINONE	12	56	14	70
BROMOHYDROQUINONE	14		28	
2,5-DICHLOROHYDROQUINONE	15	39	21	
TETRACHLOROHYDROQUINONE	31		46	

Reaction conditions:

[a]Cp_2ZrCl_2 (1.00 mmole) contained in 50 ml of water, is added
to a stirred (23,000 rpm) aqueous solution containing the diol
(1.00 mmole) in 50 ml of water with added base (2.00 mmole)
at 25°C for 1 min. stirring time.

[b]The same as a, except the Cp_2ZrCl, is contained in 50 ml of
$CHCl_3$.

[c]Portions taken from references 16 and 19.

Second, while the general trends are similar for each condensation system, the actual yields vary considerably with product
yield being greater for systems employing sodium hydroxide on
the added base. For the interfacial systems the variation of
yield may be due to triethylamine acting as a PTA, encouraging
reaction to occur away from the interface, and into the organic
phase. An alternate explanation is related to the greater ability
of sodium hydroxide to polarize the diol (3) compared to triethylamine accounting for the lesser polycondensation rate for systems
containing triethylamine.

74   C. E. CARRAHER, Jr. ET AL.

Table 3. Trends with respect to polymer reaction system and
          employed base for the synthesis of titanium polyethers.

POLYCONDENSATION SYSTEM:BASE	TREND
Sol : Et$_3$N	3≈2>5>4≈1>>6
Sol : NaOH	3>2>4>5>>1>>6
IF : Et$_3$N	2>4≈1>5≈3>6
IF : NaOH	2>4>1>3>5>6

Where 1 = hydroquinone, 2 = 2,3-dicyanohydroquinone, 3 = 2,5-
ditert-butylhydroquinone, 4 = chlorohydroquinone, 5 = 4,4'-iso-

propylidenediphenol, 6 = ethylene glycol.

Table 4. Trends with respect to polymer reaction system and
          employed base for the synthesis of zirconium polyethers.

POLYCONDENSATION SYSTEM BASE	TREND
Sol : NaOH	2>3 = 1>10>9≈8>7>4
Sol : Et$_3$N	2>3>7>9>4
IF : NaOH	1>4>10≈2>8>9≈3>7
IF : Et$_3$N	7>2>3>4

Where 1 = hydroquinone, 2 = 2,3-dicyanohydroquinone, 3 = 2,5-
ditert-butylhydroquinone, 4 = chlorohydroquinone, 7 = methylhydro-
quinone, 8 = bromohydroquinone, 9 = 2,5-dichlorohydroquinone,
10 = tetrachlorohydroquinone.

     Thus, while triethylamine may act as a PTA for the synthesis
of Group IV B polyethers, other explanations are possible. This
is not to imply that triethylamine is not a PTA for some systems,
but to suggest that each system must be investigated individually
with select studies needed that are directed specifically at
determining whether the added base is acting as a PTA.

REFERENCES

1. H. Schnell, Angew Chem., 68, 633 (1956).
2. H. Schnell, L. Bottenbruch and H. Krim, Belgian Pat. 523,543
    (1954); assigned to Farbenfabriken Bayer A.G.
3. H. Schnell, "Chemistry and Physics of Polycarbonates," Inter-
    science, New York, 1964.
4. L. Bottenbruch and H. Schnell, German Pat. 1,101,386 (1961);
    assigned to Ferbenfabriken Bayer A.G.
5. K.H. Meyer and H. Schnell, German Pat. 1,056,141 (1959);
    assigned to Farbenfabriken Bayer A.G.
6. A.J. Conix, Ind. Chim. Belge, 22, 1456 (1957).
7. A.J. Conix, Ind. Eng. Chem., 51, 147 (1959).
8. A.J. Conix, Belgium Pat. 565,478 (1958); assigned to Geraert
    Photo-Producten N.V.; Chem Abst. 55, 25356 (1961).
9. A.J. Conix and U.L. Laridon, Angew. Chem., 72, 116 (1960).
10. P.W. Morgan, SPE (Soc. Plastics Engrs.), 15, 485 (1959).
11. P.W. Morgan, "Condensation Polymers by Interfacial and Solu-
    tion Methods," Interscience, New York, 1964.
12. P.W. Morgan, "Interfacial Synthesis III. Recent Advances,"
    (Edited by C. Carraher and J. Preston), Marcel Dekker, New
    York, 1982; Chpt. 1.
13. C. Carraher, "Interfacial Synthesis II. Applications,"
    (Edited by F. Millich and C. Carraher), Marcel Dekker, New
    York, 1978.
14. C. Carraher, J. Chem. Ed., 58, 92 (1981).
15. C. Carraher, Inorganic Macromolecules Revs., 1, 287 (1972).
16. C. Carraher and L. Jambaya, Angew. Makromol. Chemie, 52,
    111 (1976) and 39, 69 (1974).
17. C. Carraher and S.T. Bajah, (Br.) Polymer, 14, 42 (1973)
    and 15, 9 (1974).
18. C. Carraher and S.T. Bajah, Br. Polymer J., 7, 155 (1975.
19. C. Carraher and L. Jambaya, J. Macromol. Sci., A8(7), 1249
    (1974).
20. C. Carraher, J. Chem. Ed., 46, 314 (1969).
21. C. Carraher and S.T. Bajah, Organic Coatings and Plastics
    Chemistry, 33 (1), 624 (1973).
22. C. Carraher and L. Jambaya, Organic Coatings and Plastics
    Chemistry, 34(2), 485 (1974).

USE OF PHASE TRANSFER AGENTS IN THE SYNTHESIS OF ANTIMONY (V) POLYAMINES

Charles E. Carraher, Jr. and Mellissa D. Naas

Department of Chemistry
Wright State University
Dayton, Ohio 45435

INTRODUCTION

We have synthesized a number of antimony, bismuth and arsenic polyesters 1 and polyoximes 2 through the condensation of the salts of the diacids and dioximes with the corresponding organometal dihalide.[1-3] Recently we extended this to include the synthesis of antimony (V) polyamines (3).[4]

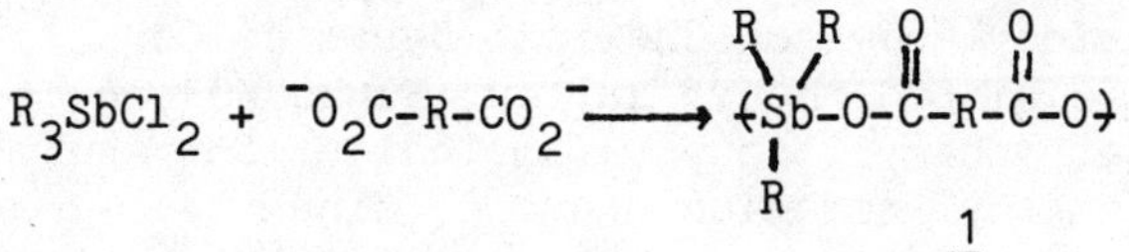

$$R_3SbCl_2 + {}^-O_2C-R-CO_2{}^- \longrightarrow \{Sb-O-\overset{\overset{R \quad R}{\diagdown \diagup}}{\underset{R}{|}}-O-\overset{O}{\overset{||}{C}}-R-\overset{O}{\overset{||}{C}}-O\} \qquad \underline{1}$$

$$R_3SbCl_2 + {}^-ON=\overset{R'}{\overset{|}{C}}-R-\overset{R'}{\overset{|}{C}}=NO^- \longrightarrow \{Sb-O-N=\overset{R'}{\overset{|}{C}}-R-\overset{R'}{\overset{|}{C}}=N-O\} \qquad \underline{2}$$

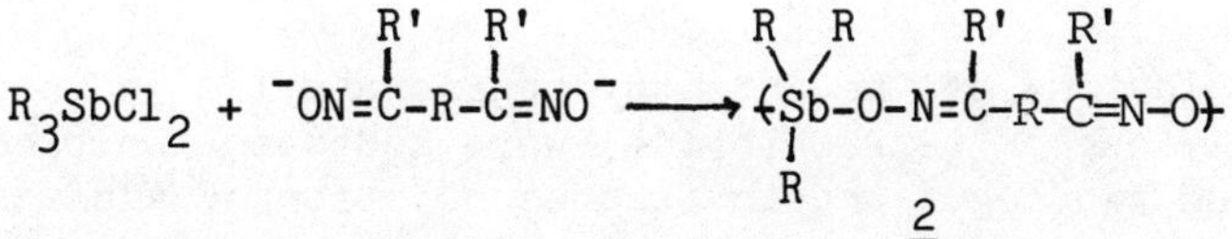

$$R_3SbCl_2 + H_2N-R-NH_2 \longrightarrow \{Sb-N-R-N\} \qquad \underline{3}$$

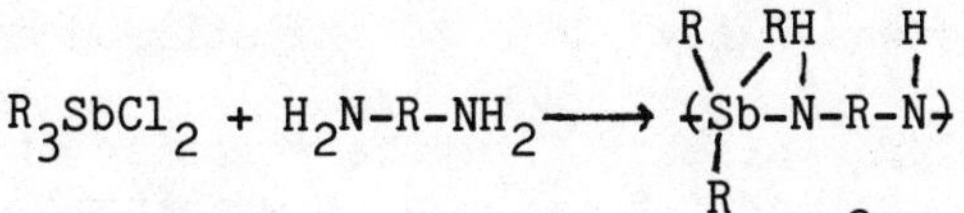

One objective is the synthesis of polymers which exhibit select biological activities for use in biomedical applications. Most antimony-containing products exhibit some biological activity. While antimony potassium and sodium tartrate, stibophen, sodium

77

antimonyl gluconate and sodium alpha,alpha-dimercaptosuccinate are utilized to control leishmaniasis, filariasis and schistosomiasis[5,6] their high toxicity is a disadvantage to human applications. Formation of a mixed chelate of antimony sodium tartrate with penicillamine gives a considerably less toxic material which retains its antiparasitic action in schistosomiasis.[7] Thus inclusion of antimony within polymers may give products which retain activities against specific organisms yet be less toxic toward the larger host. Further, organism control may occur over an extended period if a controlled release of the active moieties is involved. Further, diamines were selected which are known to offer some biological inhibition.

EXPERIMENTAL

For all the polymerization reactions conducted, a Kimax emulsifying jar was used. The jar was placed on a Waring Blendor (Model 1120) whose speed, without load, is 18,500 revolutions per minute.

The lid of the jar has one hole in it for a large-mouthed funnel through which a large volume of solution can be added in a short period of time so that the rate of addition of a solution to the jar can be held approximately constant.

The following chemicals were used as received (from Aldrich unless otherwise noted): 2,6-diaminoanthraquinone, 2,5-dichloro-p-phenylenediamine, 2,6-diamino-5-nitropyrimidine, 2,3,5,6-tetramethyl phenylenediamine, 2,6-diamino-8-purinol hemisulfate monohydrate, adenine (Chemalog Chemical Dynamics Corp.), 1,9-diaminononane, 4,4'-diaminodiphenyl sulfon (Fluka AG), 4,4'-methylenedianiline, tetraphenyl phosphonium iodide, tetrabutyl ammonium iodide, tetramethyl ammonium bromide, dibenzo-18-crown-6, 15-crown-5, triphenylantimony dichloride and triethylamine (Eastman Kodak Co.).

Solutions of the antimony organometallic in an organic solvent (usually carbon tetrachloride) were added to the blender. Rapid stirring was begun and an aqueous solution of a diamine which also contained an added base such as sodium hydroxide, NaOH, or triethylamine, Et_3N, or other agent, was added. Reaction time was begun when the organic phase was fully added to the stirred aqueous phase.

The product precipitates from the reaction mixture and is collected using suction filtration with washing by water and chloroform.

Physical determination results are reported elsewhere. Briefly analysis involving pyproprobe-chemical ionization mass spectrometry, elemental analysis, infrared spectroscopy and

analogous control reactions on products derived employing both added
bases (NaOH and Et$_3$N) and PTA's are in agreement with a product of
form 3.

Molecular weight determinations were done using a Brice-
Phoenix Model BP-3000 Universal Light Scattering Photometer.
Molecular weight determinations were carried out in the usual
manner using 1% polymer solutions with serial dilutions. Refrac-
tive Index Increments were determined using a Baush and Lombe Abbe
Refractometer Model #3-L.

RESULTS AND DISCUSSION

Antimony (V) polyamines, hereafter referred to as polyamines,
were initially synthesized utilizing sodium hydroxide as the added
base. Yields varied from poor to good but the products were mainly
oligomeric with weight average degrees of polymerization in the
order of 4 to 10. While short chains are useful for some applications, select biomedical applications require molecular weights
greater than 30,000 to prevent free drainage into the kidney, etc.
where the drug would be rapidly washed through the body.

A number of phase transfer agents (PTA's), including triethylamine, were employed in an effort to increase chain lengths.
Illustrative results appear in Table 1 for syntheses carried out
employing triethylamine. Compared with systems utilizing sodium
hydroxide as the added base, the yields and chain lengths can be
greater, equal or less. Thus the presence of an organic soluble
base does affect the reaction system either through a phase
transfer mechanism or other route. Chain lengths are still not
high.

The affect of traditional PTA's was studied. Results appear
in Table 2. Several general trends are evident. First, the effects
of the presence of PTA's is specific with regard to both the
reactants and PTA. Second, yield and chain length are generally
increased through use of PTA's and in some instances high polymers
are formed. Third, the use of PTA's allows the synthesis of some
polyamines, not synthesized by use of sodium hydroxide or triethylamine alone.

In summary, the presence of triethylamine can affect both
product yield and chain length. The use of classical PTA's typically
allows the synthesis ofpolyamines in greater yield and of longer
chain lengths. Finally, the use of classical PTA's allows the
synthesis of polyamines not synthesized utilizing analogous reaction
condition except omitting the PTA.

Table 1. Results as a function of added base

Diamine	Base	NaOH		Et_3N	
		Yield(%)	$\overline{M}_w$ (10^{-3})	Yield(%)	$\overline{M}_w$ (10^{-3})
2,6-Diaminoanthraquinone		40		37	
2,6-Diamino-5-nitropyrimidine		36		73	
2,4-Diamino-5(3,4-dimethoxy benzil)pyrimidine		21	4.8	60	
4,4-'Diaminodiphenyl sulfon		42	4.5	23	3.2
2,3,5,6-Tetramethyl-p-phenylene- diamine		8		0	
2,5-Dichloro-p-phenylenediamine		40	9.3	23	4.9
4,4'-Methylenedianiline		54		2	

Reaction conditions: Triphenylantimony dichloride (1.00 mmole) in 25 ml chloroform is added to highly stirred (18,500 rpm no load) solutions of diamine (1.00 mmole) in 25 ml of water and containing sodium hydroxide (2.00 mmole) or Et_3N (2.00 mmole) at 25°C for 30 seconds stirring time.

Molecular weights were obtained utilizing light scattering photometry in DMSO at 25°C.

Table 2. Results as a function of phase transfer agent

Diamine	NaOH		Tetrabutyl Ammonium Iodide		Tetraphenyl Phosphonium Iodide		Tetramethyl Ammonium Bromide		Dibenzo 18-Crown-6 Ether		15-Crown-5 Ether	
	Yield(%)	$\bar{M}_w$	Yield(%)	$\bar{M}_w$	Yield(%)	$\bar{M}_w$	Yield(%)	$\bar{M}_w$	Yield(%)	$\bar{M}_w$	Yield(%)	$\bar{M}_w$
2,6-Diamino-8-purinol	35	3.7	45		45	9.1			46	3.6		
2,4-Diamino-5(3,4-dimethoxybenzil)pyrimidine	21	4.8	23		45	6.0			46	2.2		
Adenine	9	2.6	0		25	7.3			28	830.0		
2,5-Dichloro-p-phenylenediamine	39	9.3	28		28	1.3	15 (28)	1.3 (2.6)	49 (36)	8.1 (17.0)		
4,4'-Diaminodiphenyl sulfon	18	4.5	23		48	4.3	23 (54)	5.9	49 (9)	5.6 (50.0)	38	9.3
1,9-Diaminononane	0	-	15		10		0		0	-	0	-
2,6-Diaminoanthraquinone	40		40		53				24			
2,6-Diamino-5-nitropyrimidine	36		36		47							
4,4'-Methylenedianiline	54						12 (6)		31 (4)			

Reaction conditions: Same as Table 1 except PTA (0.040 mmole) is added to the aqueous phase. All molecular weights are given in 10^3's. Values in ( )'s are for systems where the added base is triethylamine.

REFERENCES

1.   C. Carraher and L. Hedlund, J. Macromol. Sci.-Chem., A14(5),
     713(1980).
2.   C. Carraher and W. Venable, J. Macromol. Sci.-Chem., A14(4),
     571(1980).
3.   C. Carraher and H.S. Blaxall, Angew, Makromol. Chemie, 83,
     37(1979).
4.   C. Carraher and M. Naas, unpublished results.
5.   E. Bueding and J. Fisher, Biochem. Pharmacol., 15, 1197
     (1966).
6.   M. Pedrique and N. Freoli, Bull. World Health Org., 45,
     411 (1971).

SYNTHESIS OF AROMATIC POLYPHOSPHOANHYDRIDES BASED ON PHENYLPHOS-PHONIC DICHLORIDE EMPLOYING PHASE TRANSFER AGENTS

Charles E. Carraher, Jr., Raymond J. Linville and
Howard S. Blaxall

Department of Chemistry
Wright State University
Dayton, Ohio 45435

INTRODUCTION

We have synthesized a number of phosphorus-containing polymers utilizing the condensation of phosphorus acid chlorides and the diisocyanate with a variety of Lewis bases utilizing the inter-facial condensation technique.[1,2]

$$\underset{\underset{R}{|}}{\overset{\overset{O}{\|}}{Cl-P-Cl}} + HS-R-SH \longrightarrow \underset{\underset{R}{|}}{\overset{\overset{O}{\|}}{(P-S-R-S)}} \quad \underline{1}$$

$$\underset{\underset{R}{|}}{\overset{\overset{O}{\|}}{OCN-P-NCO}} + H_2N-R-NH_2 \longrightarrow \overset{O\ H\ O\ H\quad\ H\ O\ H}{\underset{\underset{R}{|}}{(P-N-C-N-R-N-C-N)}} \quad \underline{2}$$

We also synthesized a number of Group IV A, IV B and V A metal containing polyesters utilizing salts of dicarboxylic acids.[1,3-7]

$$R_2MX_2 + {}^-O_2C-R-CO_2^- \longrightarrow \underset{\underset{R}{|}}{\overset{\overset{R}{|}}{(M-O-\overset{\overset{O}{\|}}{C}-R\ -\overset{\overset{O}{\|}}{C}-O)}} \quad \underline{3}$$

83

We now describe the synthesis of the analogous phosphorus-containing anhydrides of form 4.

$$Cl\text{-}P\text{-}Cl \ + \ {}^{-}O_2C\text{-}R\text{-}CO_2^{-} \qquad \{P\text{-}O\text{-}C\text{-}R\text{-}C\text{-}O\}$$

4

EXPERIMENTAL

The following chemicals were utilized as received (from Aldrich unless otherwise noted): phenylphosphoric dichloride (Alfa Inorganics), terephthalic acid, nitroterephthalic acid, methylterephthalic acid, 2,5-dimethylterephthalic acid, tetramethyl-terephthalic, bromoterephthalic acid, 2,5-dichloroterephthalic acid, tetrabromoethylammonium iodide, triethylamine (Eastman Kodak Co.), tetrabutylphosphonium bromide, 18-Crown-6 and 15-Crown-5.

Polymerization procedure and apparatus is similar to that described elsewhere.[6] Briefly, the phenylphosphonic dichloride was dissolved in chloroform. The diacid was neutralized by addition of sodium hydroxide or triethylamine and was dissolved in water. Any additional phase transfer agent was also added to the aqueous phase. The phosphorus containing phase is added to the rapidly stirred aqueous phases. The reagents were used as received without further purification.

A Perkin-Elmer 457B Grating Infrared Spectrophotometer was utilized to obtain infrared spectra. Elemental analysis was carried out employing tungstic anhydride as a combustion catalyst using a Perkin-Elmer Model 240B Elemental Analyzer. Light scattering was accomplished utilizing serial dilutions and employing a modified Brice-Phoenix 2000 Universal Light Scattering Photometer. Refractive index increments were determined using a Bausch and Lomb Abbe Refractometer Model 3-L. Mass spectrometry was conducted utilizing a double-focusing Dupont 21-491 Mass Spectrometer coupled with a modified Hewlett-Packard HP-2216C computer.

RESULTS AND DISCUSSION

Structural Determination

Product structure was determined utilizing analogous reactions, control reactions, elemental analyses mass spectroscopy, light scattering photometry and infrared spectroscopy. Selected results will be briefly described.

Analogous Reactions

As noted in the Introductory section, the analogous reaction
with metal-containing acid chlorides generated the corresponding
polyester. Further, a similarity exists between metal-halide
reaction sites and phosphorus-halide reactive sites with regard
to reactions with Lewis bases under mild reaction conditions.
The condensation of phenylphosphonic dichloride with aromatic
diacid salts is considered an extension of these reactions.

Control Reactions

The reactions were repeated except omitting one of the reactants. No precipitation was formed. This is consistent with
the product containing units derived from both reactants.

Infrared Spectroscopy

An infrared spectrum illustrative of those derived from
the products is given as Figure 1.

For the product from ditriethylamine terephthalate and phenylphosphonic dichloride the following pertinent assignments are
made (all bands given in cm^{-1}): intense band at 1280 cm^{-1} in
both phenylphosphoric dichloride and product assigned to P = 0
stretching; presence of the P-phenyl moiety from the presence
of bands at 1420 (1445 in $PO\emptyset Cl_2$ itself) and 1015 (same for $PO\emptyset Cl_2$)
in the product; shift of carbonyl from 1680 for terephthalic
acid itself, and 1645 for ditriethylamine terephthalate to 1660
for the product consistent with ester or anhydride formation;
presence of new peaks at 1065 and 1110 attributed to stretching

in the P-O-$\overset{O}{\overset{\|}{C}}$ moiety; appearance of bands characteristic of the
presence of the phenyl and phenylene moieties at 525, 555, 680
(small spike), combined band with spikes at 720 and 725, and
780 ($PO\emptyset Cl_2$ exhibits bands at 790, 720 and 755; ditriethylamine
terephthalate exhibits bands at 500, 520, 550, 690, 725 and 770).

Infrared spectra of products derived employing either triethylamine or sodium hydroxide to neutralize the diacid are identical.

Pyroprobe-Mass Spectroscopy

Pyroprobe-chemical ionization mass spectroscopy was carried
out on the products. For the product derived from terephthalic
acid the following significant ion fragments were found and assignments made: unit plus P=O (334); unit minus CO_2 (238), unit (285);
unit plus CO_2 (330); $\emptyset$-P=O (123,124); $\emptyset$-O_2CPO (169,168,167,166);
O_2C-$\emptyset$-CO_2 (162); O_2C-$\emptyset$ (117); $\emptyset$ (79,73,66,60); O_2CPO (91). Thus

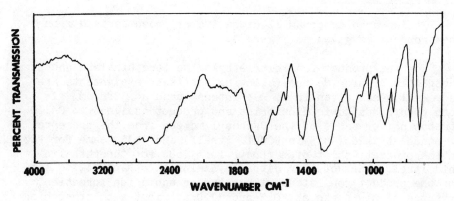

Figure 1.   Infrared spectra of the condensation product of phenylphosphonic dichloride and ditriethylamine terephthalate.

the mass spectrum agrees with a repeating unit as depicted in
form 5.

$$\{P-O-C-\bigcirc-\bigcirc-C\}$$

5

Synthesis

Though we had synthesized a number of metal-containing polyesters; previous attempts to synthesize the analogous phosphorus
anhydrides failed. We surmised that our previous failures were
due to the rapid hydrolysis of the phosphorus acid chloride since
the metal salts of dicarboxylic acids are largely insoluble in
nonpolar organic liquids as benzene, carbon tetrachloride and
chloroform. Further, reaction with the organometallic halides
occurs within the aqueous phase for interfacial systems.[7] Recent
attempts were aimed at increasing the organic solubility of the
diacid salt and use of phase transfer agents, PTA's.

The organic solubility of diacid salts is related to the
nature of the cationic portion. We found that addition of triethylamine to form the ditriethylamine salt generally yields a sparsely
soluble salt. Ditriethylamine terephthalate is partially soluble
in chloroform with a distribution coefficient (25°C, 4.00 mmoles
in 25 ml $CHCl_3$ and 25 ml H_2O) of 1.2×10^{-3} $CHCl_3/H_2O$. Thus triethylamine may act as a PTA in such systems generated as pictured
below.

$$HO_2C-R-CO_2H + 2Et_3N \longrightarrow Et_3NH^+, \ ^-O_2C-R-CO_2^-, \ Et_3NH^+$$

6

The use of ditriethylamine salts gave products from only
selected diacids. Polyphosphonanhydrides are also synthesized
when utilizing aromatic diacids neutralized by addition of sodium
hydroxide. In both cases the presence of electron withdrawing
substituents inhibits reaction (Table 1). Product yields employing
triethylamine and sodium hydroxide are approximately the same
but the systems employing the triethylamine gave products with
substantially longer chain lengths, possibly a consequence of
the greater organic solubility of the ditriethylamine acid salt,
reducing the hydrolysis of the phenylphosphonic dichloride.

Table 1. Results for aromatic diacids neutralized by addition
          of sodium hydroxide and triethylamine

Salt	Reaction Time(secs)	Yield (%)	$\overline{M}_w$
Disodium terephthalate	10	37	40,000
Disodium terephthalate	180	41	40,000
Ditriethylamine terephthalate	20	44	80,600
Ditriethylamine terephthalate	45	52	-
Ditriethylamine terephthalate	360	45	55,900
Ditriethylamine tetramethyl-terephthalate	180	45	-
Ditriethylamine 2,5-dimethyl-terephthalate	180	10	-

Reaction conditions: Phenylphosphonic dichloride (4.00 mmoles)
in 25 ml chloroform is added to rapidly stirred (18,500 rpm no
load) solutions of diacid salt (4.00 mmole) containing indicated
PTA (0.20 mmole) contained in 25 ml water at 25°C. Molecular
weights were determined in hexamethylphosphoramide at 25°C utilizing light scattering photometry. Unsuccessful - ditriethylamine
bromoterephthalate, ditriethylamine nitroterephthalate.

     Classical phase transfer agents are quantary salts and crown
ethers. The synthesis of the polyphosphoanhydrides was studied
employing a number of phase transfer agents. Select results
will be presented here illustrating the variety of results obtained. Phase transfer agents were chosen to give a broad representation of available agents. Typical results appear in Tables
2 and 3. Addition of PTA's to the triethylamine systems typically
led to lower yields and shorter chains except for 18-Crown-6
where the yield was the highest obtained, about double that obtained
without the Crown ether. For the disodium terephthalate systems
addition of a PTA typically leads to a lowering of yield and
chain length but a substantial increase in chain length when
employing 15-Crown-5.

     In summary polymeric aromatic polyphosphoanhydrides are
synthesized in moderate yields utilizing acids neutralized with

sodium hydroxide or triethylamine with those derived from triethylamine having a higher molecular weight. The use of classical
PTA's typically has a pronounced affect on the yield and/or chain
length of the product. The role of the PTA is complex and dependent
on the nature of the PTA.

Table 2. Results for systems employing triethylamine terephthalate
          and added phase transfer agents

PTA	Reaction Time(secs)	Yield (%)	$\overline{M}_w$
None	20	44	80,600
Tetrabutylammonium Iodide	180	19	42,000
Tetrabutylphosphonium Bromide	180	11	14,500
18-Crown-6	180	79	50,000
15-Crown-5	20	19	35,000

Reaction conditions: As cited for Table 1.

Table 3. Results for terephthalic acid neutralized by addition
          of sodium hydroxide as a function of added phase
          transfer agent

PTA	Reaction Time(secs)	Yield (%)	$\overline{M}_w$
None	10	37	40,000
None	180	41	40,000
Tetrabutylammonium Iodide	180	29	45,000
15-Crown-5	180	20	81,000
Tetrabutylphosphonium Bromide	180	11	14,500

Reaction conditions: As cited in Table 1.

REFERENCES

1.   C. Carraher, "Interfacial Synthesis: Polymer Applications and Technology" (F. Millich and C. Carraher, editors), Marcel Dekker, N.Y., chpt. 20, 1977.
2.   C. Carraher, Inorganic Macromolecules Reviews, 1, 287 (1972).

3.   C. Carraher and C. Deremo-Reese, J. Polymer Sci., 16, 491 (1978).
4.   C. Carraher, Inorganic Macromolecules Reviews, 1, 271 (1972).

5.   C. Carraher, J. Applied Polymer Sci., 20, 2255 (1976).
6.   C. Carraher and L. Jambaya, Angew. Makromol. Chemie, 52, 11 (1976).
7.   C. Carraher and J. Lee, J. Macromol. Sci. - Chem., A9(2), 191 (1975).

PHASE-TRANSFER POLYCONDENSATION OF

BISPHENOLATE ANIONS AND 1,6-DIBROMOHEXANE

Jung-Il Jin and Jin-Hae Chang

Department of Chemistry
Korea University
Seoul, Korea

INTRODUCTION

Phase-transfer catalysis, also often referred to as "ion pair partition" is a novel synthetic technique which has been the subject of much interest in recent years not only in the field of organic synthesis but also in polymer chemistry. The term "phase-transfer catalysis" was first introduced in 1971 by Stark[1,2] who studied kinetics in detail and the mechanism of reactions which are catalyzed by small amounts of onium salts such as quaternary ammonium or phosphonium compounds. Brändström[3-5] and Makosza[6-8] also made major initial contributions in the understanding of such reactions and the application thereof in various synthetic reactions. A generally accepted phase-transfer reaction scheme is shown in Fig. 1 for the reaction of RX + Y$^-$ ⟶ RY + X$^-$. Catalytic onium ion(Q$^+$) transfers the nucleophile Y$^-$ into the organic phase in the form of ion pair Q$^+$Y$^-$, which then reacts with RX in organic phase producing the products RY and X$^-$. X$^-$'s are removed by Q$^+$ from organic phase into aqueous phase as they are formed.

Another type of phase-transfer catalysts is synthetic macrocyclic polyethers, so-called crown ethers, and poly(ethylene glycol) derivatives. Since the discovery of crown ethers and their complexing capabilities toward metal and ammonium ions in 1967 by Pedersen[9], crown ethers and modified compounds such as cryptates[10,11] have been attracting ever increasing interest among scientists having different applications in their mind.[12] Complexes formed between these compounds and cations correspond to Q$^+$ shown in Fig. 1 and possess phase-thransfer capacity for anions. The nature of cations is known to greatly influence the complexation with a given crown either[13] or cryptate.[12]

91

```
Aqueous phase   M⁺Y⁻ + Q⁺X⁻  ⇌  Q⁺Y⁻ + M⁺X⁻
Interface      ──↓──↑── ── ── ──↓──↑── ─
Organic phase   RY + Q⁺X⁻  ⟵  Q⁺Y⁻ + RX
```

Fig. 1.   Reaction consequence of phase-transfer catalyzed reactants.

The application of phase-transfer technique in polymer chemistry
has been rapidly growing in recent years and has been explored in
development of new synthetic method of polymers, structural modification of pre-made or existing polymers, utilization of polymeric
phase-transfer catalysis, etc..[14,15] Particularly the use of phase-
transfer catalysts (PTC's) in the synthesis of condensation polymers
is now widely explored, because it can offer several advantages
over conventional single phase step-growth polymerizations especially
of nucleophilic displacement.[15] Even though there exists an apparent
similarity between "interfacial polymerization" and "phase-transfer
polymerization (PTP)," their mechanistic differences should not be
overlooked.[16]

The PTP technique has been lately utilized for the preparations
of many different polymers. Poly(phenylene oxide),[17] polycarbonates,[18] polysulfonates,[19] polysulfides,[20] polyhydroxyethers,[21]
polycarbonate/polysiloxane block copolymers,[22] polyarylethers,[23] and
arylether-sulfone copolymers[23] are some examples. Earlier Schnell,[24]
apparently without noticing the nature of reaction mechanism, could
obtain polycarbonate in good yield by reacting quaternary ammonium
salt of bisphenol A with phosgene in water-dichlorometane mixture.
Morgan[25] utilized tetraethylammonium chloride as an accelerator for
the preparation of aromatic polyesters under interfacial conditions.
In both cases it is obvious that Q^+ transferred the anions into the
organic phase with the resultant acceleration of their reactions.

In spite of the wide applicability of phase-transfer techniques
in polymer synthesis, there have not been reported many basic works.
In order to identify important reaction variables and to improve
our knowledge on the reaction mechanism of PTP technique, we have
conducted an investigation on biphasic heterogeneous polymerizations
between various bisphenolate anions and 1,6-dibromohexane in the
presence of tetrabutylammonium bromide (Q^+Br^-),[16] tetrabutylphosphonium bromide (P^+Br^-) and dibenzo-18-crown-6 (DBC)[26] (Equation 1).
Nitrobenzene was chosen as an organic phase because it was found in
preliminary experiments to be a reasonably good solvent for polymers
at the reaction temperatures.

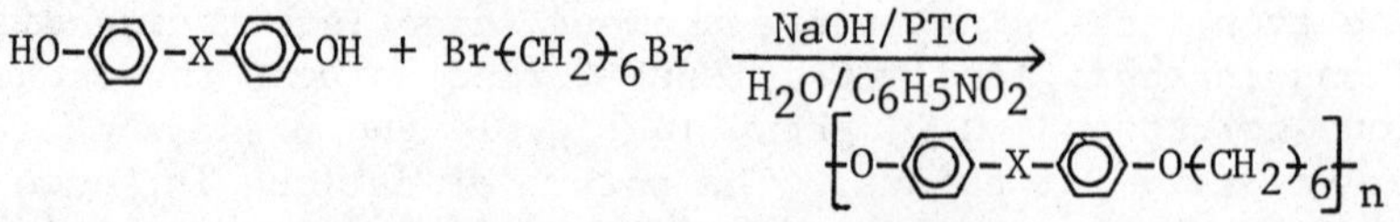

$$HO-\bigcirc-X-\bigcirc-OH + Br(CH_2)_6Br \xrightarrow[\text{H}_2\text{O}/\text{C}_6\text{H}_5\text{NO}_2]{\text{NaOH/PTC}}$$

$$\left[O-\bigcirc-X-\bigcirc-O(CH_2)_6\right]_n \qquad (1)$$

$X=C(CH_3)_2$ (BPA), CH_2 (MDP), S (TDP), SO_2 (SDP), and CO (KDP)

Preliminary experiments also showed that the PTC's employed were able to solubilize to a satisfactory extent bisphenol anions in nitrobenzene.

EXPERIMENTAL

Chemicals

All of the bisphenols were obtained from commercial sources and were purified by recrystallization. 1,6-Dibromohexane (DBH) and nitrobenzene were purified by fractional distillation. Q^+Br^- obtained from Aldrich Chemical was used as received. P^+Br^- and DBC were prepared according to the literature procedures. Other chemicals used in this investigation were of reagent grade and were used as received.

Synthetic Reactions

A general procedure for the synthetic reactions employed in this study was as follows: Bisphenol (5.0m mole) and alkali hydroxide (12.0m mole) were dissolved in 20mℓ of deionized distilled water. DBH (5.0m mole) and PTC (0.1m mole) were dissolved in 20mℓ of nitrobenzene. These two solutions were placed in a 250mℓ round bottom flask equipped with a nitrogen inlet, a water-cooled condenser, a thermometer, and a stirrer.

The reaction mixture was vigorously stirred at $78^{\circ}C$ under a nitrogen atmosphere. At the end of the reaction, the nitrobenzene layer was separated and poured into 100mℓ of acidified (pH=1) methanol. The precipitated polymers were thorougly washed with dilute hydrochloric acid, water, and then methanol.

Characterization of Polymers

Intrinsic viscosities of polymer solutions were measured using a Cannon Ubbelohde type viscometer. Number average molecular weights of polymers were determined by vapor pressure osmometry (Knaner Co.). Thermal properties of polymers were studied using a DSC (Perkin Elmer, Model IB) and a TGA (Perkin Elmer, Model TGA-1) instruments. IR spectra of polymers were obtained by a diffraction grating infrared spectrophotometer (Jasco, Model DS-701G).

Study of Phase-Transfer Phenomenon

The fraction of bisphenolate anions distributed between aqueous and nitrobenzene phases in the presence of a PTC was estimated by the determination of the concentration of bisphenolate anions dissolved in nitrobenzene phase using acid-base titration method or from the amount of K^+ dissolved in nitrobenzene which was determined by flame emission spectrometry (Pye Unicam).

A mixture formulated with the same amount of each of the
ingredients as described in "Synthetic Reactions", excluding DBH,
was placed in a 250mℓ separatory funnel and agitated vigorously for
15 minutes. The mixture was then kept in a constant temperature
bath at 78°C or 88°C for 24 hours. Nitrobenzene layer was separated
and titrated against a standardized hydrochloric acid solution.
When the concentration of K^+ in nitrobenzene was to be determined,
one mℓ of nitrobenzene solution was taken by a pipet and transferred
to a platinum crucible. The solution in the crucible was combusted
in an electrical furnace. The combustion products were then dissolved
in measured amount of deionized distilled water. The solution was
subjected to flame emission spectrophotometry using cesium as an
ionization retarder.

RESULTS AND DISCUSSION

Catalytic Effect of Q^+Br^-, P^+Br^-, and DBC

When the polymerization between bisphenol A (BPA) and DBH was
conducted in the absence of any catalyst, only a trace amount of
polyethers was obtained even after 7 hours at 78°C and 88°C. The
addition of a small quantity of catalysts, however, greatly enhanced
the polymerization rate as shown in Fig. 2. Since we earlier
observed that such biphasic polymerizations depend on the stirring
speed up to a certain rpm,[16] in order to avoid any effect of its
variations, we maintained a very high stirring speed throughout the
present study. Since DBC catalyzed reactions were significantly
slower than others at 78°C, they were conducted at a higher temperature of 88°C throughout this study.

By increasing the amount of the catalyst, regardless of its
structure, the reaction rate was raised but only up to a certain
level above which increase in the quantity of the catalyst used,
did not enhance the polymerization rate to any appreciable extent,
see Fig. 3. At this threshold level of catalyst the supply of
bisphenolate anions by the catalyst to reaction site apparently
becomes equal to or exceeds their consumption rate through polymer-
zation. The results shown in Fig. 2 and 3 imply that the catalytic
activity, being judged by polymer yield, was $P^+Br^- > Q^+Br^- > DBC$.
Further discussion on this point will be made in the later part of
the discussion.

The relationship between inherent viscosities of the resulting
polymers and reaction time is shown in Fig. 4. Molecular weight of
polymers, as reflected by their solution viscosities, increased
with reaction time, which indicates that, as expected, step-growth
mechanism is operating for this type of polymerization. Overall

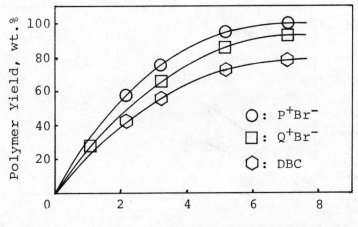

Fig. 2.   Phase-transfer polymerization between bisphenol A and
          1,6-dibromohexane in the presence of 2 mole % catalyst.
          Excess (20 mole %) NaOH or KOH was used.

reactivity of the bisphenolate anions examined was BPA>TDP>KDP>SDP,
as shown in Fig. 5, when Q^+Br^- or DBC was used as a catalyst.   The
same order of reactivity was also observed when P^+Br^- was used
(Table 1).   The order of reactivity of bisphenols parallels the
basicity of corresponding anions or inductive effect of the

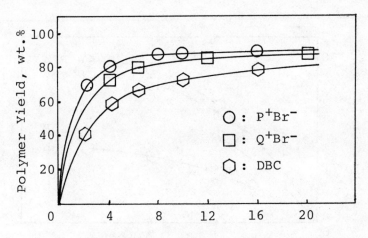

Fig. 3.   Effect of the concentration of PTC on the polymerization
          of thiodiphenol and 1,6-dibromohexane.   Reactions were
          conducted for 3 hours in the presence of 20 mole % excess
          NaOH or KOH.

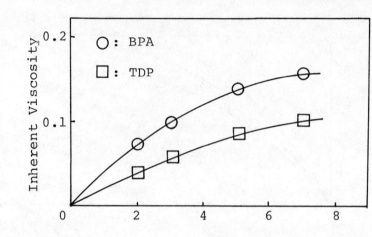

Fig. 4.   Dependence of solution viscosities of polymers on the
           reaction time. Polymerizations were conducted at 78°C
           using 2 mole % P⁺Br⁻ in the presence of 20 mole % excess
           NaOH.

substituents $(C(CH_3)_2, S, CO, and SO_2)$ between the phenolic rings.
Additional factors to be taken into consideration are their transferability into nitrobenzene in the presence of PTC's, which will
be dealt with later in the discussion.

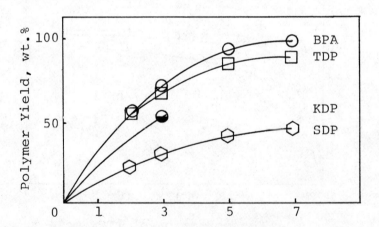

Fig. 5.   Dependence of reaction rate on the structure of bisphenols.
           Polymerization was conducted at 78°C using 2 mole %
           P⁺Br⁻ and 20 mole % excess NaOH.

Table 1. Dependence of Polymer Yield on the Structure
of Bisphenols*

Bisphenol	BPA	TDP	MDP	KDP	SDP
Polymer yield, wt.%	73.2	69.9	64.2	50.0	31.6

*Data were obtained after 3 hrs. of reaction at 78°C
using 2 mole % P^+Br^-.

Effect of Alkali Metal Cations and Concentration of Hydroxides

Since it is well known that stability of a complex between a
crown ether and an alkali metal cation greatly depends on the size
of the inner cavity of the former and of the latter and that of all
the alkali metal ions dibenzo-18-crown-6 forms the most stable complex with K^+,[27] it was interesting to see whether these facts in a
parallel way influence the present polymerization reactions. We
conducted a series of polymerizations under the same condition as
described in "Synthetic Reactions", using various alkali hydroxides
(LiOH, NaOH, KOH, and CsOH); the results are tabulated in Table 2.
The results strongly imply that phase transfer of the nucleophiles,
bisphenolate anions, from the aqueous to the nitrobenzene phase was
most efficient when the metal ion was K^+. This, in turn, indicates
that the crown ether-K^+ complex is the most stable among the complexes expected to be formed with the alkali metals examined, as
has been amply demonstrated by others. For example, Pedersen[9]

Table 2. Dependence of Polymerization Rate on Alkali
Metal Cations*

Alkali metal hydroxide	BPA polyether yield, wt.%
LiOH	No reaction
NaOH	5.3
KOH	60.9
CsOH	1.2

*Polymerization of BPA and 1,6-dibromohexane at 88°C for
3 hours. The amounts of reactants and solvent were the
same as described in "Experimental" section. Excess
(20 mole %) LiOH, CsOH, NaOH and KOH were used.

Table 3.   The Effect of Hydroxide Ion Concentration on the
           Reaction Rate*

Salt concentration	KOH, mole			
	0.020	0.025	0.030	0.035
BPA polymer yield, wt.%	49.5	77.5	77.9	79.0

*Data obtained from the polymerization of BPA and 1,6-dibro-
mohexane at 88°C for 3 hrs. in the presence of 4 mole % DBC.

studied solvent extraction by DBC of picrate salts from the aqueous
phase to methylene chloride and found that Li^+, Na^+, K^+, and Cs^+
salt was phase-transferred into the organic phase to the extent of
0%, 1.6%, 25.2%, and 5.8%, respectively. This trend is quite similar
to that shown in Table 2. UV study of complexation between DBC and
various alkali cations lead to the same conclusion.[9] The tremendous
increase in polymer yield with K^+ strongly emphasizes the importance
of selecting the proper complexing cation when a crown ether is employed as a phase-transfer catalyst. As can be seen from Table 3
and 4, polymerization was very much accelerated by the use of the
excess amount of KOH. The presence of higher concentration of K^+
in the aqueous phase must have favored the formation of complexes
between K^+ and the crown ether at the interface, increasing the
transfer of bisphenolate anions from the aqueous to the nitrobenzene

Table 4.   Dependence of Polymer Yield on the Amount
           of Potassium Hydroxide Used*

Ratio of KOH to reactants	Polymer yield, wt. %			
	BPA	TDP	KDP	SDP
1 : 1	49.5	38.9	30.7	18.8
1.2 : 1	79.4	68.1	51.2	27.2

*Polymerization of bisphenolate anions and 1,6-dibro-
mohexane at 88°C for 3 hours. Amount of the DBC used
was 6 mole % of diphenoxide.

Table 5. Dependence of Polymer Yields on the Amount of Sodium
          Hydroxide Used

Mole ratio of NaOH to reactant	Polymn. with Q$^+$Br^{-}[1]			Polymn. with DBC[2]			
	BPA	TDP	SDP	BPA	TDP	KDP	SDP
1 : 1	40.2	24.1	11.6	49.5	38.9	30.7	18.8
1.2 : 1	64.5	53.1	16.8	79.4	68.1	51.2	27.2

[1] Polymerized for 3hrs. at 78°C in the presence of 2 mole %
    catalyst.
[2] Polymerized for 3hrs. at 88°C in the presence of 6 mole %
    catalyst.

phase. This point will be discussed more fully in later part of
this paper. We have observed the same phenomena in the polymerizations with Q$^+$Br$^-$ and P$^+$Br$^-$ catalysts. The results tabulated in
Table 5 clearly demonstrate that the use of excess amounts of sodium
hydroxide resulted in higher yields of polymers. The presence of a
higher concentration of hydroxide ions in the aqueous phase apparently favored more facile transfer of bisphenolate anions into
nitrobenzene, which was confirmed by a distribution study to be
discussed in the following section.

Distribution of Bisphenolate Anions between Aqueous Phase and Nitrobenzene in the Presence of PTC

Catalytic activity of a PTC in the reactions under investigation should depend on its ability to bring selectively bisphenolate
anions to the reaction site and to remove the competitive by-product,
bromide ions, from the site, tightness of ion pairs between the
catalyst and reacting species transferred to the organic phase, and
others such as polarity of the organic phase and solubility of
resulting polymers. In order to compare the phase-transfer capacity
of PTC's employed in this study, we examined partition of bisphenolate anions between the aqueous phase and nitrobenzene. The results
are tabulated is Table 6.

Solubilization of anions by PTC's regardless of their stucture
was BPA>TDP>KDP>SDP. This seems to be in accord with the lyophilicity
of bisphenolate anions. Among the three PTC's used P$^+$Br$^-$ was the most
efficient in transferability of bispholate anions from the aqueous
phase into nitrobenzene and DBC was the least. The use of an excessive amount of hydroxide favored a higher degree of partition of

Table 6.   Solubilization of Bisphenolate Anions by PTC's
           into Nitrobenzene*

Bisphenol	P^+Br^-		Q^+Br^-		DBC	
	1:1	1:1.2	1:1	1:1.2	1:1	1:1.2
BPA	78	88	70	82	75	81
TDP	71	78	60	64	63	65
KDP	64	68	–	–	24	31
SDP	43	61	27	47	10	14

*Data presented are mole % bisphenolate anions dissolved
into nitrobenzene and expressed as mole % of catalysts
used, i.e., 2 x moles of bisphenolate in nitrobenzene
divided by moles of catalyst initially added. Distribution was studied at 78°C (P^+Br^- and Q^+Br^-) and 88°C (DBC)
using 2 mole % catalysts of bisphenols. The ratios of
1:1 and 1:1.2 designate the mole ratios of bisphenols to
NaOH (P^+Br^- and Q^+Br^-) and KOH(DBC).

the anions by PTC's in the organic phase. All of these phenomena,
at least partially, can explain the observations made in polymerizations such as its order of catalytic activity of PTC's, enhanced
reaction rate when excess amount of hydroxide was used, and the
order of reactivity of bisphenol anions. The combination of the
two factors, greatest nucleophility and highest transferability by
PTC into the nitrobenzene phase, makes the BPA anion the most reactive among the anions examined. The opposite is true for SDP.

    The lowest catalytic activity of DBC is rather perplexing con-
sidering its high transfer capacity of anions into the organic
phase, which was comparable to that of Q^+Br^-, see Table 6. Some
possible reasons among many can be surmised: slower anion transfer
by DBC into nitrobenzene due to its insolubility in the aqueous
phase, less efficient removal of bromide anions, a reaction product,
from the organic phase because of high lyophilicity of DBC, and a
formation of tighter ion pairs between K^+-DBC complex and bisphenolate anions. We do not have at the present moment any experimental
evidence to approve or disapprove the above mentioned reasons. We
could, however, qualitatively observe that DBC could carry KBr from
the aqueous phase into nitrobengene in the presence of bisphenol
anions to a much greater extent than Q^+Br^- and P^+Br^- did.

Mechanism of Polymerization

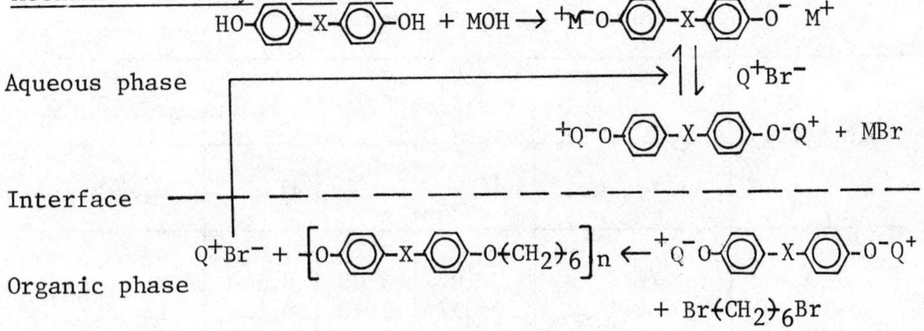

$$HO\text{-}\bigcirc\text{-}X\text{-}\bigcirc\text{-}OH + MOH \rightarrow {}^+M\text{-}O\text{-}\bigcirc\text{-}X\text{-}\bigcirc\text{-}O^- M^+$$

Aqueous phase

$$Q^+Br^-$$

$$^+Q^-O\text{-}\bigcirc\text{-}X\text{-}\bigcirc\text{-}O^-Q^+ + MBr$$

Interface ——

Organic phase

$$Q^+Br^- + \left[O\text{-}\bigcirc\text{-}X\text{-}\bigcirc\text{-}O\text{-}(CH_2)_6\right]_n \leftarrow {}^+Q^-O\text{-}\bigcirc\text{-}X\text{-}\bigcirc\text{-}O^-Q^+$$

$$+ Br(CH_2)_6Br$$

Fig. 6. Phase-transfer polymerization mechanism between bisphenols and 1,6-dibromohexane in H_2O/nitrobenzene mixture.

All of the observations described so far imply that the widely accepted mechanism is operative in the present polymerizations. Bisphenolate anions originally present in the aqueous phase are tranferred by Q^+ or P^+ or K^+-DBC complex into nitrobenzene, where they react with DBH producing polyethers. The side product, bromide ion, generated in the organic phase during reaction is brought by the catalysts into the aqueous phase. This cycling of catalyzing species between the two phases continues to supply the reactant to the reaction site and to transport the bromide back into aqueous phase. Schematic representation of the mechanism is shown in Fig. 6. When DBC is used as a catalyst, its complex with K^+ corresponds to the onium ion Q^+ in the scheme shown above. Rather poor catalytic efficiency and high lyophilic character of DBC, however, suggest that a modified mechanism may be applicable in the case of DBC catalyst. The mechanism to be proposed here is the one in which DBC-K^+ complex, once formed at the interface between water and nitrobenzene, resides mostly in organic phase and exchange its counter anions mainly at the interface. As shown in Fig. 6, when onium salts are employed as a catalyst, they cycle between the two phases crossing the border exchanging their counter anions. In a separate experiment, we observed that DBC was not solubilized into aqueous KBr or KOH solutions to any measurable extent in the absence as well as presence of bisphenols. This supports the assumption that DBC mainly, if not completely, stays in the nitrobenzene phase under the present polymerization conditions. In spite of such a mechanistic difference, it is obvious that PTC's make actual polycondensation to occur homogeneously in the organic phase, not at the interface, which is a basic distinction between phase-transfer polymerizations and conventional interfacial reactions.

The above shown mechanism is expected to bring about two important consequences as far as the structure of resulting polymers are concerned. When the reaction is conducted at low concentration

Table 7.   Properties of Polyethers Prepared by PTP of
           Bisphenols and 1,6-Dibromohexane

Bisphenol	Polymers I[*1]			Polymers II[*2]	
	Yield, wt.%	$[\eta]$	$\overline{Mn}$	ηinh	Tm, ^{o}C
BPA	79.4	0.102	3500	0.340	113
TDP	68.1	0.097	2540	0.338	150
MDP	–	–	–	0.339	144
KDP	51.2	0.091	2270	0.329	219
SDP	27.2	0.081	1860	0.338	167

[*1] Polymers I were obtained after 3 hours of reaction at
     88°C using 6 mole % DBC. Intrinsic viscosities of
     polyethers from BPA, TDP, and SDP were measured at 25°C
     and that of KDP at 50°C using chloroform, chlorobenzene,
     1,2-dichlorobenzene, and DMF, respectively.
[*2] Polymers II were obtained after 24 hours of reaction at
     100°C using 10 mole % P^{+}Br^{-}. All of the inherent
     viscosities were measured on 0.4g/100mℓ solution of the
     samples in 1,1,2,2-tetrachloroethane at 70°C, with the
     exception that ηinh of KDP polymer was measured for
     trifluoroacetic acid solution at 50°C.

of PTC, in other words, below the threshold level mentioned earlier,
the supply of bisphenolate ions by PTC to the organic phase would
be slower than their consumption by reaction and the transport would
become rate-controlling. Under this condition the growing chains
in the organic phase would be of ω-bromohexyl group, and resulting
polymers would have bromoalkyl end groups. IR spectra of the polymers prepared at low concentration (for example, 2 mole %) of PTC's
showed absorption bands around 660 cm^{-1} corresponding to C-Br
stretching and no phenolic O-H absorptions. We could also observe
the formation of AgBr precipitates when DMF solution of polymers
was treated with AgNO$_3$. This is in agreement with the observations
made by N'Guyen and Boileau[28].

On the contrary, polymers having phenolic as well as bromoalkyl
end groups will be obtained when the amount of PTC used is higher
than the threshold level, since supply of bisphenolate anions by
PTC to nitrobenzene would exceed their disappearance through reaction. Even though IR spectra of the polymers prepared using 10 mole
% P^{+}Br^{-} (Table 7) showed very weak absorptions both of C-Br and
phenolic O-H, concrete conclusions can not be drawn yet.

Properties of Polymers

General properties of polymers prepared under two different reaction conditions are summarized in Table 7. Molecular weights of polymers were not high even after prolonged reaction using rather high concentration of catalyst. All of the polymers showed rather sharp melting endotherms on DSC thermograms indicating semicrystalline nature. Polymers derived from BPA, TDP, and SDP started to lose their weights around 350°C when examined by TGA. DSC analysis showed that they underwent endothermic thermal degradation at the corresponding temperatures. Polymers of higher molecular weight prepared from BPA, TDP. MDP and SDP were soluble in hot 1,1,2,2,-tetrachloroethane, but those derived from KDP and BDP were not soluble. Lower molecular weight polymers were more readily soluble in various solvents as indicated in Table 7.

ACKNOWLEDGEMENT

Financial support of this work by the Ministry of Education of Korea is acknowledged. Thanks are also due to the American Chemical Society for the partial finacial support to J.-I. Jin to participate in the ACS symposium on "Crown Ethers and Phase Transfer Catalysis in Polymer Chemistry" held at Las Vegas on March 29th through April 2nd, 1982.

REFERENCES

1. C.M. Starks, J. Amer. Chem. Soc., 93(1), 195 (1971).
2. C.M. Starks and R.M. Owens, J. Amer. Chem. Soc., 95(11), 3613 (1973).
3. A. Brändström and U. Junggren, Acta Chem. Scand., 23, 2203, 2204, 2536, 3585 (1969); ibid, 25, 1469 (1971).
4. A. Brändström and K. Gustavii, Acta Chem. Scand., 23, 1215 (1969).
5. A. Brändström, K. Gustavii, and S. Allanson, Acta Chem. Scand., 25, 77 (1971).
6. M. Makosza, Tetrahedron Lett., 673, 677 (1969).
7. M. Makosza and W. Wawrzyniewicz, Tetrahedron Lett., 4659 (1969).
8. M. Makosza and E. Bialecka, Tetrahedron Lett., 4517 (1971).
9. C.J. Pedersen, J. Amer. Chem. Soc., 89, 7017 (1967).
10. B. Dietrich, J.M. Lehn, and J.P. Sauvage, Tetrahedron Lett, 2885, 2889 (1969).
11. J.M. Lehn, U.S. Pat., 3,888,877 (1975).
12. R.M. Izatt and J.J. Christensen, Eds., "Synthetic Multidentate Macrocyclic Compounds," Academic Press, New York, 1978.
13. H.K. Frensdorff, J. Amer. Chem. Soc., 93(3), 600 (1971).
14. L.J. Mathias, J. Macromol. Sci.-Chem., A15(5), 853 (1981).
15. Y. Imai, J. Macromol. Sci.-Chem., A15(5), 833 (1981).

16. J.-I. Jin, Y.-W. Jung, K.-S. Lee, and K.-W. Chung, J. Korean
    Chem. Sec., 24(3), 259 (1979).
17. M. Hirose and Y. Immamura, Nippon Kagaku Zasshi, 1, 113 (1977).
18. K. Soga, S. Hosoda, and S. Ikeda, J. Polym. Sci.: Polym. Chem.
    Ed., 17, 517 (1979).
19. Y. Imai, M. Ueda, and M. Ii, Makromol, Chem., 179, 2085 (1978).
20. Y. Imai, A. Kato, M. Ii, and M. Ueda, J. Polym. Sci.: Polym.
    Lett. Ed., 17, 579 (1979).
21. A.K. Banthia, D. Lunsford, D.C. Webster, and J.E. McGrath, J.
    Macromol. Sci.-Chem., A15(5), 946 (1981).
22. J.S. Riffle, R.G. Freelin, A.K. Banthia, and J.E. McGrath, J.
    Makromol. Sci.-Chem., A15(5), 967 (1981).
23. D.J. Gerbi, R.F. Williams, R. Kellman, and J.L. Morgan, Polym.
    Preprints, 22(2), 385 (1981).
24. H. Schnell, Angew. Chem., 68, 633 (1954).
25. P.W. Morgan, Macromolecules, 5, 536 (1970).
26. J.-I. Jin, K.-S. Lee, J.-H. Chang and S.-J. Kim, Polymer (Korea),
    6(1), 60 (1982).
27. C.J. Pedersen and H.K. Frensdorff, Angew. Chem. Int. Ed., 11(1),
    16 (1972).
28. T.D. N'Guyen and S. Boileau, Polym. Preprints, 23(1), 154 (1982).

PHASE TRANSFER FREE RADICAL REACTIONS:[1] POLYMERIZATION OF

ACRYLIC MONOMERS

Jerald K. Rasmussen and Howell K. Smith, II

Central Research Laboratories
3M, 3M Center
St. Paul, Minnesota 55144

INTRODUCTION

The use of phase transfer catalysis (PTC) for the purpose
of carrying out ionic reactions has become commonplace in recent
years.[2] Although the techniques of PTC have been concerned for
the most part with the transfer of anionic reagents between two
immiscible phases for subsequent nucleophilic reactions, the
concept is believed to be much more general.[3] Recently, for
instance, a few examples describing extension of the process to
cationic (electrophilic) reagents have been reported.[4,5]

While PTC has become a powerful tool for the synthetic
organic chemist, it has also had tremendous impact in the field
of polymer science.[6] Numerous examples of polymer modification
and functionalization reactions employing phase transfer catalysts
have been described. Even more striking, however, has been the
role of PTC in actual anionic polymerization reactions, where
dramatic effects on polymerization rates, yields, and microstructure can be attributed to the catalyst. Condensation polymerizations have also been facilitated in the presence of phase
transfer catalysts. Only recently we reported the first examples
of phase transfer initiated free radical polymerization.[7,8] The
present article will detail the features of phase transfer free
radical polymerizations and will also describe some of the characteristics of the polymers formed.

EXPERIMENTAL

Unless indicated otherwise, all materials were obtained

105

commercially and used without further purification. Notably, to
mimic commercial processes, monomers were utilized without the
removal of inhibitors. Typically, polymerization reactions were
conducted under an inert atmosphere at 40% solids, 55°C for 24
hours. Following reaction, conversion of monomer to polymer was
determined by a simple gravimetric technique. Inherent viscosities
were measured at 30°C using a Canon viscometer at a concentration
of 0.1g/100 ml. tetrahydrofuran. Gel-permeation chromatography
was performed using a Waters gel-permeation chromatograph, model
200, with columns containing 10^3-10^7 angstrom pore sizes. In the
kinetic studies, disappearance of peroxydisulfate (initially 0.01$\underline{M}$)
was measured by an indirect coulometric technique, using ferrous
ion as an intermediate reagent.

RESULTS AND DISCUSSION

Crown Ethers As Phase Transfer Agents

     Our investigations into the possible use of phase transfer
catalysis to accomplish free radical polymerizations began in the
latter part of 1976. In our initial experiment, we attempted to
polymerize ethyl acrylate in acetone solvent utilizing potassium
peroxydisulfate and 1,4,7,10,13,16-hexaoxacyclooctadecane(18-crown-
6, $\underline{1}$) at the 1 and 2 mole percent levels respectively based on
monomer as the catalyst system. The major question at the outset
seemed to be whether peroxydisulfate could be effectively phase
transferred, a notoriously difficult process with divalent anions.[9]
To circumvent this potential problem, water was added to the
reaction medium until homogeneity was obtained. This occurred at
a 93.5/6.5(w/w) acetone/water ratio. The stirred reaction mixture, under Argon, was heated at reflux for thirty minutes to
give a quantitative conversion of monomer to polymer. The exo-
therm associated with the polymerization was observed to the
exceedingly mild, particularly since the initiator level was
10-25 times that normally utilized for acrylic solution polymerizations. The mildness and facility of this reaction prompted our
subsequent investigations. Table 1 compares the results obtained
with several initiator systems under similar reaction conditions.
The results indicated that the $K_2S_2O_8 \cdot 2$(18-crown-6) intiator ($\underline{2}$)

$\underline{1}$

TABLE 1. Polymerization[a] Of Ethyl Acrylate

Catalyst	Reaction Time, hrs.	Conversion, %
$K_2S_2O_8 \cdot 2$(18-crown-6)	1.5	73
AIBN	1.5	58
Benzoyl peroxide	3.5	62
$K_2S_2O_8$(control)	6.0	36[b]

[a]Reaction conditions: ethyl acrylate(50g),distilled water(3g),
acetone(73g),catalyst(0.1mole% based on monomer),70°C bath
temperature under argon
[b]Nonhomogeneous reaction mixture

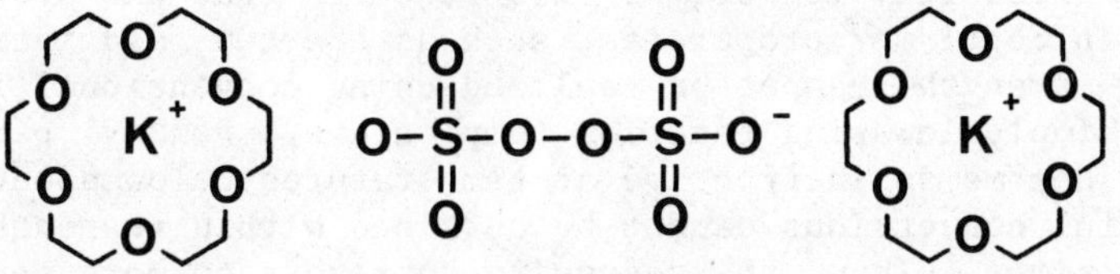

2

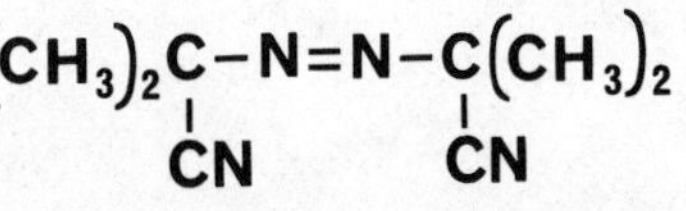

3

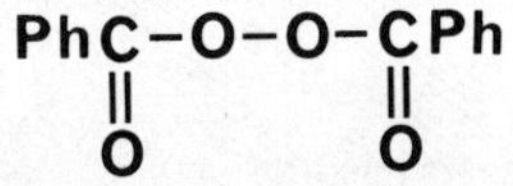

4

was far more efficient than the more conventional initiators,
azobisisobutyronitrile (AIBN, 3) and benzoylperoxide (4), despite
the fact that all three were expected to have similar activation
energies for decomposition ($K_2S_2O_8$,[10] 28-31 kcal mol^{-1} in neutral
aqueous solution; AIBN,[11] 31 kcal mol^{-1}; benzoyl peroxide,[12] 32.7
kcal mol^{-1}.

This lead us to speculate that complexation with a crown ether might be lowering the activation energy for persulfate decomposition. This thought was quite intriguing since it suggested the possibility for in situ generation of highly reactive initiators from quite stable precursors.

To test the practical utility of this concept, we decided to study the influence of polymerization temperature on polymer properties. We chose a monomer composition of 80 mole % isooctylacrylate[13] (IOA)/20 mole % acrylic acid (AA). This was polymerized in acetone, 0.1 mole % $K_2S_2O_8 \cdot 2$(18-crown-6) as initiator, for 24 hours at temperatures ranging from 35°C to 60°C. Conversions were excellent in all cases except that at 35°C, which was slightly lower. The polymer solutions were coated on polyester film and dried to enable testing as pressure sensitive adhesives.[14] Polymer and tape properties are listed in Table 2. As expected, as the polymerization temperature was lowered, the molecular weight of the resultant polymer increased. Relative shear strength, a measure of the cohesion of the polymer,[14] is plotted in Figure 1 and compared with $\overline{M}_w$ as measured by gel-permeation chromatography (GPC).

The noteworthy feature of this study is that the dramatic changes in copolymer properties, such as the ten-fold increase in shear strength, cannot be realized using conventional initiators by simply lowering reaction temperature. Benzoyl peroxide and AIBN become so ineffective at temperatures below about 50-55°C that useful conversions cannot be obtained within reasonable reaction times. Thus, the necessity to resort to more active and correspondingly hazardous initiators can be obviated.

TABLE 2.   Polymerization of 80:20 IOA/AA

Catalyst	Temp., °C	Conversion, %	$10^{-6}\overline{M}_w$	Shear,[a] min
AIBN	60	96.0	2.21	43
$K_2S_2O_8$/18-crown-6	60	95.0	2.09	30
$K_2S_2O_8$/18-crown-6	50	95.4	4.53	105
$K_2S_2O_8$/18-crown-6	45	96.0	6.06	174
$K_2S_2O_8$/18-crown-6	40	94.7	8.12	286
$K_2S_2O_8$/18-crown-6	35	92.0	7.67	286

[a]Measured according to the procedure cited in reference 14 using a 1000 gram load and an adhesive contact area of $1.61cm^2$.

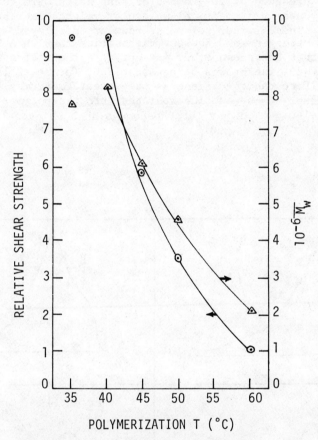

Fig. 1.   Comparison of relative shear strength and molecular
          weight by GPC of IOA/AA copolymers as a function of
          polymerization temperature.

The molecular weight distributions of the polymers turned out to be quite unusual. Whereas polydispersities were normal to a bit broader than normal, the distributions often showed a high molecular weight shoulder or were nearly bimodal (Figure 2). At the lower polymerization temperatures, this high molecular weight fraction approached an astonishing 2.5×10^7 in molecular weight. This phenomenon appears to be fairly specific to the phase transfer initiated polymers, since polymers initiated with AIBN in the presence of 18-crown-6 or quat salts were found to have normal, nearly symmetrical molecular weight distributions (Figure 3). These latter polymers, however, displayed somewhat lower shear strength and inherent viscosities than an AIBN control, indicating that the crown ether may act as a chain transfer agent. With persulfate, the crowns being associated (at least loosely) with the sulfate radical anions as they are formed, may have a greater tendency to become incorporated into the polymer as initiation sites, then may later become branch or crosslink sites (Scheme 1).

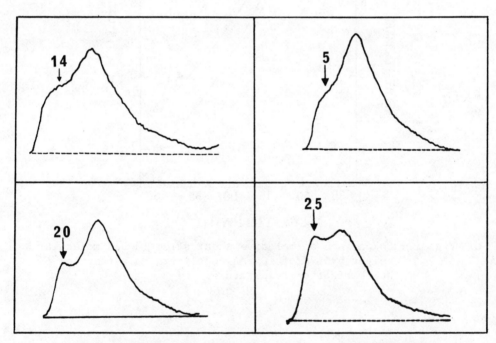

Fig. 2 Representative GPC traces of IOA/AA copolymers using
K$_2$S$_2$O$_8 \cdot 2$(18-crown-6) as initiator. Peak molecular weights
(x10^{-6}) of the high molecular weight fractions are
indicated by the arrows.

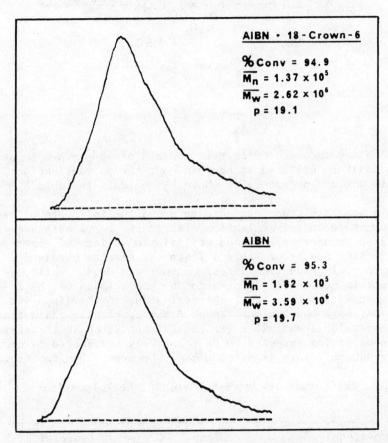

Fig. 3. GPC traces of IOA/AA copolymers polymerized in acetone, 55°C, 24 hours, in the presence and absence of 18-crown-6, using AIBN as initiator.

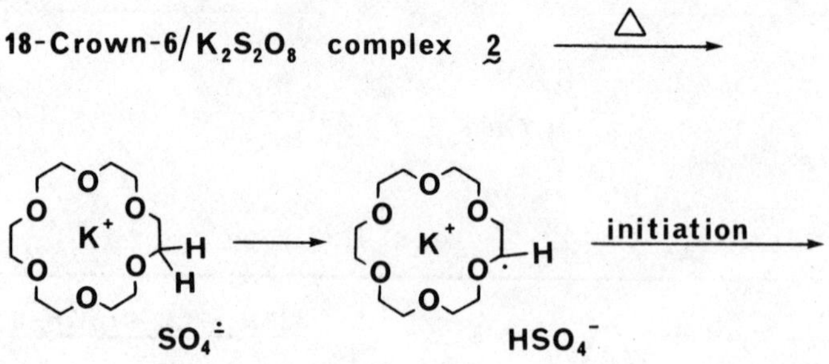

18-Crown-6/$K_2S_2O_8$ complex 2 $\xrightarrow{\Delta}$

Scheme 1

        To this point, stoichiometric amounts of 18-crown-6 to per-
sulfate had been utilized in carrying out the polymerizations.
That this was not necessary is shown by the data in Table 3.

        In a separate study designed to probe the influence of crown
ether structure on polymerization efficiency, butyl acrylate was
polymerized in acetone solution utilizing a variety of crowns as
phase transfer agents.[7] Table 4 lists the results obtained
when commercial grade (inhibited) monomer was used. While one
cannot obtain meaningful information by comparision of the inherent viscosities throughout the series due to the disparity in
conversion, some conclusions can be drawn concerning chain transfer.
Where conversion are similar (Entries 1 and 2), chain transfer,
if it is occurring, appears to be relatively unaffected by crown
ether structure. This is not the case, however, when the crown

TABLE 3.   Effect of 18-Crown-6 Level In The Polymerization[a] Of
           80:20 IOA/AA

18-crown-6,equiv.[b]	Temp., $^\circ$C	Conversion, %
0.5	50	92
0.25	55	94
0.1	55	94
0.01[c]	55	71

[a] Identical conditions to those described for Table 2.
[b] Based on $K_2S_2O_8$ charged.
[c] Monomer used was isooctylacrylate only.

TABLE 4. Poly(Butylacrylate) via $K_2S_2O_8$/Crown Ethers

Entry	Crown	Conversion, %	ηinh, dlg^{-1}
1	15-crown-5	63	3.10
2	Dibenzo-18-crown-6	65	3.08
3	Dicyclohexano-18-crown-6	85	2.86
4	18-crown-6	94	2.58
5	[2.2.2]-cryptand	95	0.86
6	[a]	91	2.24
7	None	14	--

[a]Threo-2,5-poly(tetrahydrofurandiyl); W.J. Schultz, M.C. Etter, A.V. Pocius, S. Smith, J. Amer. Chem. Soc., 102, 7981 (1980).

ether contains heteroatoms other than oxygen (compare entries 4 and 5). The presence of nitrogen atoms in the cryptand apparently dramatically enhance the chain transfer ability of that catalyst.

In addition to information concerning chain transfer ability, the data in Table 4 indicated a relationship between conversion and the complexing ability of the particular crown ether utilized as phase transfer catalyst. In fact, when percent conversion was plotted vs the log of the binding constants (log K) of the respective crowns for potassium cation, an apparently linear correlation was obtained. On the basis of some simple assumptions and free radical addition polymerization theory, however, it may be deduced that conversion should correlate with K rather than log K (Equation 1). This anomaly was resolved by using inhibitor-free monomer (Table 5), whereupon a good correlation (R=0.948)[7] between the variables was obtained.

$$\frac{-d[\text{Monomer}]}{dt} = \frac{kK[\text{Crown}][\text{Monomer}]}{[\text{Crown} \cdot \text{K}^+]} \qquad (1)$$

TABLE 5. Polymerization[a] Of Butyl Acrylate by $K_2S_2O_8$/Crown Ethers

Crown	Conversion, %	log K_{CH_3OH}
18-crown-6	73.4 ± 0.6[b]	6.06[c]
dicyclohexyl-18-crown-6	49.2	5.70[c,e]
21-crown-7	41.1	4.22[c]
dibenzo-18-crown-6	23.6	4.36[d]
15-crown-5	10.7	3.77[c]
cyclohexyl-15-crown-5	13.7	3.58[d]
dibenzo-24-crown-8	1.04	3.49[d]
1,10-diaza-18-crown-6	3.10	2.04[d]
1,10-dithia-18-crown-6	0.25	1.15[d]
12-crown-4	0.16	f
none(control)	0.10	

[a] Polymerization conditions: Monomer(45.5g),acetone(90ml), $K_2S_2O_8$(0.5mmol),crown(1mmol),55°C, Argon,3 hrs.

[b] Average of two trials.

[c] J.D. Lamb, R.M. Izatt, C.S. Swain, J.J. Christensen, J. Am. Chem. Soc., 102, 475-479 (1980).

[d] J.J. Christensen, D.J. Eatough, R.M. Izatt, Chem. Rev., 74, 351-384(1974).

[e] The crown used was a mixture of isomers: isomer A(log K=6.01) and isomer B(log K=5.38).

[f] Not available.

In an attempt to get a more basic understanding of the initiator system, we undertook a study of the kinetics of decomposition of aqueous potassium persulfate.[15] In 0.1N potassium hydroxide buffer, the apparent activation energy for decomposition was found to decrease from 33.5 kcal/mole[16] to 19.9 kcal/mole upon the addition of 18-crown-6. A more detailed investigation has now shown[17] this to be a radical chain decomposition in which crown is being oxidized. The accelerated decomposition seems to be due to the reaction of the crown cation radical and persulfate dianion (Scheme 2). Addition of a free radical trap, in this case methacrylonitrile, suppressed the rate of persulfate disappearance to that normally observed in the absence of crown. Obviously the chemistry observed in aqueous media in the absence of monomer cannot be directly correlated with the polymerization behavior in organic media.

Scheme 2

In an independent study, Voronkov and coworkers have reported[18] the preparation of a 1:2 $K_2S_2O_8$/18-crown-6 complex characterized as being soluble in methanol, dimethyl sulfoxide, and dimethylformamide. Results obtained using this complex for the polymerization of styrene and methyl methacrylate (MMA) in methanol at 60°C are listed in Table 6. The obviously reduced efficiency of this sytem as compared with results obtained under our conditions certainly can be traced in part to the use of less reactive monomers. However, we can speculate that the choice of solvent, protic vs aprotic, may play a major role as well.

TABLE 6. Polymerizations[a] Using $K_2S_2O_8 \cdot 2$(18-Crown-6) in Methanol

| Monomer[b] | Concentration,wt% | | Time, | Conversion | |
	Monomer	Initiator	hrs	%	$10^{-3}\overline{M}_w$
Styrene	80	3	24	34	25
Styrene	50	3	24	25	22
MMA	80	3	4	80	160
MMA	50	3	6	60	201

[a] Reference 18.
[b] Inhibitor free monomer used.

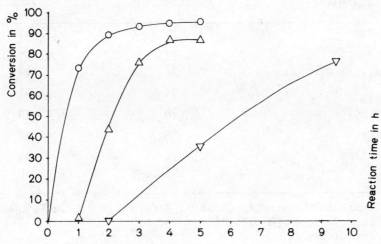

Figure 4.   Polymerization of Butylacrylate at 55°C Using Various Initiators

Quaternary Salts As Phase Transfer Agents

        Subsequent to our initial discovery that 18-crown-6 could
effectively phase transfer peroxydisulfate, we began a study aimed
at determining the scope and generality of the process of phase
transfer free radical polymerization.[19] Besides the crown ethers
several other classes of phase transfer agents were found to be
useful including quaternary onium salts, linear polyethers, di-
and polyamines, and amphoteric or zwitterionic compounds. Of
these, the quaternary ammonium salts were found to be extremely
efficient,[8] even more so than 18-crown-6. Results obtained using
several quat salts (Table 7) verify the phase transfer nature of
the system. As has been found with other phase transfer catalyzed
reactions, small quaternary ions are less effective than larger
ones and micellar type ions are less effective than the more
symmetrical ones. This latter point is more dramatically illustrated in Figure 4, which also compares the activity of AIBN.

TABLE 7. Poly(Butylacrylate)[a] Via $K_2S_2O_8$/Quat Salts[b]

Phase Transfer Agent	Reaction Time, hrs.	Conversion, %
$\underline{n}\text{-Bu}_4N^+$ HSO_4^-	21	97
Aliquat 336[c]	21	97
$CH_3(CH_2)_{15}\overset{+}{N}C_5H_5$ Br^-	24	92
$CH_3(CH_2)_{15}\overset{+}{N}(CH_3)_3$ Br^-	24	91
$(CH_3)_4N^+$ Cl^-	24	43
$\underline{n}\text{-Bu}_4P^+$ Br^-	18.5	88
18-crown-6	24	93
None	21	10

[a]Conditions: Butyl Acrylate(45.5g), EtOAc(67g), H_2O(3g), $K_2S_2O_8$
(0.5mol), P.T. Agent(1mmol),55°C.
[b]Data from Reference 8.
[c]Methyltricaprylylammonium chloride, Henkel Corp., Mpls, MN.

In many instances, particularly when utilizing the quat salts as phase transfer agents, we have added sufficient amounts of water to dissolve the persulfate. This produces a polymerization mixture consisting largely of an organic phase containing solvent, monomer and phase transfer agent, and a minor amount of an aqueous phase containing the persulfate. Dissolution of the persulfate in water appears to be completely optional, not only in the case of crown ethers, but with the quat salts as well. In general, however, we have found that addition of up to about 5% by weight of water based on the total reaction mixture imparts improved homogeneity to the resultant polymer solutions.[19]

In conjunction with our studies on quaternary ammonium salts as phase transfer agents for persulfate, we discovered that it was quite easy to prepare pure quaternary ammonium persulfates from organically soluble quat salts by a simple ion-pair extraction technique (Equation 2). With Aliquat 336 one simply shakes a pentane solution of the quat salt with a saturated aqueous solution of excess ammonium persulfate for about five minutes, draws off the aqueous layer and removes the pentane under reduced pressure to achieve quantitative conversion to the quat persulfate. This enabled us to investigate the decomposition of persulfate in organic media (Table 8). Based upon the somewhat limited data, two points can be made. First of all, rate of decomposition seems to be little influenced by the structure of the cation. Secondly, the activation energy for decomposition of persulfate does indeed appear to be lowered by 2-4 kcal/mole over that observed in neutral aqueous media. This is enough to change the half life at $60^{\circ}C$ from 18 hours to approximately 49 minutes, and may in large part be responsible for the initiator efficiency.

TABLE 8. Decomposition of Quaternary Ammonium Persulfates $(Q^+)_2S_2O_8$

Q^+	Temp., $^{\circ}C$	Solvent	k, sec^{-1}	E_a, kcal mol^{-1}
$(\underline{n}\text{-}C_4H_9)_4\overset{+}{N}$ [a]	45	Acetone	4.49×10^{-5}	
$(C_8\text{-}C_{12})_3\overset{+}{N}CH_3$ [b]	45	Acetone	4.49×10^{-5}	
"	50	Acetone	6.94×10^{-5}	
"	50	MEK[c]	6.43×10^{-4}	
"	55	MEK[c]	1.21×10^{-4}	25.8
"	60	MEK[c]	2.37×10^{-4}	

[a] Tetrabutylammonium.
[b] Methyltricaprylylammonium.
[c] Methyl ethyl ketone.

$$2 \left[R^1 R^2 R^3 R^4 N^+ \ X^- \right]_{org} + \left[M_2 S_2 O_8 \right]_{aq} \rightleftharpoons$$

$$(2)$$

$$\left[(R^1 R^2 R^3 R^4 N^+)_2 S_2 O_8^{2-} \right]_{org} + 2 \left[MX \right]_{aq}$$

SUMMARY AND CONCLUSIONS

The process of phase transfer free radical polymerization extends the use of typically organic-insoluble free radical initiators into solution and bulk processes, whereas previously they could only be used in water-based systems such as in emulsion polymerizations. This is advantageous since the initiators in question are generally more stable and therefore present fewer storage and handling problems as opposed to typical organic-soluble initiators, many of which require refrigeration, etc. In the course of this work, we discovered that the process surprisingly allows polymerizations to be carried out at substantially lower temperatures and/or in much shorter reaction times than those previously possible using common organic-soluble initiators. This has tremendous advantages, especially in bulk or high solids solution polymerizations, in reducing problems associated with heat dissipation. Shorter reaction times also have their obvious economic advantages.

A number of questions still must be answered before a complete understanding of this polymerization system can be formulated. Foremost among these is the apparent contradiction that even though the phase transfer initiators are substantially more reactive than conventional ones, polymerizations utilizing them exhibit lower or more controllable exotherms. This observation has been experimentally verified in pilot plant scale-up through 2000-gallon batches. A second question relates to the superiority of the quat salts over crown ethers in this system. In fact, the quat salts work as well with inhibited monomer as the crowns do in the absence of inhibitor. A part of this may be due simply to better solubilization by the quat salts. Finally, at this point we have been unable to observe, at least qualitatively, any redox polymerizations in organic media utilizing persulfate, a phase transfer agent, and any of a number of reducing agents which commonly work for polymerizations in aqueous media. In contrast to this, Takeishi and coworkers have recently reported[20] that redox polymerization of methyl methacrylate does occur in the absence of solvent with the system potassium persulfate/sodium bisulfate/18-crown-6. These and other questions are the subjects of continuing investigations.

REFERENCES

1.   Chemistry of Naked Persulfate. 6.
2.   C.M. Starks, C. Liotta "Phase Transfer Catalysis:
     Principles and Techniques," Academic Press, New York, 1978.
3.   Reference 2, page 3.
4.   D. Landini, F. Rolla, J. Org. Chem., $\underline{47}$ 154(1982) and
     references therein.
5.   D.W. Armstrong, M. Godat, J. Am. Chem. Soc., $\underline{101}$ 2489(1979).
6.   L.J. Mathias, J. Macromol. Sci.-Chem., A15, 853(1981).
7.   J.K. Rasmussen, H.K. Smith II, J. Am. Chem. Soc., $\underline{103}$, 730
     (1981).
8.   J.K. Rasmussen, H.K. Smith II, Makromol. Chem., $\underline{182}$, 701(1981).
9.   Reference 2, p. 162.
10. D.A. House, Chem. Rev., $\underline{62}$, 185(1962).
11. W.A. Pryor, "Introduction to Free-Radical Chemistry,"
     Prentice-Hall, Inc., Englewood Cliffs, N.J., 1966.
12. R. Curci, J.O. Edwards, in "Organic Peroxides," Vol. 1,
     D. Swern, Editor, Wiley-Interscience, New York, 1970, p.281.
13. L.F. Hatch, "Higher Oxo Alcohols," Wiley, New York, 1957,
     p. 33.
14. S.M. Heilmann, H.K. Smith II, J. Appl. Polym. Sci., $\underline{24}$, 1551
     (1979).
15. J.K. Rasmussen, S.M. Heilmann, P.E. Toren, presented at the
     Second Symposium on Macrocylic Compounds, Provo, Utah,
     August 14-16, 1978, abstract #II.7.
16. I.M. Kolthoff, I.K. Miller, J. Am. Chem. Soc., $\underline{73}$, 3055(1951).
17. J.K. Rasmussen, S.M. Heilmann, P.E. Toren, A.V. Pocius,
     T.A. Kotnour, submitted for publication.
18. T.N. Rakhmatulina, E.N. Baiborodina, A.V. Rzhepka, V.A.
     Lopyrev, M.G. Voronkov, Vysokomol. Soedin., Ser. B., $\underline{21}$, 229
     (1979); Chem. Abstr., $\underline{90}$, 187436v (1979).
19. J.K. Rasmussen, U.S. Patent 4,326,049, April 20, 1982.
20. M. Takeishi, H. Ohkawa, S. Hayama, Makromol. Chem., Rapid
     Commun., $\underline{2}$ 457(1981).

SYNTHESES OF CARBON-CARBON CHAIN POLYMERS AND POLYSULFIDES

BY PHASE TRANSFER CATALYZED POLYCONDENSATION

Yoshio Imai

Department of Textile and
Polymeric Materials
Tokyo Institute of Technology
Meguro-ku, Tokyo 152, Japan

and

Mitsuru Ueda

Department of Polymer Chemistry
Yamagata University
Yonezawa, Yamagata, Japan

INTRODUCTION

Recently, phase transfer catalysis has been effectively exploited in the field of synthetic organic chemistry.[1-3] However, there are scant reports on the syntheses of condensation polymers by phase transfer methods. Five years ago, we started a broad investigation into the use of phase transfer catalysis (PTC) to effect polycondensation. Since then, we have successfully synthesized various types of condensation polymers with high molecular weights. For example, aromatic polysulfonate III from aromatic disulfonyl chloride I and bisphenol II [Eq.(1)],[4,5] aromatic polyphosphonate V from phenylphosphonic dichloride IV and II [Eq.(2)],[6] and aromatic polyether VII from activated aromatic dichloride VI and II [Eq.(3)].[7]

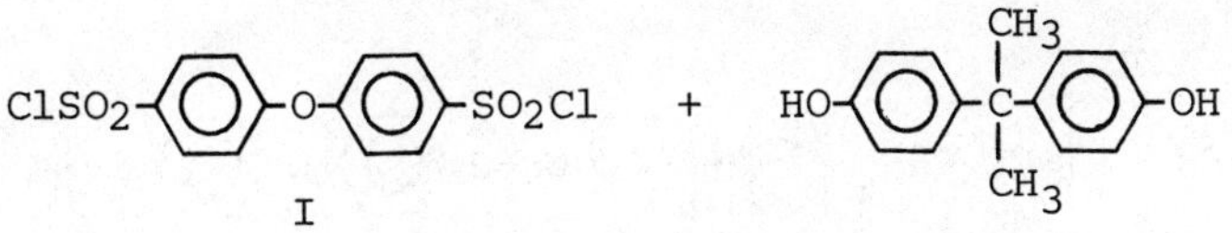

$$ClSO_2-\text{①}-O-\text{①}-SO_2Cl \quad + \quad HO-\text{①}-\overset{\overset{\displaystyle CH_3}{|}}{\underset{\underset{\displaystyle CH_3}{|}}{C}}-\text{①}-OH$$

I

II

$$\xrightarrow{\text{PTC}} \left[-SO_2-\bigcirc-O-\bigcirc-SO_2O-\bigcirc-\underset{\underset{CH_3}{|}}{\overset{\overset{CH_3}{|}}{C}}-\bigcirc-O- \right]_n \quad (1)$$

III

$$\underset{\underset{Ph}{|}}{\overset{\overset{O}{||}}{Cl-P-Cl}} \quad + \quad II \quad \xrightarrow{\text{PTC}} \left[-\underset{\underset{Ph}{|}}{\overset{\overset{O}{||}}{P}}-O-\bigcirc-\underset{\underset{CH_3}{|}}{\overset{\overset{CH_3}{|}}{C}}-\bigcirc-O- \right]_n \quad (2)$$

IV   V

$$\underset{O_2N}{}\underset{}{Cl-\bigcirc-SO_2-\overset{\overset{NO_2}{}}{\bigcirc}-Cl} \quad + \quad II$$

VI

$$\xrightarrow{\text{PTC}} \left[\underset{O_2N}{}-\bigcirc-SO_2-\overset{\overset{NO_2}{}}{\bigcirc}-O-\bigcirc-\underset{\underset{CH_3}{|}}{\overset{\overset{CH_3}{|}}{C}}-\bigcirc-O- \right]_n \quad (3)$$

VII

Phase transfer catalyzed polycondensation was also applicable to the syntheses of aliphatic and aromatic polysulfides. For example, aliphatic polysulfide X from aliphatic dibromide VIII and aliphatic dithiol IX [Eq.(4)],[8] aromatic polysulfide XII from activated aromatic dichloride VI and aromatic dithiol XI [Eq.(5)],[9] and another type of polysulfide XIV from activated dichloride XIII and XI [Eq.(6)].[10]

$$Br(CH_2)_4Br \quad + \quad HS(CH_2)_6SH \quad \xrightarrow{\text{PTC}} \left[-(CH_2)_4-S-(CH_2)_6-S- \right]_n \quad (4)$$

VIII   IX   X

$$VI \quad + \quad HS-\bigcirc-O-\bigcirc-SH$$

XI

$$\xrightarrow{\text{PTC}} \left[\underset{O_2N}{}-\bigcirc-SO_2-\overset{\overset{NO_2}{}}{\bigcirc}-S-\bigcirc-O-\bigcirc-S- \right]_n \quad (5)$$

XII

$$Cl-CH=CH-\underset{\underset{O}{||}}{C}-\bigcirc-\underset{\underset{O}{||}}{C}-CH=CH-Cl \quad + \quad XI$$

XIII

$$\xrightarrow{\text{PTC}} \left[-CH=CH-\underset{\underset{O}{||}}{C}-\bigcirc-\underset{\underset{O}{||}}{C}-CH=CH-S-\bigcirc-O-\bigcirc-S- \right]_n \quad (6)$$

XIV

All of these results were already presented at the ACS Houston Meeting on March, 1980.[11,12] Since that time, further efforts have been made to synthesize novel condensation polymers by phase transfer catalyzed polycondensation. The present article deals with our recent works on the syntheses of carbon-carbon chain polymers and new types of polysulfides.[13] The following abbreviations of phase transfer catalysts have been used throughout this article: tetramethylammonium chloride (TMAC), tetraethylammonium chloride (TEAC), tetrabutylammonium chloride (TBAC), benzyltriethylammonium chloride (BTEAC), cetyltrimethylammonium chloride (CTMAC), cetyltrimethylammonium bromide (CTMAB), benzyltriphenylphosphonium chloride (BTPPC), cetyltributylphosphonium bromide (CTBPB), 15-crown-5 (15-C-5), 18-crown-6 (18-C-6), dibenzo-18-crown-6 (DB-18-C-6), dicyclohexyl-18-crown-6 (DC-18-C-6), dibenzo-24-crown-8 (DB-24-C-8), and dicyclohexyl-24-crown-8 (DC-24-C-8).

RESULTS AND DISCUSSION

Syntheses of Carbon-Carbon Chain Polymers[14,15]

Active methylene compounds are known to undergo condensation with alkyl halides to give excellent yields of dialkylated products under phase transfer conditions.[1-3] We have extended the phase transfer catalyzed alkylation to polycondensation of active methylene compounds such as t-butyl cyanoacetate and phenylacetonitrile with activated dichlorides, and successfully synthesized carbon-carbon chain polymers with high molecular weights. Other types of carbon-carbon chain polymers of rather low inherent viscosities were already prepared by the solution polycondensation of malononitrile with bischloromethyl aromatic compounds in dimethyl sulfoxide using sodium hydride[16] or triethylamine[17] as a base.

The phase transfer catalyzed polycondensation that led to the formation of polymer XVII was first carried out with t-butyl cyanoacetate XV and α,α'-dichloro-p-xylene XVI in a benzene-aqueous alkaline solution system [Eq. (7)].[14]

$$NC-CH_2-\underset{\underset{O}{\|}}{C}-O-\underset{\underset{CH_3}{|}}{\overset{\overset{CH_3}{|}}{C}}-CH_3 \quad + \quad ClCH_2-\bigcirc-CH_2Cl$$

XV   XVI

$$\xrightarrow{PTC} \left[-CH_2-\bigcirc-CH_2-\underset{\underset{COOC(CH_3)_3}{|}}{\overset{\overset{CN}{|}}{C}}-\right]_n \qquad (7)$$

XVII

The effect of reaction temperature on inherent viscosity of
polymer XVII in the two-phase polycondensation with BTEAC is shown
in Figure 1.   The optimum range of reaction temperature was 20-50°C
for the preparation of high molecular weight polymer XVII.   The
reaction product prepared at 80°C contained hydrolyzed structure to
some extent, which was evidenced by the decrease in intensity of
ester carbonyl absorption in the infrared spectrum.

Figure 2 shows the effect of sodium hydroxide concentration in
the aqueous phase on the polycondensation.   High sodium hydroxide
concentration of 50 wt%, which was conveniently used for the general
phase transfer catalyzed alkylation,[1-3] was necessary to prepare
polymer XVII of high molecular weight.

An essential feature of this type of polycondensation is that
relatively large amounts of the phase transfer catalyst, at least
50 mol%, based on each monomer were required to yield the high
molecular weight polymer, as shown in Figure 3.   This presents a
striking contrast to the fact that only catalytic amounts of the
catalyst are necessary in the usual phase transfer catalyzed polycondensation reported previously.[12]   It may be conveniently explained
by the fact that the catalyst is actually occluded in a polymeric
mass which is separated during the polymerization.

Table 1 shows the effect of solvents on the two-phase polycondensation.   In addition to benzene, some aromatic solvents such as
anisole and benzonitrile could be used effectively for the synthesis
of polymer XVII with high inherent viscosity.

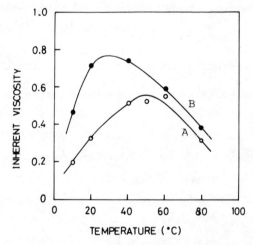

Fig. 1.   Effect of reaction temperature on inherent viscosity of
          polymer XVII formed in polycondensation with BTEAC (50
          mol%) in benzene-50 wt% aqueous NaOH system. (A) Polymeri-
          zation for 2 h; (B) polymerization for 6 h.

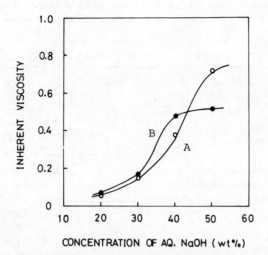

Fig. 2. Effect of concentration of aqueous NaOH on inherent viscosity
         of polymer XVII formed in polycondensation with BTEAC (50
         mol%) in benzene-water system. (A) Polymerization at 20°C
         for 6 h; (B) polymerization at 50°C for 2 h.

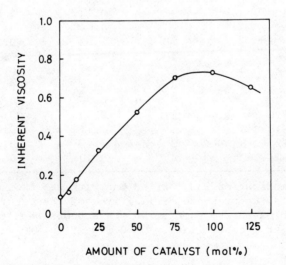

Fig. 3. Effect of amount of BTEAC on inherent viscosity of polymer
         XVII formed in polycondensation in benzene-50 wt% aqueous
         NaOH system at 50°C for 2 h.

Table 1. Synthesis of Polymer XVII in Various Aromatic
          Solvent-Water systems with BTEAC Catalyst[a]

Solvent	Polymer	
	Yield (%)	η_{inh}[b]
Benzene	87	0.73
Toluene	87	0.36
p-Xylene	89	0.32
Anisole	85	1.60
Benzonitrile	85	0.77

[a]The polymerization was carried out with 2.5 mmol of each
monomer in the presence of 1.25 mmol of BTEAC in 1 mL of
solvent and 5 mL of 50 wt% aqueous NaOH at 20°C for 6 h.
[b]Measured at a concentration of 0.5 g/dL in NMP at 30°C.

Table 2. Synthesis of Polymer XVII with Various Phase
          Transfer Catalysts in Benzene-Water System[a]

Catalyst	Reaction time (h)	η_{inh} of polymer[b]	
		Alkaline component	
		NaOH	KOH
TMAC	2	0.14	0.13
TEAC	2	0.68	0.21
TBAC	2	0.93	0.22
BTEAC	2	0.52	0.30
CTMAC	2	0.47	0.16
15-C-5	4	0.24	0.21
18-C-6	5	0.37	0.27
DB-18-C-6	2	0.20	0.05
DC-18-C-6	3	0.17	0.06
DB-24-C-8	3	0.17	0.04
DC-24-C-8	4	0.19	0.17

[a]The polymerization was carried out with 2.5 mmol of each
monomer in the presence of 1.25 mmol of the catalyst in
1 mL of benzene and 5 mL of 50 wt% aqueous NaOH at 50°C.
[b]Measured at a concentration of 0.5 g/dL in NMP at 30°C.

The influence of catalysts and alkaline components on the polycondensation is summarized in Table 2. The two-phase polymerization
was strongly catalyzed by some quaternary ammonium salts such as
TEAC, TBAC, BTEAC, and CTMAC, leading to the formation of high
molecular weight polymer XVII. All of the crown ethers including
15-C-5, 18-C-6, and others were found to be less effective than the

quaternary ammonium salts. In the two-phase system, sodium hydroxide as alkaline component in the aqueous phase was generally more effective than potassium hydroxide for producing polymer XVII of high molecular weight.

The time dependence of inherent viscosity of polymer XVII is shown in Figure 4. The phase transfer catalyzed polycondensation proceeded much more rapidly with BTEAC than that with 18-C-6 at 50°C.

The two-phase polycondensation of phenylacetonitrile XVIII with activated dichloride XVI leading to polymer XIX was then conducted under similar phase transfer conditions [Eq.(8)].[15]

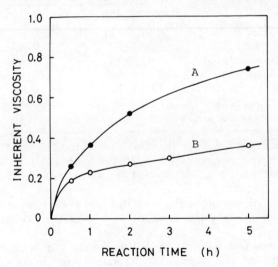

$$\langle\bigcirc\rangle - CH_2 - CN \quad + \quad XVI \quad \xrightarrow{PTC} \quad \left[-CH_2 - \bigcirc - CH_2 - \underset{\underset{\bigcirc}{|}}{\overset{\overset{CN}{|}}{C}} - \right]_n \qquad (8)$$

XVIII   XIX

The results of the polycondensation are summarized in Table 3. In the absence of catalyst, no polymeric product was obtained even after reaction of 40°C for 18 h, whereas all of the catalysts used were effective for the preparation of polymer XIX. In general, the quaternary ammonium and phosphonium salts including TEAC, TBAC, BTEAC, and CTBPB appeared to be more efficient than the crown ethers such as 18-C-6 and DB-24-C-8. In the crown ether catalyzed system, potassium hydroxide could favorably be substituted for sodium hydroxide as an alkaline component in the aqueous phase.

Fig. 4. Time dependence of inherent viscosity of polymer XVII formed in polycondensation with various catalysts (50 mol%) in benzene-50 wt% aqueous NaOH system at 50°C. (A) Polymerization with BTEAC; (B) polymerization with 18-C-6.

Table 3. Synthesis of Polymer XIX with Various Phase
          Transfer Catalysts in Benzonitrile-Water System[a]

Catalyst	Alkaline component	Polymer	
		Yield (%)	η_{inh}[b]
None	NaOH	0	-
TEAC	NaOH	89	0.58
TBAC	NaOH	90	0.47
BTEAC	NaOH	87	0.38
CTBPB	NaOH	87	0.39
18-C-6	NaOH	85	0.31
18-C-6	KOH	84	0.43
DB-24-C-8	NaOH	94	0.25
DB-24-C-8	KOH	83	0.27

[a]The polymerization was carried out with 2.5 mmol of each
monomer in the presence of 1.25 mmol of the catalyst in
2 mL of benzonitrile and 5 mL of 50 wt% aqueous alkaline
solution at 50°C for 2 h.
[b]Measured at a concentration of 0.5 g/dL in NMP at 30°C.

Table 4. Synthesis of Polymer XIX in Various Aromatic
          Solvent-Water Systems with TBAC Catalyst[a]

Solvent[b]	Polymer	
	Yield (%)	η_{inh}[c]
Benzene (2.2)	80	0.49
Anisole (4.3)	95	0.65
Chlorobenzene (6.6)	84	0.42
o-Dichlorobenzene (9.9)	78	0.62
Benzonitrile (25.2)	90	0.47
Nitrobenzene (34.8)	87	0.40

[a]The polymerization was carried out with 2.5 mmol of each
monomer in the presence of 1.25 mmol of TBAC in 2 mL of
the solvent and 5 mL of 50 wt% aqueous NaOH at 50°C for
2 h.
[b]The values in parentheses are dielectric constants.
[c]Measured at a concentration of 0.5 g/dL in NMP at 30°C.

     Table 4 shows the effect of solvents on inherent viscosity of
polymer XIX. All of the aromatic solvents such as benzene, anisole,
and others were used efficiently as an organic layer to give high
molecular weight polymer XIX. The dielectric constant of the
solvent had almost no influence on attained molecular weight of
polymer XIX.

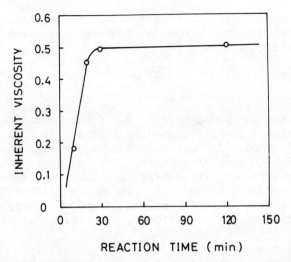

Fig. 5. Time dependence of inherent viscosity of polymer XIX formed in polycondensation with TBAC (50 mol%) in benzonitrile-50 wt% aqueous NaOH system at 50°C.

The course of the polycondensation in terms of inherent viscosity of polymer XIX is shown in Figure 5. The polymerization proceeded fairly rapidly at 50°C, with separation of the polymeric mass during the reaction, and was almost completed in 30 min.

Polymer XVII thus formed from t-butyl cyanoacetate was readily soluble at room temperature in chloroform, THF, and pyridine, as well as in polar aprotic solvents such as DMF, DMAc, and DMSO, and a transparent and tough film could be cast from THF solution of the polymer. On the other hand, polymer XIX from phenylacetonitrile was less soluble than polymer XVII; this polymer dissolved only in hot THF and hot polar aprotic solvents.

Polymer XIX had a melting temperature around 220°C, and a clear but somewhat brittle film could be molded by hot-pressing. Galss transition temperature of the polymer was 167°C. On the other hand, polymer XVII was rather thermally unstable; it began to decompose around 120°C without melting.

Syntheses of Polysulfides[18],[19]

Previously we reported the syntheses of aliphatic polysulfides with high molecular weights by the phase transfer catalyzed polycondensation of dibromoalkanes with aliphatic dithiols in aqueous potassium hydroxide.[8] Sodium sulfide is an attractive monomer for the preparation of polysulfides, since it is readily available and inexpensive. The two-phase polycondensation of dibromooctane XX

with aqueous sodium sulfide that led to the formation of polysulfide
XXI was performed in the presence of quaternary ammonium and phosphonium salts [Eq. (9)].[18]

$$Br(CH_2)_8Br \quad + \quad Na_2S \quad \xrightarrow{PTC} \quad \left[-(CH_2)_8-S-\right]_n \qquad (9)$$

$$\quad\quad XX \qquad\qquad\qquad\qquad\qquad\qquad\qquad XXI$$

The results are summarized in Table 5. Since the S^{2-} anion,
a divalent anion derived from sodium sulfide, is strongly hydrated,
it is assumed to be difficult to transfer the S^{2-} anion from the
aqueous phase to the organic phase. This problem could be overcome
by use of highly lipophilic catalysts. Such catalysts as CTMAC,
CTMAB, and CTBPB were found to be particularly effective for the
preparation of high molecular weight polysulfide XXI, whereas TBAC,
BTEAC, and BTPPC were all ineffective for this type of polymerization.

Figure 6 shows the effect of amount of catalyst on the two-phase
polycondensation. Inherent viscosity of polymer XXI was
highest with 10 mol% of CTMAC based on each monomer.

The effect of reaction temperature on the polycondensation was
examined in a temperature range of 40-110°C with CTMAC or CTBPB.
As shown in Figure 7, the inherent viscosity increased with increasing
temperature. The polycondensation catalyzed by CTMAC gave polymer
XXI of high molecular weight over all the temperature range compared with the polymerization catalyzed by CTBPB.

Table 5.   Synthesis of Polysulfide XXI with Various Phase
           Transfer Catalysts in Water System[a]

Catalyst	Polymer	
	Yield (%)	η_{inh}[b]
None	94	0.08
TBAC	99	0.09
CTMAC	91	0.71
CTMAB	86	0.70
CTBPB	99	0.42
BTEAC	99	0.19
BTPPC	96	0.08

[a]The polymerization was carried out with 5 mmol of XX and
5 mL of 1M aqueous Na_2S in the presence of the catalyst
at 100°C for 1 d.
[b]Measured at a concentration of 0.5 g/dL in conc. sulfuric
acid at 30°C.

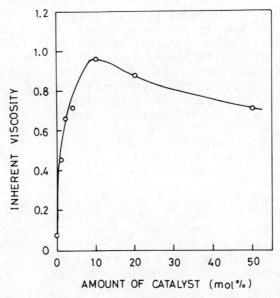

Fig. 6. Effect of amount of CTMAC on inherent viscosity of polysulfide XXI formed in polycondensation in aqueous system
at 100°C for 1 d.

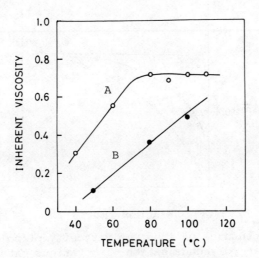

Fig. 7. Effect of reaction temperature on inherent viscosity of
polysulfide XXI formed in polycondensation with various
catalysts (4 mol%) in aqueous system for 1 d. (A) Polymerization with CTMAC; (B) polymerization with CTBPB.

This difference may be attributed to the different state of polycondensation. Below 90°C, the polymerization catalyzed by CTBPB proceeded with polymer precipitation, which limited the increase in molecular weights. Above that temperature, the polymer became melt state, thus permitting the polycondensation in a water-melt two-phase system. Accordingly, CTBPB became soluble in polymer-melt phase and effectively catalyzed the polycondensation as a phase transfer catalyst. On the other hand, the polymerization with CTMAC proceeded in emulsion even at a low temperature. This observation suggests that a substantial fraction of the growing molecules is solubilized in micelles by CTMAC which acts as cationic surfactant. Therefore, both phase transfer catalyzed and micelle catalyzed systems are probable in the polycondensation with CTMAC.

Figure 8 shows the course of the polycondensation in terms of inherent viscosity of polymer XXI. The polymerization proceeded fairly rapidly at 100°C with CTMAC and gave the polysulfide having inherent viscosity of 0.6 dL/g in 1 h. On the other hand, the poly-condensation with CTBPB at 100°C proceeded much more slowly, compared with the polymerization with CTMAC at 60°C.

The results of the polycondensation of sodium sulfide with various dibromoalkanes are shown in Table 6. Aliphatic polysulfides of high molecular weights were obtained readily with CTMAC or CTBPB.

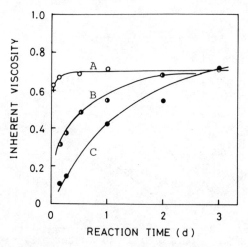

Fig. 8. Time dependence of inherent viscosity of polysulfide XXI formed in polycondensation with various catalysts (4 mol%) in aqueous system. (A) Polymerization with CTMAC at 100°C; (B) polymerization with CTMAC at 60°C; (C) polymerization with CTBPB at 100°C.

Table 6.   Syntheses of Aliphatic Polysulfides from
Dibromoalkanes and Sodium Sulfide[a]

$$Br-R-Br \quad + \quad Na_2S \quad \xrightarrow{\text{Catalyst}} \quad \left[-R-S-\right]_n$$

R	Catalyst	Polymer η_{inh}[b]
$-(CH_2)_6-$	CTMAC	0.64
$-(CH_2)_6-$	CTBPB	0.68
$-(CH_2)_8-$	CTMAC	0.71
$-(CH_2)_8-$	CTBPB	0.71
$-(CH_2)_{10}-$	CTMAC	0.83
$-(CH_2)_{10}-$	CTBPB	0.60

[a]The polymerization was carried out with 5 mmol of the
dibromoalkane and 5 mL of 1M aqueous Na_2S in the presence
of the catalyst (4 mol%) at 100°C for 3 d.
[b]Measured at a concentration of 0.5 g/dL in conc. sulfuric
acid at 30°C.

These aliphatic polysulfides (Table 6) were readily soluble in
chloroform at room temperature, and in hot THF and hot m-cresol.
All the polysulfides were highly crystalline and had melting temperatures below 100°C. They gave tough films by casting from the
chloroform solutions or by hot-pressing.

In addition to various types of polysulfides reported previously,
8-10 a new polysulfide has been synthesized using bischloroacetyl
aromatic compounds, which are known to be highly reactive dichlorides.
The two-phase polycondensation of bis(4-chloroacetylphenyl) ether
XXII with 4,4'-oxybisbenzenethiol XI leading to polysulfide XXIII
was conducted at 30°C in a 1,2-dichloroethane-aqueous potassium
hydroxide system [Eq.(10)].[19]

Table 7. Synthesis of Polysulfide XXIII with Various Phase
          Transfer Catalysts in Dichloroethane-Water System[a]

Catalyst	Polymer	
	Yield (%)	η_{inh}[b]
None	98	0.24
TEAC	78	0.19
TBAC	58	0.29
BTEAC	96	0.38
CTMAC	80	0.45
BTPPC	69	0.20
CTBPB	69	0.29
18-C-6	76	0.17
DC-18-C-6	73	0.15

[a]The polymerization was carried out with 1.25 mmol of each
monomer in the presence of the catalyst (10 mol%) in 3 mL
of 1,2-dichloroethane and 2.5 mL of 1M aqueous KOH at 20°C
for 24 h.
[b]Measured at a concentration of 0.5 g/dL in DMAc at 30°C.

     Table 7 shows the effect of catalysts on the polycondensation.
In this system, the polymerization proceeded even without use of
catalyst leading to polysulfide XXIII with inherent viscosity of
0.2 dL/g. Among the quaternary onium salt catalysts employed, BTEAC
and CTMAC were found to be effective for producing high molecular
weight polymer XXIII. Crown ethers such as 18-C-6 and DC-18-C-6
were quite ineffective for this type of polycondensation.

     The effect of solvents on the two-phase polycondensation is
shown in Table 8. Chlorinated hydrocarbons such as dichloromethane,
chloroform, and 1,2-dichloroethane, and aromatic solvents like benzene
and nitrobenzene were used favorably to yield polysulfide XXIII with
inherent viscosities of 0.3-0.7 dL/g.

     Figure 9 shows the course of the polycondensation in terms of
inherent viscosity of polysulfide XIII. The polymerization with
BTEAC proceeded rapidly at 30°C in the two-phase system and gave
the polymer having inherent viscosity of 0.6 dL/g in 1 h. On the
other hand, the polycondensation with CTMAC proceeded much more
slowly in an emulsion or in a micelle system.

     Polysulfide XXIII thus formed was only soluble in DMAc and di-
chloroacetic acid at room temperature. A yellow but somewhat brittle
film could be cast from the DMAc solution. The polymer did not melt
below 300°C.

Table 8.   Synthesis of Polysulfide XXⅢ in Various Solvent-
Water Systems with BTEAC Catalyst[a]

Solvent	Polymer	
	Yield (%)	η_{inh}[b]
Dichloromethane	87	0.36
Chloroform	83	0.69
1,2-Dichloroethane	79	0.40
sym-Tetrachloroethane	96	0.38
Benzene	88	0.48
Anisole	82	0.32
Nitrobenzene	86	0.38

[a]The polymerization was carried out with 1.25 mmol of each
monomer in the presence of BTEAC (10 mol%) in 3 mL of the
solvent and 2.5 mL of 1M aqueous KOH at 30°C for 12 h.
[b]Measured at a concentration of 0.5 g/dL in DMAc at 30°C.

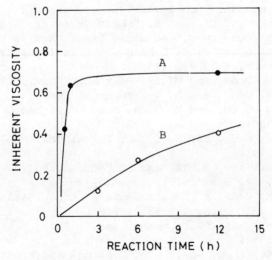

Fig. 9.   Time dependence of inherent viscosity of polysulfide XXⅢ
formed in polycondensation with various catalysts (10 mol%)
in 1,2-dichloroethane-aqueous KOH system at 30°C. (A)
Polymerization with BTEAC; (B) polymerization with CTMAC.

CONCLUSION

It is clear from these results that the application of phase transfer catalyzed synthesis to polycondensation has high potential value. With the proper choice of conditions, activated dihalides react readily with divalent nucleophiles including bisphenols, dithiols, and active methylene compounds in organic solvent-aqueous alkaline solution system to yield various types of condensation polymers such as polysulfonates,[4,5] polyphosphonates,[6] polyethers,[7] polysulfides,[8-10,18,19] and carbon-carbon chain polymers.[14,15]

$$X-R-X \quad + \quad HY-R'-YH \quad \xrightarrow{\text{PTC}} \quad \left[-R-Y-R'-Y-\right]_n \qquad (11)$$

-X : -Cl, -Br

$-YH$: -OH, -SH, $-\overset{|}{\underset{|}{C}}H$

Phase transfer catalyzed polycondensation [Eq. (11)] provides advantages of the simplicity of reaction and work-up for obtaining condensation polymers, and so this method should enjoy enormous growth over the next few years.

REFERENCES

1. W. P. Weber and G. W. Gokel, "Phase Transfer Catalysis in Organic Synthesis," Springer-Verlag, New York (1977).
2. C. M. Starks and C. Liotta, "Phase Transfer Catalysis, Principles and Techniques," Academic Press, New York (1978).
3. E. V. Dehmlow and S. S. Dehmlow, "Phase Transfer Catalysis," Verlag Chemie, Weinheim (1980).
4. Y. Imai, M. Ueda, and M. Ii, Makromol. Chem., 179, 2085 (1978).
5. Y. Imai, M. Ueda, and M. Ii, Kobunshi Ronbunshu, 35, 807 (1978).
6. Y. Imai, N. Sato, and M. Ueda, Makromol. Chem., Rapid Commun., 1, 419 (1980).
7. Y. Imai, M. Ueda, and M. Ii, J. Polym. Sci., Polym. Lett, Ed., 17, 85 (1979).
8. Y. Imai, A. Kato, M. Ii, and M. Ueda, J. Polym. Sci., Polym. Lett. Ed., 17, 579 (1979).
9. Y. Imai, M. Ueda, M. Komatsu, and H. Urushibata, Makromol. Chem., Rapid Commun., 1, 681 (1980).
10. M. Ueda, N. Sakai, M. Komatsu, and Y. Imai, Makromol. Chem., 183, 65 (1982).
11. Y. Imai, Am. Chem. Soc., Org. Coatings and Plastics Chem. Preprints, 42, 257 (1980).
12. Y. Imai, J. Macromol. Sci.-Chem., A15, 833 (1981).
13. Y. Imai and M. Ueda, Am. Chem. Soc., Polym. Preprints, 23 [1], 164 (1982).
14. Y. Imai, T.-Q. Nguyen, and M. Ueda, J. Polym. Sci., Polym. Lett. Ed., 19, 205 (1981).

15. Y. Imai, A. Kameyama, T.-Q. Nguyen, and M. Ueda, J. Polym. Sci.,
     Polym. Chem. Ed., 19, 2997 (1981).
16. D. Brown, M. E. B. Jones, and W. R. Maltman, J. Polym. Sci.,
     B, 6, 635 (1968).
17. N. Kawabata, K. Matsubara, and S. Yamashita, Bull. Chem. Soc.
     Jpn., 46, 3225 (1973).
18. M. Ueda, Y. Oishi, N. Sakai, and Y. Imai, Macromolecules, 15,
     000 (1982).
19. Y. Imai, R. Takasawa, and M. Ueda, Polym. Preprints, Jpn., 30,
     884 (1981).

ANIONIC POLYMERIZATIONS WITH COMPLEX BASES

AND THEIR DERIVATIVES

P. Caubere

Université de Nancy I
Laboratoire de Chimie Organique I
Nancy, France

INTRODUCTION

Organometallics as well as metals such as sodium or
potassium are well known as initiators for anionic polymerizations.
Moreover, in order to increase or modulate the reactivity of these
classical initiators, the influence of additives (such as tertiary
amines, linear and macrocyclic polyethers, cryptands, alcohols and
alkoxides) has been the subject of many works[1-11].

However the drawback of these initiators for industrial
purposes is generally their high cost and difficulty in handling.

In order to strongly reduce these inconveniences, we
decided to investigate the possibilities offered by a new class of
cheap and easily handled potential initiators first investigated in
our laboratory as synthetic organic reagents. Indeed the concept of
base synergy, we developed a few years ago, led to the preparation
of new reagents $NaNH_2$-RONa called Complex Bases (C.B.). RONa (called
an activating agent) could be, for example, an alkoxide or a ketone
enolate. We have fully demonstrated (for a review see ref. 12) that
C.B. have properties different from those of $NaNH_2$ or RONa taken
separately, and may be powerful bases even in weakly polar solvents
such as tetrahydrofuran (THF) or benzene. Thus it might be expected
that the enhanced reactivity of the amide anion could be used for
anionic polymerizations.

Before describing the results obtained in the anionic
polymerization field, we shall give some information about C.B.
themselves.

Complex Bases

A large part of the studies concerning C.B. were performed with $NaNH_2$. Moreover sodium alkoxides were the main activating
agents used to perform anionic polymerizations. So we shall essentially report briefly here the results concerning C.B. $NaNH_2$-RONa
obtained from sodamide and alcohols.

C.B. are simply prepared by addition of an alcohol to a
calculated excess of $NaNH_2$[12]. The main factors influencing their
reactivity are the ratio $NaNH_2$/RONa, the nature of the solvent and
of the activating agent. To classify the activity of C.B. we used[12]
the test reactions given in Scheme 1.

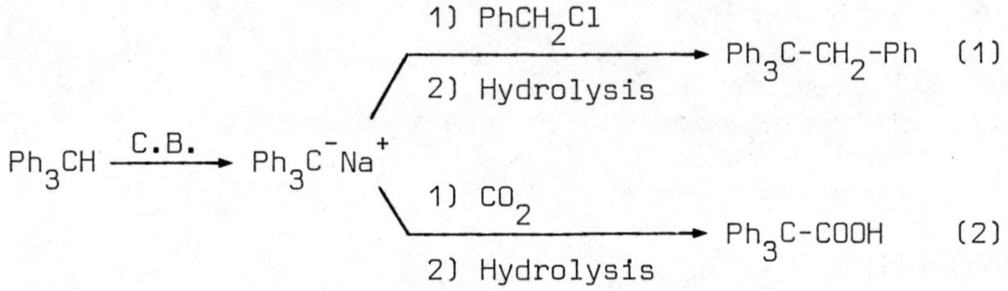

$$Ph_3CH \xrightarrow{\text{C.B.}} Ph_3C^- \; Na^+$$

1) $PhCH_2Cl$
2) Hydrolysis $\longrightarrow Ph_3C-CH_2-Ph$ (1)

1) CO_2
2) Hydrolysis $\longrightarrow Ph_3C-COOH$ (2)

Scheme 1

Generally even in the presence of an excess of C.B.
triphenyl methane is not converted quantitatively to the corresponding carbanion. Thus after addition of a certain amount of benzyl
chloride (eq. 1) the red colour disappears. If addition of the
halogeno compound is stopped, the coloration appears again and the
condensation may be pursued as a coloured titration. The time needed
for condensation of a given amount of benzyl chloride on a given
amount of triphenyl methane represents the ability of C.B. to
regenerate a carbanion (Basic Renewal Power : B.R.P.). On the other
hand, the yield of acid formed by carbonatation (eq. 2) indicates
the ability of C.B. to form a carbanion (Carbanionic Generation
Power : C.G.P.). B.R.P. and C.G.P. are easily and rapidly determined.
They conveniently give excellent information about the properties
of C.B.

Thus we drew the following important conclusions :

- the best ratio $NaNH_2$/RONa is 2.

- C.B. are much more efficient in polar aprotic solvents than in
  nonpolar ones.

- the role of the activating agent is very important.

Briefly summarized it may be said that in THF the best
B.R.P. may be expected from diethylenglycol monoalkyl ether
alkoxide. In the same solvent good B.R.P. may also be obtained from
primary and secondary alkoxides(provided that they contain a light
branched chain) as well as from tertiary alkoxides. DME has a levelling effect on B.R.P. values with numerous inversions relative to
the order defined in THF. Concerning C.G.P. values, THF as well as
DME orders are similar to the B.R.P. order in THF, again with a
levelling effect of DME. It is worth noting that in nonpolar
solvents such as benzene or cyclohexane efficient bases may be
obtained using di- or triethylene glycol monoalkyl ether alkoxides
as activating agents.

Taking into account that C.B. are heterogeneous reagents
constituted of some remaining solid sodamide, a white suspension
and a limpid liquid we interpreted the phenomena as follows
(Scheme 2) :

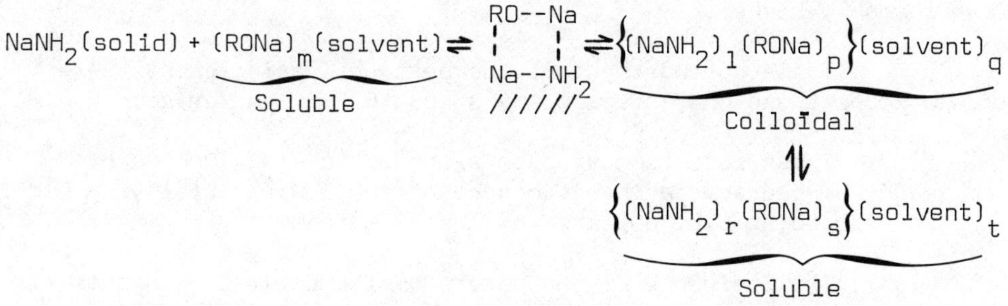

$$NaNH_2(solid) + \underbrace{(RONa)_m (solvent)}_{Soluble} \rightleftharpoons \underbrace{\begin{array}{c} RO--Na \\ | \quad | \\ Na--NH_2 \\ /////// \end{array}}_{} \rightleftharpoons \underbrace{\left\{(NaNH_2)_1 (RONa)_p\right\}(solvent)_q}_{Colloidal}$$

$$\Big\Downarrow$$

$$\underbrace{\left\{(NaNH_2)_r (RONa)_s\right\}(solvent)_t}_{Soluble}$$

Scheme 2

The alkoxide interacts with $NaNH_2$ on the crystal surface
and transfers it from the solid phase to the colloidal liquid phase
and (certainly to a small extent only) a completely soluble complexed
form. Aggregates are, of course, responsible for the reactivities
observed. Moreover a series of experiments[14] led us to conclude that
during a reaction between aggregates and an organic substrate, free
alkoxide is released which must go back to complex some more solid
sodamide. Finally these results have been extended with some restric-
tions to $LiNH_2$ and KNH_2[13].

STUDY OF VINYL MONOMERS

C.B. initiated anionic polymerization of vinyl monomers in
solution[13-16]

With these data in hand we first undertook the study of
the polymerization of vinyl monomers (styrene, 2-vinyl pyridine,
methyl methacrylate, methacrylonitrile and acrylonitrile) in THF
and toluene.

The influences of different parameters were studied with
C.B. taken from among sodamide containing C.B.

Interestingly it was found that the highest yields and
lowest polydispersity $\overline{Mw}/\overline{Mn}$ were obtained when the ratio $NaNH_2/RONa$
was 2, the value leading to the best B.R.P. and C.G.P. (vide supra).
Concerning the solvent, it has been shown that changing from a low
polar solvent (THF) to a nonpolar one (cyclohexane) the yields
decreased and the polydispersity increased. Finally we showed that
the activating agents played an important role in these polymerizations. In order to illustrate this influence we have reported in
Table 1 some of the results obtained with styrene and methyl methacrylate. For comparison the polymerizations of a given monomer in
a given solvent were performed using the same conditions. Control
experiments showed, as expected, that C.B. behaviour was completely
different from the behaviour of separate constituents. With these
latter, yields were generally much lower and polydispersities
considerably higher.

It is noteworthy that the best activating agents were
found among those characterized by a good C.G.P. and/or good B.R.P.

These results have been extended to $LiNH_2$-ROLi and
KNH_2-ROK. Of course, taking into account the low reactivity of $LiNH_2$
the corresponding C.B. appeared of less interest.

The influence of the nature of the activating agents
deserves some brief comments. We saw above that the activation of
alkali amides by alkoxides must be due to aggregate formations.
Of course the structure of the aggregates and their solubility
depend on the nature of the activating agent and of the solvent.
Some portion of these aggregates constitute the active initiating
centers. Thus for a given amide, the number -and the reactivity- of
active centers depend on the nature of the activating agent and of
the solvent.

In other words, and this constitutes one of the interesting properties of C.B. as polymerization initiators, it is possible
for a given solvent to modulate the initiation and thus the properties of the polymer, by a simple change in the nature of the activating agent.

Now, what has been determined concerning the mechanism
of the C.B. initiated polymerizations ?

We have established that the NH_2^- part of the C.B.
initiated the polymerizations.

Table 1 : Polymerization of vinyl monomers (88 mM) by Complex Base NaNH$_2$ (16.7 mM) - RONa (8.3 mM)

Monomer	Styrene (88)[a]								Methyl methacrylate (100)[a]							
Solvent	THF (40)				PhCH$_3$ (40)				THF (40)				PhCH$_3$ (40)			
Activating agent R-	Yield %	Mn[b]	Mw[b]	Mw/Mn	Yield %	Mn[b]	Mw[b]	Mw/Mn	Yield %	Mn[b]	Mw[b]	Mw/Mn	Yield %	Mn[b]	Mw[b]	Mw/Mn
nBu-	35	5100	9200	1.8	15	7950	38150	4.8	40	3200	5450	1.7	37	1500	2700	1.8
C$_5$H$_{11}$-	50	10300	20600	2.0	15	10650	57500	5.4	50	5000	9000	1.8	43	2600	5450	2.1
C$_6$H$_{13}$-	52	12750	34400	2.7	10	9000	47700	5.3	43	4500	8600	1.9	45	2950	6500	2.2
C$_7$H$_{15}$-	47	16000	46400	2.9	15	17250	105250	6.1	55	6350	10800	1.7	47	3550	8500	2.4
C$_8$H$_{17}$-	53	18650	61550	3.3	19	24400	153700	6.3	75	9000	16200	1.8	55	4700	10350	2.2
C$_{10}$H$_{21}$-	22	11200	50400	4.5	25	45000	342000	7.6	70	10500	15750	1.5	70	7000	13300	1.9
tBuCH$_2$-	15	1200	3400	2.8	30	19500	159900	8.2	60	4500	6750	1.5	50	2650	5300	2.0
CH$_2$=CH-CH$_2$-	87	53500	85600	1.6	37	100000	600000	6.0	60	6600	9900	1.5	100	10000	17000	1.7
iPr-	72	37500	97500	2.6	16	16500	75900	4.6	100	15000	24000	1.6	72	3950	6700	1.7
Et$_2$CH-	58	13350	37400	2.8	10	11000	53900	4.9	100	20000	30000	1.5	60	4200	7550	1.8
tBu-	62	18600	39100	2.1	35	45500	163800	3.6	100	25000	40000	1.6	83	5400	9700	1.8
nBu$_3$C-	40	5900	14700	2.5	22	18700	71050	3.8	100	16000	28800	1.8	65	3900	7400	1.9
Et(OCH$_2$CH$_2$)-	40	6250	11900	1.9	63	123000	430500	3.5	100	10000	17000	1.7	70	5600	8950	1.6
Et(OCH$_2$CH$_2$)$_2$-	100	60500	90750	1.5	78	245700	786250	3.2	100	30000	48000	1.6	100	12500	18750	1.5
Et(OCH$_2$CH$_2$)$_3$-	50	25350	53200	2.1	72	158400	586050	3.7	100	25000	45000	1.8	100	10500	16800	1.6
Et(OCH$_2$CH$_2$)$_6$-	27	5500	12650	2.3	51	61200	250950	4.1	90	12000	19200	1.6	90	6300	10700	1.7
Me(OCH$_2$CH$_2$)-	38	3800	6450	1.7	50	70000	245000	3.5	100	4300	6450	1.5	88	5050	8050	1.6
Bu(OCH$_2$CH$_2$)$_2$-	95	59300	100800	1.7	80	290400	1132550	3.9	100	35000	56000	1.6	100	18700	29900	1.6
Bu(OCH$_2$CH$_2$)$_3$-	47	24100	53050	2.2	64	179200	824300	4.6	80	21600	38900	1.8	100	16000	28800	1.8
Furfuryl	100	72200	137200	1.9	22	30800	194050	6.3	100	8000	13600	1.7	95	5700	10250	1.8

[a]Styrene : temperature 30°C, polymerization time 3 h ; Methyl methacrylate : temperature 40°C, polymerization time 2 h in THF and 6 h in PhCH$_3$

[b]Measured by g.p.c. in THF at 30°C

Moreover the study of the variation of $\overline{DPn}$ (degree of polymerization) versus the monomer concentration in THF as well as in toluene showed a linear relation as long as the monomer concentrations were not too high. These observations strongly support a living nature of the polymers. This hypothesis was confirmed by the preparation of block copolymers from styrene and methyl methacrylate. These experiments showed that the polymer chains formed from styrene carried active centers able to initiate polymerization of other monomers.

On the other hand spectroscopic study of the stability of the polystyryl carbanion leading to the propagation of the polymerization of styrene initiated by a C.B. confirmed the living polymer hypothesis. Moreover the rate of disappearence of the carbanion in our conditions, compared to literature data, led us to the conclusion that alkoxide complexes the carbanion which promotes polymerization.

Finally we confirmed the living nature of the polymers and demonstrated the possibility of preparing polymers with high $\overline{Mn}$ values by performing polymerizations with different monomer/ initiator ratios as illustrated in Table 2.

Table 2 : Polymerization of acrylonitrile and methacrylonitrile at different monomer/initiator ratios

Acrylonitrile + NaNH$_2$ + Et(OCH$_2$CH$_2$)$_2$ONa in THF (40 ml) at 25°				
Monomer mM	NaNH$_2$ mM	RONa mM	Yield %	$\overline{Mn}$
100	16.7	8.3	100	15000
100	8	4	95	35000
150	4	2	90	72000
Methacrylonitrile + NaNH$_2$ + Bu^{t}ONa in toluene (50 ml) at 30°				
150	16.7	8.3	100	12600
150	8	4	100	35000
200	4	2	100	84600

Solid Complex Bases and the bulk polymerizations of vinyl monomers[14,16,17]

The results obtained with C.B. initiated polymerization in solution led us to ponder on the potentialities offered by C.B. as bulk polymerization initiators.

As a preliminary finding it was of paramount interest to know if it was possible to remove the solvent from a C.B. without loss of activity.

The studies performed on two C.B. taken from among the best led to the following information :

- Vacuum removal of the solvent from C.B. prepared in low polar solvents such as THF, led to very reactive yellow powders (moisture and air sensitive). Addition of melted triphenyl methane led to the instantaneous formation of the red carbanion.

- Yellow powders obtained from C.B. prepared in a nonpolar solvent such as benzene or cyclohexane, showed low reactivity.

- Exploratory experiments performed with styrene showed that these yellow powders were able to initiate the polymerization after a latent period.

From these observations we concluded that it was possible to obtain solid active reagents by simple removal of the solvent from a C.B., that some residual solvent could be included into the solid and that they are able to initiate polymerizations.

They were called Solid Complex Bases (S.C.B.).

Note : During the preliminary experiments dealing with styrene polymerization it was found that addition of alcohol, precursor of RONa to solid $NaNH_2$ (without solvent) also led to a S.C.B. able to initiate bulk polymerizations. However the latent period was longer than for S.C.B. prepared from a C.B. Moreover we shall see later that generally the S.C.B. prepared without solvent led to less satisfactory results.

Turning now toward bulk polymerizations themselves the influences of different parameters were studied, as for solution polymerizations initiated with C.B. The conclusions drawn from these works performed essentially with sodamide containing S.C.B. will be briefly given below.

First of all it must be emphasized that with S.C.B. (as for C.B.) the optimum $NaNH_2$/RONa ratio was found to be two, whether the S.C.B. were prepared from C.B. or from addition of an alcohol to $NaNH_2$ without solvent !

The nature of the solvent used to prepare the C.B. from which S.C.B. were obtained as well as the absence of solvent during the S.C.B. preparation play a role in the bulk polymerization results. This is well illustrated in Table 3.

Table 3 : Polymerization of styrene (88 mM) by Solid Complex Base
         NaNH$_2$ (16.7 mM) - C$_2$H$_5$(OCH$_2$CH$_2$)$_2$ONa (8.3 mM) at 45°C
         for 2 h

Solvent of Complex Base preparation	Yield %	$\overline{Mn}$ [a]	$\overline{Mw}$ [a]	$\overline{Mw}/\overline{Mn}$
None	100	157400	928650	5.9
THF	100	121600	522900	4.3
DME	100	117400	493100	4.2
Diglyme	97	136100	680500	5.0
Toluene	75	95750	450050	4.7
Benzene	76	90800	491250	5.3
Cyclohexane	70	87650	447000	5.1

[a]Measured by g.p.c. in THF at 30°C

Finally the nature of the activating agent is very
important as examplified in Table 4.

Broadly speaking, the best activating agents for polymerization in solution were also the best ones for S.C.B. initiated
bulk polymerizations. Of course control experiments showed that
S.C.B. behaviour was completely different from the behaviour of
their constituents taken separately. Note that $\overline{Mw}/\overline{Mn}$ ratios are
rather low for bulk polymerizations.

The above results were successfully extended to other
vinyl monomers such as 2-vinyl pyridine, methacrylonitrile and
acrylonitrile.

Study of LiNH$_2$ and KNH$_2$ containing S.C.B. showed that
these bases were able to initiate bulk polymerizations. As expected
LiNH$_2$ containing S.C.B. was not very reactive. Those of KNH$_2$ were
comparable to NaNH$_2$ containing S.C.B.

Finally study of the variations of $\overline{DPn}$ versus monomer
concentration during the bulk polymerization of styrene initiated
by a sodamide S.C.B., showed the living nature of the polymers. This
was confirmed by copolymerization performed between styrene and
methyl methacrylate.

Salt Complex Bases and the solution or bulk polymerizations of vinyl
monomers[14,16,18]

A few yeras after we had discovered sodamide containing
C.B., Biehl and coworkers[19] observed the activation of NaNH$_2$ by
inorganic sodium salts such as NaSCN or NaNO$_2$.

Table 4 : Polymerization of vinyl monomers by Solid Complex Bases NaNH$_2$ (16.7 mM) - RONa (8.3 mM)

| Monomer (mM) | Styrene (88)[a] | | | | | | | | Methyl methacrylate (100)[a] | | | | | | | |
| Solvent | THF | | | | - | | | | THF | | | | - | | | |
Activating agent R-	Yield %	$\overline{Mn}$[b]	$\overline{Mw}$[b]	$\overline{Mw}/\overline{Mn}$	Yield %	$\overline{Mn}$[b]	$\overline{Mw}$[b]	$\overline{Mw}/\overline{Mn}$	Yield %	$\overline{Mn}$[b]	$\overline{Mw}$[b]	$\overline{Mw}/\overline{Mn}$	Yield %	$\overline{Mn}$[b]	$\overline{Mw}$[b]	$\overline{Mw}/\overline{Mn}$
nBu-	35	39200	121500	3.1	28	45100	157850	3.5	37	5200	11950	2.3	30	5600	19050	3.4
C$_5$H$_{11}$-	53	74200	252300	3.4	50	99500	358200	3.6	40	6400	13450	2.1	35	8050	25750	3.2
C$_6$H$_{13}$-	50	75000	247500	3.3	47	97750	361650	3.7	38	6550	16350	2.5	32	7400	26600	3.6
C$_7$H$_{15}$-	51	77000	269500	3.5	45	94500	330750	3.5	50	8700	20900	2.4	40	10400	36400	3.5
C$_8$H$_{17}$-	55	99100	485600	4.9	53	120850	640500	5.3	45	7650	19900	2.6	37	10450	31350	2.0
C$_{10}$H$_{21}$-	39	74100	377900	5.1	40	103000	535600	5.2	40	7850	23500	3.0	35	10500	43050	4.1
tBuCH$_2$-	26	18200	56400	3.1	20	18000	73800	4.1	60	5900	17100	2.9	49	5900	19500	3.3
CH$_2$=CH-CH$_2$-	100	310000	837000	2.7	80	396000	1168000	3.0	47	23950	57500	2.4	75	55500	160950	2.9
iPr-	89	135500	400500	3.0	83	178450	588900	3.3	80	17600	38700	2.2	70	26250	65500	2.5
Et$_2$CH-	80	124000	409200	3.3	57	127400	407700	3.2	72	21600	51650	2.4	50	20500	55350	2.7
tBu-	90	153000	443700	2.9	78	175500	508950	2.9	88	30800	64700	2.1	85	40800	97900	2.4
nBu$_3$C-	52	53350	198200	3.7	61	92400	378850	4.1	78	23000	57500	2.5	40	16000	41600	2.6
Et(OCH$_2$CH$_2$)-	78	304200	821350	2.7	70	308000	924000	3.0	90	54000	102600	1.9	90	72000	158400	2.2
Et(OCH$_2$CH$_2$)$_2$-	100	480000	1152000	2.4	95	570000	1482000	2.6	95	190000	361000	1.9	92	184000	386400	2.1
Et(OCH$_2$CH$_2$)$_3$-	87	382800	1110100	2.9	80	416000	1331200	3.2	87	69600	167053	2.4	83	81350	195200	2.4
Et(OCH$_2$CH$_2$)$_6$-	53	212000	890400	4.2	45	216000	950400	4.4	80	48250	120600	2.5	78	62400	156000	2.5
Me(OCH$_2$CH$_2$)-	66	277200	859300	3.1	40	160000	512000	3.2	90	48300	111100	2.3	84	61900	167150	2.7
Bu(OCH$_2$CH$_2$)$_2$-	100	495000	1732500	3.5	87	539400	2211550	4.1	85	225250	585653	2.6	80	163200	424300	2.6
Bu(OCH$_2$CH$_2$)$_3$-	87	391500	1291950	3.3	62	325500	1236900	3.8	87	76550	191400	2.5	70	64750	148900	2.3
Furfuryl	100	375000	1087500	2.9	87	346250	1108000	3.2	83	59350	136500	2.3	60	48000	96000	2.0

[a]Temperature : 45°C, polymerization time : Styrene 1 h, Methyl methacrylate 0.5 h

[b]Measured by g.p.c. in THF at 30°C

In a comparative study (see 12) we showed that, generally speaking, inorganic salts had lower activating properties than sodium alkoxides or ketone enolates.

However we decided to further study the activation of alkali amides by inorganic salts and to determine whether the best of them (called Salt Complex Bases : Sa.C.B.) could be used as initiators of solution as well as bulk polymerization of vinyl monomers.

From experiments conducted as for C.B. and S.C.B. (vide supra) on styrene we obtained the following important information :

- Efficient Sa.C.B. M^1NH_2-M^2X may be obtained with $M^1 = M^2$ as well as $M^1 \neq M^2$. Whatever the cation may be, counteranions of the activating part can be classified as follows :

$$NO_2^- \gg SCN^- > {}^-OCN \gg {}^-CN$$

As usual we have verified that, except for KNH_2, neither alkali amide nor alkali salts taken separately were able to initiate polymerizations.

- From studies performed with $NaNH_2$ and $NaNO_2$ it appeared that the most powerful initiator was obtained when the ratio $NaNH_2/NaNO_2$ was 2, whether the polymerizations were performed in solution (THF, toluene) or as bulk polymerizations.

- Polymers, here too, were living ones.

These results were used to perform polymerizations of other vinyl monomers. A good idea of the possibilities offered by Sa.C.B. may be obtained from data given in Table 5.

Some remarks about the microstructures of the polymers

Stereoregularity of the polymers obtained during the work described above, depends on several parameters. However some features emerge. For example concerning polystyrenes[20] obtained by solution polymerizations there was no large influence of the nature of the solvent or of the nature of the activating agent as illustrated from some data gathered in Table 6.

It clearly appears that the polymers obtained are mostly syndiotactic. The interpretation[20] of these results is that during the propagation step, the alkoxides, by complexation of the cation, maintained a sufficient electron density to propagate the polymerization ; however, they are too far from the active site to have enough steric influence on the pathway of polymerization.

Table 5 : Polymerization of vinyl monomers by Salt Complex Bases
NaNH$_2$ (16.7 mM) - ZNa (8.3 mM)

		Solution polymerization							
Solvent (ml)		THF (40)				PhCH$_3$ (30)			
Monomer (mM)	Activating agent	NaNO$_2$	NaSCN	NaCNO	NaCN	NaNO$_2$	NaSCN	NaCNO	NaCN
2VP (80)[c]	Yield (%)	60	55	40	47	50	52	50	30
	$\overline{Mn}$ [a]	4900	3600	2000	2400	4600	3900	2800	1700
MMA (100)[d]	Yield (%)	100	100	96	90	100	100	90	85
	$\overline{Mn}$ [a]	43500	22000	19200	16200	41000	19000	15300	12800
	$\overline{Mw}$	56500	46200	48000	38900	195600	87000	62800	71400
	$\overline{Mw}/\overline{Mn}$ [a]	1.3	2.1	2.5	2.4	4.8	4.6	4.1	5.6
MAN (80)[e]	Yield (%)	100	75	63	50	100	95	74	57
	$\overline{Mn}$ [b]	20000	11250	8500	6500	19500	10000	6800	4800
AN (80)[f]	Yield (%)	100	100	70	65	90	70	60	50
	$\overline{Mn}$ [b]	6300	4200	2800	2300	4000	2500	2100	1400
		Bulk polymerization							
Solvent (ml)		THF [g]							
2VP (80)[c]	Yield (%)	50	55	40	35	55	45	35	45
	$\overline{Mn}$ [a]	55000	55000	36000	29800	88000	54000	33300	38700
MMA (100)[d]	Yield (%)	70	65	75	55	60	75	55	60
	$\overline{Mn}$ [a]	63000	53900	51000	34100	72000	66000	38500	40800
	$\overline{Mw}$ [a]	151200	140300	158100	143200	172800	191400	134800	195900
	$\overline{Mw}/\overline{Mn}$ [a]	2.4	2.6	3.1	4.2	2.4	2.9	3.5	4.8
MAN (80)[e]	Yield (%)	85	80	75	60	85	87	80	75
	$\overline{Mn}$ [b]	63800	50400	34500	28200	72300	56600	42000	37500
AN (80)[f]	Yield (%)	90	95	80	75	80	85	80	60
	$\overline{Mn}$ [b]	52000	36100	27200	24800	49600	34000	32000	21000

[a] Measured by g.p.c. at 30°C in THF

[b] Measured by v.p.o. at 130°C in DMF

[c] Solution polymerization temperature 40°C, polymerization time 4 h,
bulk polymerization temperature 45°C, polymerization time 2 h

[d] Solution and bulk polymerization : temperature 35°C, polymerization
time 2 h

[e] Solution polymerization : temperature 35°C, polymerization time
20 min ; bulk polymerization : temperature 45°C, polymerization
time 30 min

[f] Solution polymerization : temperature 40°C, polymerization time
30 min ; bulk polymerization : temperature 45°C, polymerization
time 30 min

[g] Solvent of Complex Bases preparation

Table 6 : Tacticity of polystyrenes obtained in THF with $NaNH_2$-RONa

Activating agent (R)	Triads		
	Iso	Hetero	Syndio
nBu-	0.09	0.35	0.56
Et_2CH-	0.08	0.35	0.57
tBu-	0.11	0.39	0.50
$Et(OCH_2CH_2)_2-$	0.13	0.34	0.53
$Et(OCH_2CH_2)_6-$	0.12	0.43	0.45
C_6H_5	0.09	0.44	0.47
(o-tolyl, CH_3)	0.13	0.34	0.43
$C_6H_5-C=CH_2$	0.14	0.40	0.46

With methyl methacrylates it has also been found[16] that the nature of the solvent as well as that of the activating agent does not have much influence on the microstructure of the polymers which were generally mostly heterotactic.

STUDY OF DIENE MONOMERS[14,16,21]

Taking account of the results obtained with vinyl monomers, we decided to get an insight into the polymerization of butadiene and isoprene.

As an illustration of the results obtained we have reported in Table 7 data concerning experiment performed on buta- diene with C.B. (solution) and S.C.B. (bulk) as polymerization initiators.

It appeared that only the best bases for polymerization of vinyl monomers led to acceptable results with diene monomers although rates and yields were lower. Moreover only oligomers could be obtained even in bulk conditions. The rather low weights have been explained by the fact that oligomers contain an NH_2 group which is well known[22] as favoring transfer reactions which terminate the propagation reactions.

With isoprene, the reactions seemed more complicated since initiation of the polymerizations was due to C.B. as well as to the allylic carbanion formed by abstraction of the proton of the

Table 7 : Polymerization of butadiene (100 mM) by Complex Bases MNH_2 (16.7 mM) - ROM (8.3 mM) at 35°C for 3 h

Solvent (ml)	THF (20)				PhCH$_3$ (20)				−			
Amide	NaNH$_2$		KNH$_2$		NaNH$_2$		KNH$_2$		NaNH$_2$		KNH$_2$	
Activating agent R-	Yield (%)	$\overline{Mn}$ a	Yield (%)	$\overline{Mn}$ a	Yield (%)	$\overline{Mn}$ a	Yield (%)	$\overline{Mn}$ a	Yield (%)	$\overline{Mn}$ a	Yield (%)	$\overline{Mn}$ a
None	0	-	20	900	0	-	10	1100	0	-	0	-
$CH_2=CH-CH_2-$	45	1150	40	1500	35	1500	70	2750	30	4100	65	5250
iPr-	60	1500	50	1700	30	1800	75	3100	20	3950	40	4300
tBu-	65	1950	60	2100	35	2000	65	2500	25	5000	35	2950
$Et(OCH_2CH_2)-$	60	1200	65	1800	30	1000	70	3250	40	4700	60	5200
$Et(OCH_2CH_2)_2-$	80	2500	85	3700	45	2000	85	4900	65	20000	80	15000
$Et(OCH_2CH_2)_3-$	40	1200	50	2900	40	1600	70	3500	30	13000	60	10200
$Et(OCH_2CH_2)_6-$	40	800	50	1500	35	900	60	3400	45	7400	50	5400
$MeOCH_2CH_2-$	50	1000	60	1700	40	1400	55	2750	50	4500	40	7100
$Bu(OCH_2CH_2)_2-$	40	1450	70	2000	42	2000	65	3100	60	15300	55	12300
$Bu(OCH_2CH_2)_3-$	50	1500	35	1450	35	1600	60	2100	20	7100	35	4000
Furfuryl	60	1000	40	1850	39	1800	55	1500	30	8200	30	6000

a Measured by v.p.o. at 37°C in toluene

proton of the methyl group of the isoprene molecule. Note that control experiments showed that, except for KNH_2, none of the constituing reagent taken separately was able to initiate oligomerizations.

Finally we examined the structure of the oligomers. As may be seen from Table 8, where the results obtained have been briefly summarized, here too we do not observe a large influence of the nature of the solvent or activating agent.

The influence of the nature of the cation was more important.

The apparent contradiction of the alkoxide influences on the initiation and propagation rates compared to the lack of influence on the nature of the oligomer formed has been interpreted as given above for vinyl monomers.

CONCLUSION

In conclusion, I have not given all the results obtained since the goal of this short review is only to sensitize the reader to the possibilities offered by Complex Bases and their derivatives.

The following adventages presented by these new initiators must be emphasized :

- They are much less air or humidity sensitive than classical anionic initiators.

- They may be prepared from inexpensive commercial grade reagents on small as well as large scales. Moreover they are very easily handled.

- By simply varying the nature of the activating agent it is possible to modulate the initiator activity and thus the characteristics of the polymers formed.

- They lead to living polymers thus allowing the preparation of block copolymers.

Finally, although being mainly an organic chemist specialized in synthetic reagents, I hope to have shown how Complex Bases, Solid Complex Bases and Salt Complex Bases could be of interest in the anionic polymerizations or oligomerizations fields and thus to have opened a new way for carrying out this research.

Table 8 : Structure of polybutadiene and polyisoprene

Polymer	Polybutadiene						Polyisoprene								
Solvent	THF		PhCH₃		None		THF			PhCH₃			None		
Initiator	1,4	1,2	1,4	1,2	1,4	1,2	1,4	1,2	3,4	1,4	1,2	3,4	1,4	1,2	3,4
LiNH₂-ROLi [a]	70	30	75	25	80	20	75	5	20	75	5	20	80	5	15
NaNH₂-RONa [a]	60	40	65	35	65	35	50	15	35	55	15	30	60	15	25
KNH₂	75	25	70	30	-	-	80	5	15	65	15	20	60	15	25
KNH₂-ROK [a]	50	50	50	50	50	50	50	15	35	50	15	35	55	15	30

[a] Experiments were performed with all the activating agents given in Table 7

ACKNOWLEDGEMENTS

        I express my sincer appreciation to my coworkers
S. RAYNAL and G. NDEBEKA. I am grateful to S. LECOLIER (S.N.P.E.)
and also to Mrs G. ROQUES (Université Nancy I) for fruitful
discussions. I thank the Société Nationale des Poudres et Explosifs
and the Centre National de la Recherche Scientifique for financial
support.

REFERENCES

1. M. Szwarc, "Carbanions living polymers and electron transfer
      processes", Interscience, New York (1968)
2. A.W. Langer, Jr., Trans. N.Y. Acad. Sci., 27:741 (1965)
3. M. Shinohara, J. Smid, M. Szwarc, J. Amer. Chem. Soc., 90:2175
      (1968)
4. S. Kopolow, T.E. Hogen-Esch, J. Smid, Macromolecules, 66:133
      (1973)
5. S. Boileau, B. Kaempf, J.M. Lehn, F. Schue, J. Polym. Sci. B,
      12:203 (1974)
6. M. Mamhashi, H. Takida, Makromol. Chem., 124:172 (1969)
7. D.M. Wiles, S. Bywater, J. Phys. Chem., 68:1983 (1964)
8. H.L. Hsich, C.F. Wofford, J. Polym. Sci. A, 1:449 (1969)
9. T. Narita, T. Yasumara, T. Tsunita, Polym. J., 4:421 (1973)
10. T.C. Cheng, A.F. Halasa, D.P. Tate, J. Polym. Sci. Polym. Chem.
      Ed. 11:253 (1973) ; ibid 14:573 (1976)
11. L. Lochmann, D. Doskocilova, J. Trekoval, Coll. Czech. Chem.
      Commun., 42:1355 (1977) and references cited therein
12. For reviews see : P. Caubère, Acc. Chem. Res., 7:301 (1974)
      and references cited therein ; P. Caubère, Topics in Current
      Chemistry, 73:50 (1978) and references cited therein
13. G. Ndebeka, P. Caubère, S. Raynal, S. Lecolier, Polymer, 22:347
      (1981) and references cited therein
14. G. Ndebeka, Thesis, Nancy (1979)
15. S. Raynal, S. Lecolier, G. Ndebeka, P. Caubère, J. Polym. Sci.
      Polym. Lett. Ed., 18:13 (1980)
16. S. Raynal, Thesis, Nancy (1979)
17. S. Raynal, S. Lecolier, G. Ndebeka, P. Caubère, Polymer, 22:356
      (1981)
18. S. Raynal, S. Lecolier, G. Ndebeka, P. Caubère, Polymer, 23:283
      (1982)
19. E.R. Biehl, E. Nieh, K.C. Hsu, J. Org. Chem., 34:3595 (1969) ;
      E.R. Biehl, K. Hsu, E. Nieh, J. Org. Chem., 35:2454 (1970)
20. S. Raynal, G. Ndebeka, P. Caubère, S. Suparno, J. Sledz, F.Schue
      J. Macromol. Sci. Chem. A, 17:667 (1982)
21. S. Raynal, S. Lecolier, G. Ndebeka, P. Caubère, Polymer, 22:1425
      (1981)
22. D.M. French, Rubber Chem. Technol., 42:90 (1969)

ANIONIC POLYMERIZATION IX: A REVIEW OF THE
USE OF CROWN ETHER AS A MODIFIER IN THE ANIONIC
POLYMERIZATION AND COPOLYMERIZATION OF DIENE

Tai Chun Cheng

Corporate R&D
Raychem Corporation
Menlo Park, California   94025

and

Central Research Laboratory
Firestone Tire and Rubber Company
Akron, Ohio   44317

ABSTRACT

     Ever since Petersen reported the complexing ability of the
crown ether with alkali, alkaline earth and other cations,
crown ether became a major subject for researchers. This is
due to the fact that crown ether possesses the ability to form
complexes with a variety of inorganic salts and also the ability
to solubilize these salts in aprotic solvents. The complexation
between metal cation and crown ether is believed to involve
ion-dipole interactions and therefore is similar in nature to
ordinary solvation. In addition, crown ether/metal cation
complexes can serve as catalysts in reactions involving ionic
intermediates. Polymerization of diene with crown ether/metal
cation complexes is a typical example of this subject, since this
reaction involves ionic intermediates. The detailed information
including a brief history, chemical properties of crown ether and
its application in the anionic polymerization and copolymerization
have been discussed.

INTRODUCTION

     It is generally known that the polymerization of butadiene
with organosodium in aliphatic hydrocarbon solvent gives a

polymer with a relatively low molecular weight and low conversion.[1]
Furthermore, the rate of polymerization of butadiene with organosodium is slow. At 30°C, it requires several days in order to
obtain a reasonable conversion. If the polymerization temperature is raised, to increase the rate, a spontaneous termination
mechanism becomes important. Based on our previous study,[2] it is
believed that an allylic sodium would be responsible for the
instability of the live end. We can consider that allylic sodium
is able to disproportionate rapidly in hexane by two different
pathways, one giving NaH and the other giving sodium metal. The
pathway to NaH is a dead end because NaH is inactive for diene
polymerization under the conditions employed. The sodium metal,
on the other hand, is known to initiate polymerization of butadiene
very slowly in hydrocarbon solvent by an electron-transfer
mechanism.[29] Thus, sodium metal formed rapidly by disproportionation of allylsodium could reinitiate diene polymerization in a
slow process. The net result would be a dead polymer of broad
molecular weight distribution and NaH or Na metal.

By introducing either lithium or potassium t-butoxide, as
described by us, in the early papers,[2] the entire story of the
n-butylsodium system is changed. The lithium t-butoxide is more
effective than potassium t-butoxide with regard to retention of
the stability of the live end. It is not clear whether this is
due to an inherently greater ability of lithium to stabilize the
live end or to its greater solubility in hexane, since both
complexes are insoluble in hexane.

It is known that crown ether possesses the ability to form
stable complexes with alkali cation and also the ability to
solubilize these salts in hydrocarbon solvents.[3] Furthermore,
crown ether/metal cation complexes can serve as catalysts in
reactions involving ionic intermediates.[4] It is felt that such
a system would provide us valuable information regarding the
stability of allylic sodium. Our early publication has achieved
this goal. This paper is a review of crown ether and its application in the areas of anionic polymerization and copolymerization
of diene.

HISTORY

The first cyclic polyether was reported by Ruggli in 1912.[5]
This compound was prepared through the self-condensation of a

bifunctional molecule in a very dilute solution. As expected, the resulting product from this reaction contained no more than two oxygen atoms in the polyether ring.

This method has since been improved by Luttringhaus and Ziegler. They reported the successful synthesis of the first cyclic polyether containing four or more oxygen atoms in the polyether ring. This compound was prepared from resorcinal.[6] Subsequently, Lutteringhaus also prepared other cyclic polyethers derived from hydroquinnone and 1,5-and 2,6-dihydroxynaphthalenes,[7] from 4,4'-hydroxydiphenyl and 4,4'-dihydroxydiphenyl-methane,[8] and from 4,4'-dihydroxydiphenyl ether.[9]

In 1941, Adams and Whitehill reported that 1,1',4,4'-bis (hexamethylenedioxy)-dibenzene (a cyclic polyether containing four oxygen atoms in the ring) was successfully synthesized by the direct condensation of bis (w-bromo-n-hexyl) ether of hydroquinone and hydroquinone in amyl alcohol with dry potassium carbonate. The yield of the product was about 15 percent. Since then, numerous papers prior to 1967 have been published concerning the synthesis of cyclic polyethers. Table I lists several of the papers and authors that have contributed to the synthesis of this product.[10,11,12,13,14]

The macrocyclic ethers synthesized by Petersen[15,16,17] have drawn considerable interest due to their ability to form a stable complex with metal salts[18,19,20] and serve as catalysts in chemical reactions. Because of these reasons, new monocyclic analogues have been synthesized. It is now becoming one of the hottest subjects in today's chemical fields.

SYNTHESIS

Macrocyclic polyethers, in general, can be prepared by five different methods. They are prepared by the reaction of organic diols with organic dihalides in the presence of base (e.g., NaOH). The reaction mixture is usually allowed to reflux with good agitation for 23 to 30 hours in a solvent. Isolation of a pure product is followed by filtration and recrystallization techniques. The five different methods for the synthesis of cyclic polyethers are shown in the following equations where R represents aromatic groups, R' represents either aromatic or aliphatic groups, and X represents halides. In addition, we will include an example for each method described in this section.

TABLE I

HISTORY OF CROWN ETHER

METHOD	REFERENCE
1. RESORCINOL CYCLIC POLYETHER	LUTTRINGHAUS AND ZIEGLER, ANN. 528, 155 (1937)
2. HYDROQUINONE OR 1,5 AND 1,6 DIHYDROXYNAPHTHALENE CYCLIC POLYETHER	LUTTRINGHAUS, ANN. 528, 181, (1937)
3. 4,4′ DIHYDROXYDIPHENYL OR 4,4′ DIHYDROXYDIPHENYL METHANE CYCLIC POLYETHER	LUTTRINGHAUS, ANN. 528, 211 (1937) LUTTRINGHAUS, ANN. 528, 211 (1937)
4. 4,4′ DIHYDROXYDIPHENYL ETHER CYCLIC POLYETHER	LUTTRINGHAUS, ANN. 528, 223 (1937)
5. HYDROQUINONE + $Br(CH_2)_6 OPhO(CH_2)_6Br$ 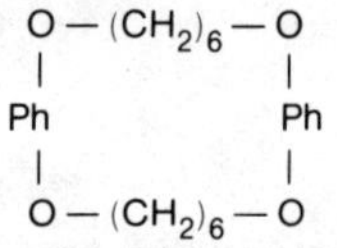	ADAMS AND WHITEHILL, JACS, 63, 2073 (1941)

$$O-(CH_2)_6-O$$
$$|\qquad\qquad |$$
$$Ph\qquad\qquad Ph \quad (15\%)$$
$$|\qquad\qquad |$$
$$O-(CH_2)_6-O$$

6. ACETONE + FURAN $\longrightarrow$

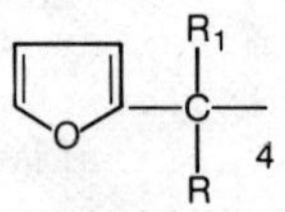

(18%)

ACKMAN, BROWN & WRIGHT, J.O.C., 20, 1147 (1955)

7.

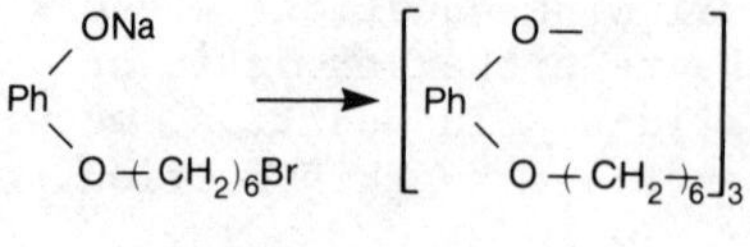

LUTTRINGHAUS AND SICHERT-MODROW, MAKROMOL. CHEM. 18-19, 511 (1956)

HISTORY OF CROWN ETHER

METHOD	REFERENCE
8. CYCLIC TETRAMER OF PROPYLENE	DOWN, LEWIS, MOORE & WILKINSON PROC. CHEM. SOC., 209 (1957) J. CHEM. SOC., 3767 (1959)
9. SYNTHESIS OF CYCLIC POLYMERS AND THEIR COMPLEXES WITH METAL SALTS	PETERSEN, JACS, 89, 7017 (1967)
10. NEW MACROCYCLIC POLYETHERS	PEDERSEN, JACS, 92, 391 (1970)

METHOD	REFERENCE
11. STABILITY CONSTANTS OF CYCLIC POLYETHER COMPLEXES WITH UNIVERLENT CATIONS	FRENSDORFF, JACS, 93, 600 (1971)

(continued)

TABLE I (Continued)

HISTORY OF CROWN ETHER

METHOD	REFERENCE
12. STUDY OF UNI- AND BIVALENT METAL IONS WITH CROWN ETHER	IZATT, NELSON, RYTTING, HAYMORE AND CHRISTENSEN, JACS, 93 1619 (1971)
13. STABILITY CONSTANTS OF CYCLIC POLYETHER COMPLEX WITH UNIVALENT CATIONS	FRENSDORFF, JACS, 93, 4684 (1971)
14. SALT COMPLEXES OF CYCLIC POLYETHERS	SMID, et al., MACROMOLES, 6, 133, (1973)
15.	TAKEKOSHI & WEBB, U.S. PATENT 3,824215 (1974)

HISTORY OF CROWN ETHER

METHOD	REFERENCE

16. $Li^+ < Na^+ < K^+ < Rb \leq Cs^+$
    $Be^{2+} < Mg^{2+} < Ca^{2+} < Sr^{2+} \leq Ba^{2+}$

    BLASIUS, ADRIAN, JANZEN AND KLAUTKE,
    J. OF CHROMATOGRAPHY, 96, 89 (1974)

17. CYCLIC POLYETHERS

    STEWART, WADDAN AND BORROWS
    BRITISH PAT. 785,229 (1977)

18. POLYMER-MULTIHETEROMACROCYCLES

    CRAM, U.S. PATENT 4,043,979 (1977)

19.

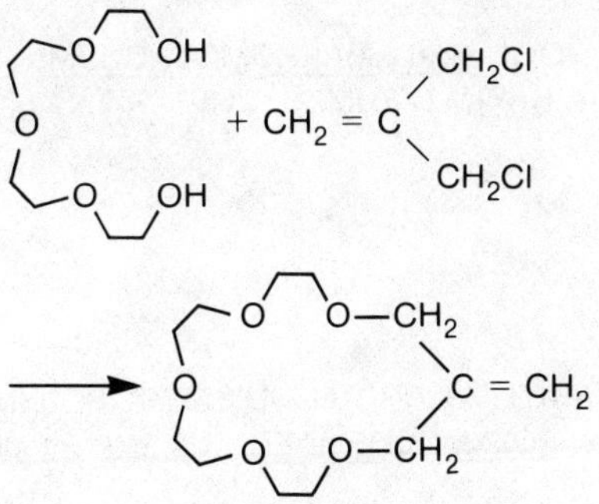

    TOMOI, et al., TETRAHEDRAN LETTERS,
    33, 3031 (1978)

(Continued)

TABLE I (Continued)

HISTORY OF CROWN ETHER

METHOD	REFERENCE
20. TACTIC POLY (CROWN ETHERS)	KIMURA, et al., POLYM. BULL., 1, 403 (1979)
21. POLYMERIC PSEUDOCROWN ETHERS: SYNTHESIS AND COMPLEXATION WITH TRANSITION METAL ANIONS	WARSHAWSKY, et al., JACS, 101, 4249 (1979)

22.

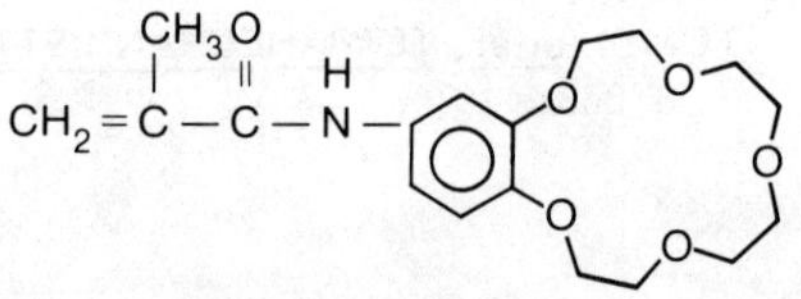

AND CATION BINDING PROPERTIES

YOGI, et al., MAKROMOL. CHEM., RAPID COMM., 1, 263 (1980)

23.

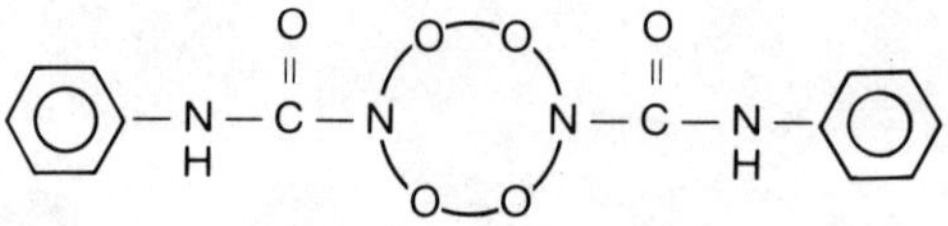

MATHIAS AND AL-JUMAH, J. POLYM. SCI., POLYM. CHEM. ED., 18, 2911 (1980)

1) $HOROH + 2NaOH + XR'X \longrightarrow$ $+ 2NaCl + H_2O$

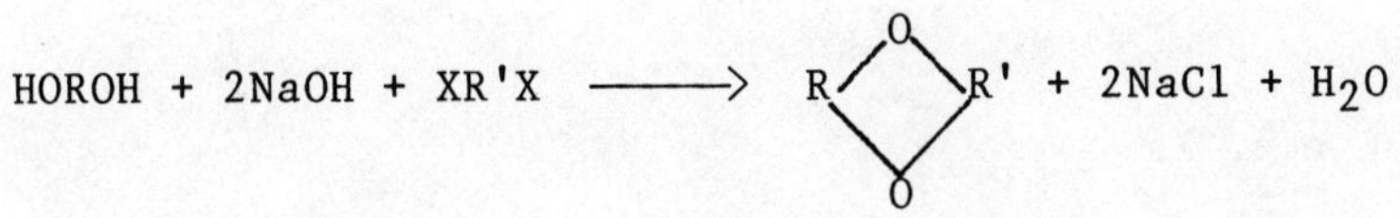

e.g. 15   (4-62%)

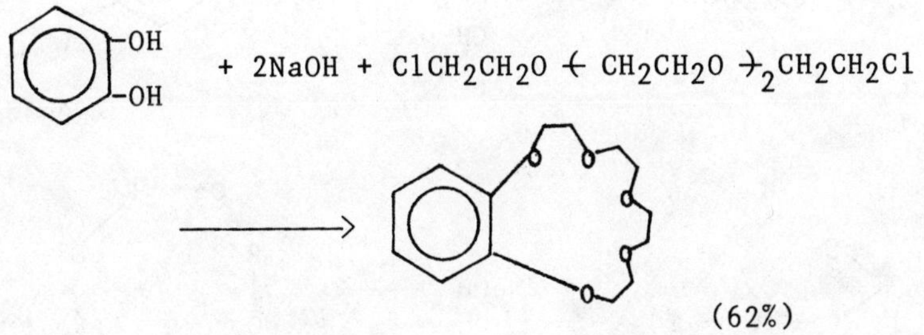

$+ 2NaOH + ClCH_2CH_2O \left(\!CH_2CH_2O\!\right)_2 CH_2CH_2Cl$

$\longrightarrow$

(62%)

2) $+ 2\ ROH + XR'X \longrightarrow$

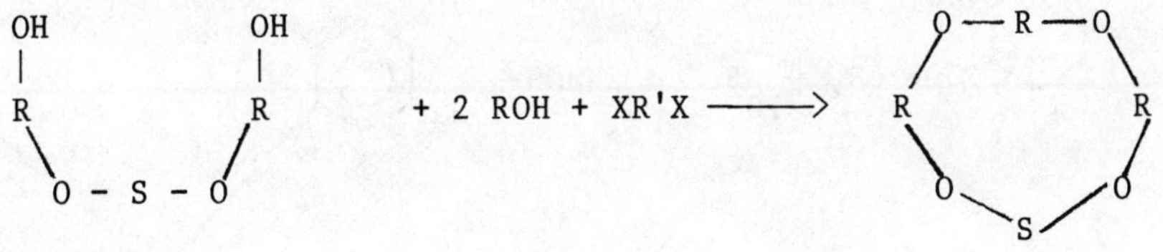

e.g. 15   (4-80%)

$+ 2NaOH +$ $\longrightarrow$

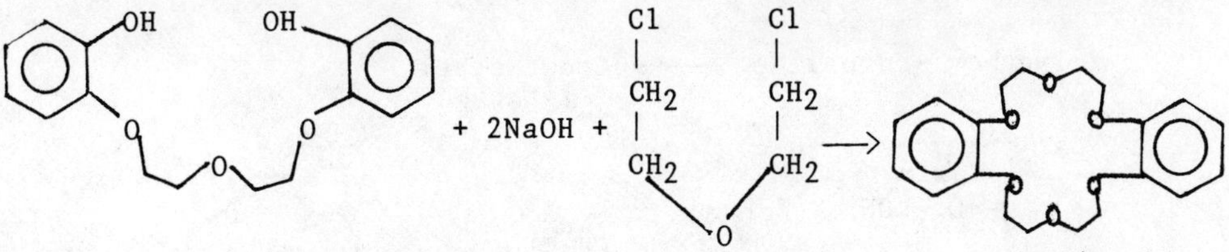

(80%)

3) 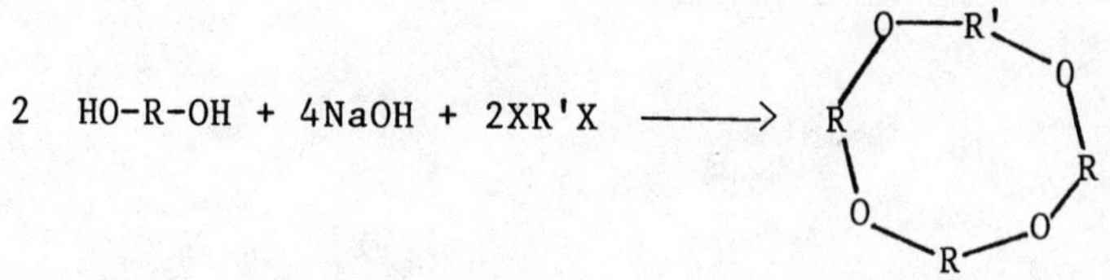   2 HO-R-OH + 4NaOH + 2XR'X ———⟶

e.g. 15

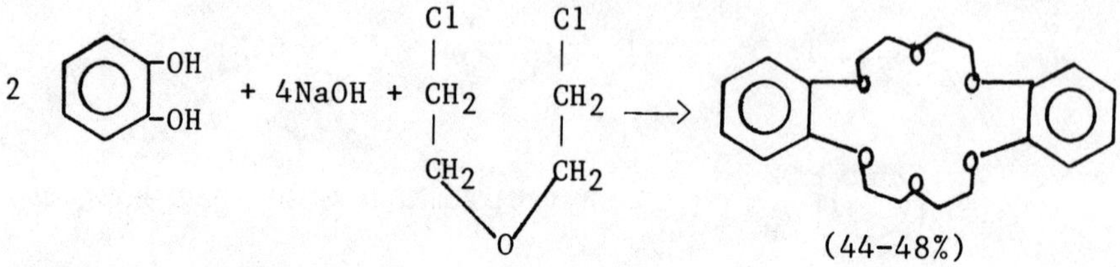

(44-48%)

4) 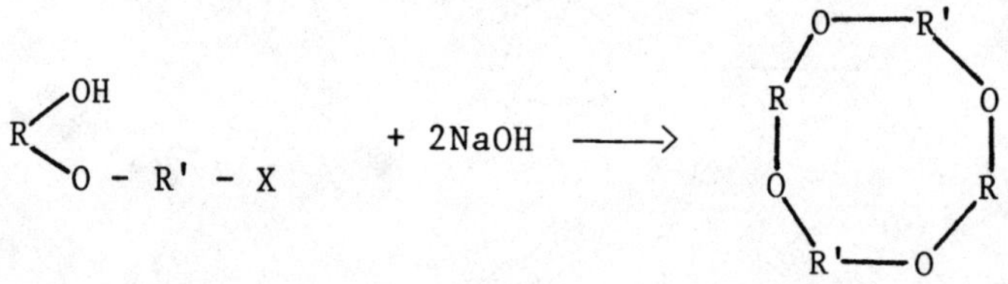   + 2NaOH ———⟶

e.g. 15

$(CH_2)_{10}$

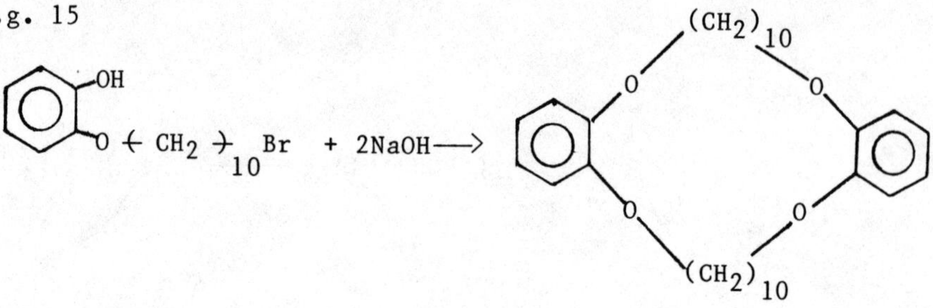

(3%)

5)   Miscellaneous

Polyether + dihalide ———⟶ Crown ether
                              (35-98%)
e.g. 21, 22

(a) 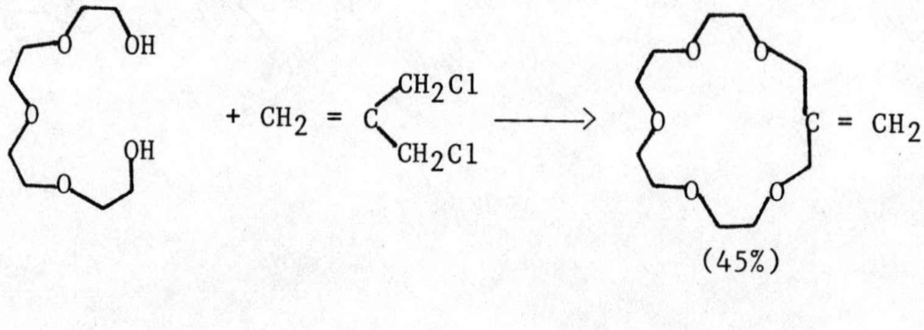   + $CH_2 = C \begin{matrix} CH_2Cl \\ CH_2Cl \end{matrix}$ ———⟶ $C = CH_2$

(45%)

(b)

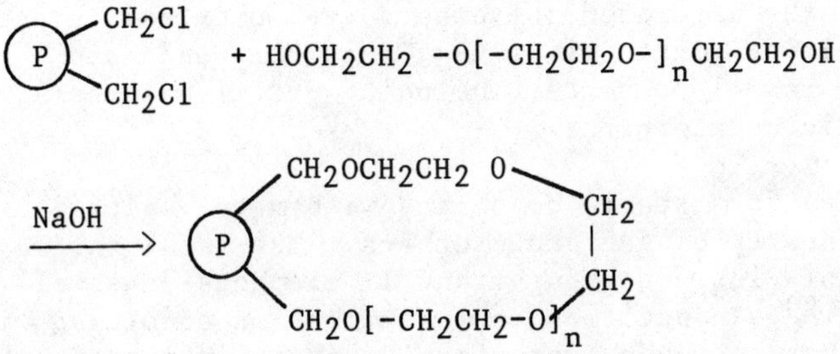

$$\underset{P}{\bigcirc}\begin{array}{l}-CH_2Cl \\ \\ -CH_2Cl\end{array} \quad + \quad HOCH_2CH_2 -O[-CH_2CH_2O-]_n CH_2CH_2OH$$

$$\xrightarrow{\text{NaOH}} \quad \underset{P}{\bigcirc}\begin{array}{l}-CH_2OCH_2CH_2 \ O \\ \hspace{5em}CH_2 \\ \hspace{5em}| \\ \hspace{5em}CH_2 \\ -CH_2O[-CH_2CH_2-O]_n\end{array}$$

(c)

$$\bigcirc\!\!\!\bigcirc\begin{array}{l}-CH_2Br \\ \\ -CH_2Br\end{array} \quad + \quad HO \ (\ CH_2CH_2O \ )_{n+2} H$$

$$\xrightarrow{\text{2KotBu}}$$

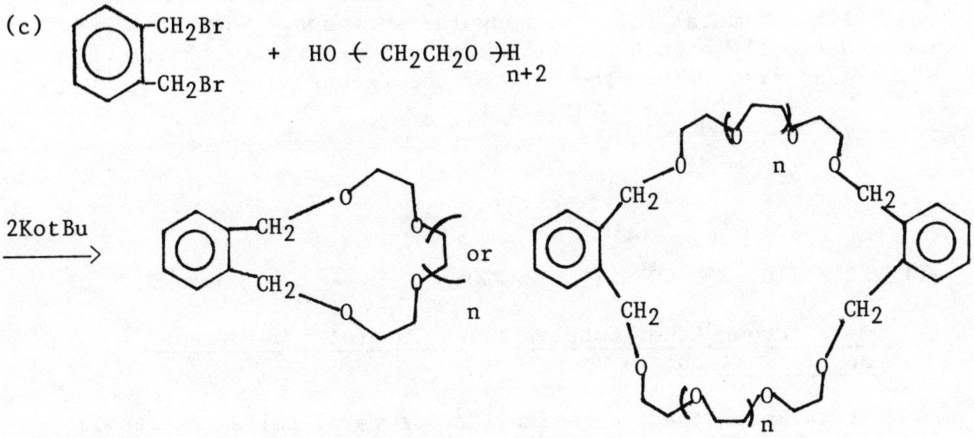

1,2-cyclohexyl derivatives of crown ether are usually prepared by the hydrogenation of the benzo compounds in the presence of ruthenium dioxide as a catalyst. Isolation of this product is done by using a column chromatography technique.

PROPERTIES

The cyclic polyethers, in general, are known as colorless crystalline compounds when the polyethers contain aromatic side rings. The melting point of the polyethers increases as the number of benzo groups increase (e.g., m.p. 39-40°C for [18] crown-6, 43-44° for benzo [18] crown-6, 164°C for dibenzo [18] crown-6, and 190-192°C for tribenzo [18] crown-6).

The solubility of cyclic polyethers is also dependent on the number of aromatic rings. The compounds containing more than one benzo group are generally insoluble in protic solvents at room temperature. However, they are readily soluble in chlorinated solvents such as methylene chloride and chloroform.[15]

In contrast, the saturated polyethers are colorless, viscous liquids or low melting point solids.[18] These compounds are much more soluble in all solvents (including petroleum ether) than their aromatic counterparts.

The ability to form stable complexes with metal salts is another unique property of the crown ethers. The metal salts described previously include Group I and II elements[15] as well as Al^{3+} in group III.[20] In addition, ammonium cation complexed with cyclic polyether has also been reported.[18] It is interesting to note that the presence of cyclic polyethers strongly increase the solubility of metal salts in nonpolar solvents. This property can drastically affect the structure and reactivity of ion pairs [24, 25] and also affect the rate of the reactions.[26, 27]

POLYMERIZATION AND COPOLYMERIZATION

(A) Metal-Crown Ether Complex as an Initiator for Diene Polymerization

It is known that the solubility of metal salts in nonpolar media is drastically increased if small amounts of crown ethers (cyclic polyethers) are used as complexation agents. Such a concept has been demonstrated in various areas of chemistry. For example, they are used as phase-transfer catalysts in organic synthesis. Moreover, Cheng[27] and Schue[28] have expanded this idea in the areas of anionic polymerization.

It is generally recognized that sodium metal produces a low molecular weight polymer at low conversion when used as an initiator for the anionic polymerization of 1,3-butadiene in a hydrocarbon solvent.[29] The modification of sodium initiator with lithium t-butoxide or K-t-butoxide produces a polymer with a high molecular weight.[30] This is presumably due to the formation of a complex (I) between allylic sodium and metal alkoxide.

$$\text{———CH}_2 \underset{M}{\overset{Na}{\diamond}} \text{O-t-Bu}$$

(I)

Such a delocalized active complex might be expected to be more stable and give a diene polymer having a high molecular weight. But the range of solvents in which stable sodium-metal alkoxide complex solutions could be obtained was severely limited by solubility and decomposition problems, especially in hydrocarbon media. The application of dicyclohexyl-18-crown-6 or DCHE (II) to complex the sodium metal as reported by Schire[28] has

(II)

proved to be very successful for increasing the solubility in hydrocarbon solvents and enhancing the stability of carbon-sodium bonds. As a result, this system gives a polymer with high molecular weight (DSV 2-3 g/dl) at high conversion (92 percent). In addition, the vinyl (or 1,2) content also increases from 62 percent to 75 percent at the expense of 1,4 content.

Other group I metals, such as K, Rb, Cs in the DCHE system also showed similar results in terms of the conversion, the rate of polymerization, and microstructure.[28] This is shown in Table II.

The polymerization of isoprene with crown ether (DCHE) modified group I metal initiator was also reported by Schue.[28] Their data suggested that an enhancement in the rate of polymerization and a change in the geometry of the polymer chain were observed as compared to the system without using crown ether as a modifier. This is shown in Table III.

The reasons for this are believed to be associated with solubilization of the polar salts in non-polar solvents and enhanced ion separation. Tight complexation of the cation by crown ethers in general will enhance anion activity. As a result, a different polymerization mechanism is observed.

(B) Organosodium-Crown Ether Complex as an Initiator

(i) Homopolymerization of Butadiene. The polymerization of butadiene with organosodium in a hydrocarbon solvent is known to produce a polymer with low molecular weight and low conversion.[31] The addition of polar modifiers to the organosodium, such as aliphatic ethers and amines, initiates butadiene polyermization

TABLE II

POLYMERIZATION OF 1,3-BUTADIENE

INTIATOR	MODIFIER	Pzn TEMP °C	% CONVERSION	MICROSTRUCTURE		REFERENCES
				1,2%	1,4%	
Na	—	30	30	62	38	(28)
Na	DCHE	30	92	75	25	(29)
K	—	10	—	35	65	(30)
K	DCHE	10	90	69	31	(26)
Rb	—	10	—	62	38	(31)
Rb	DCHE	10	98	63	37	(26)
Cs	—	10	—	53.6	46.4	(32)
Cs	DCHE	10	100	58	42	(26)

TABLE III

POLYMERIZATION OF ISOPRENE IN BENZENE AT 10° C

INITIATOR METAL		YIELD %	MICROSTRUCTURE		
			1,2 %	3,4 %	1,4 %
Na	—		10	58	32
	+ DCHE	70	24	44	32
K	—		6	41	53
	+ DCHE	95	19	60	21
Rb	—		8	39	53
	+ DCHE	92	21	45	34
Cs	—		10	31	59
	+ DCHE	100	17	41	42

F. SCHUE, et al., POLYM. LETTERS. ED., 13, 397 (1975)

in a hydrocarbon solvent and also produces low molecular polymer
in low conversion.[35] However, the situation was changed when
crown ethers were used as modifiers in the above polymerization.
Polybutadiene was obtained in high conversion and high molecular
weight. This is true whether we chose tricyclohexyl-18-crown-6
(III) or bicyclohexyl-18-crown-6 as modifiers (II).

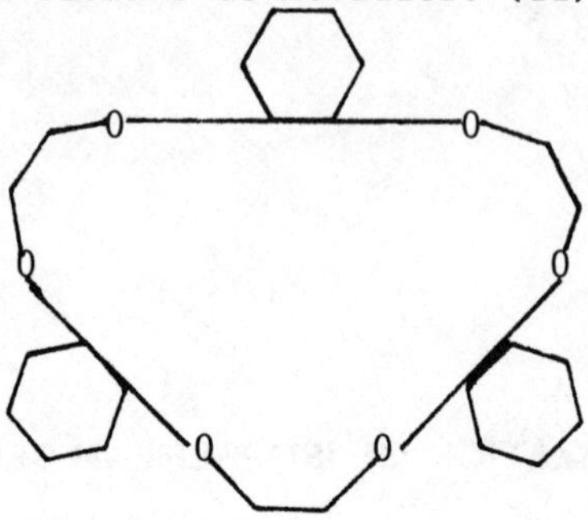

(III)

As can be seen in Table IV, the conversion of the polymer
increases from 40 percent to 94.5 percent as the ratio of n-butylsodium to bicyclohexyl-18-crown-6 changes from 25.4 to 5.3. The
molecular weight of the polymer was also found to be high as indicated by Dilute Solution Viscosity (DSV) (1.60 and 5.2 g/dl).

The effect of crown ethers on conversion cannot be explained
by the change of polarity of the solvent since aliphatic ethers
do not give the same result as crown ethers do. It is believed
that the crown ethers stabilize the growing allylic anion through
metal complexation. The complexes (III) between alkali metal and
crown ether have been observed by other workers and have been
isolated in certain cases.[15]

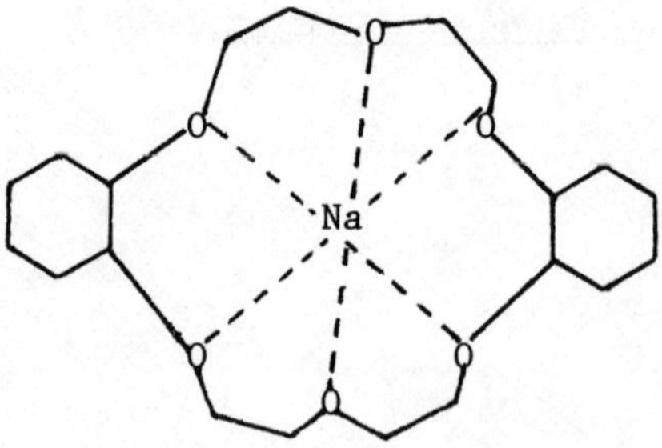

(IV)

TABLE IV

POLYMERIZATION OF 1,3-BUTADIENE WITH n-BUTYLSODIUM MODIFIED
WITH CROWN ETHER[a] AT 30°C IN HEXANE

| Na/CROWN | DSV, dl/g[b] | INFRARED ANALYSIS | | | CONVER-SION, % | Tg,°C |
		cis-1, 4%	trans-1, 4%	1,2, %		
—	∨	8.9	39.1	52.0	< 20	—
19[c]	0.81	1.2	28.1	70.7	76.0	-38
6.35[d]	3.72	4.6	20.8	74.6	92.0	-38
5.30[d]	5.20	5.2	19.5	75.3	94.5	-32.5
25.4[c]	1.68	1.1	30.6	68.4	41.0	-41
4.60[d]	4.60	1.6	18.5	80.0	93.0	-30

[a]MONOMER CONCENTRATION, 2.10 mole/l.; CATALYST CONCENTRATION, 0.006-0.009 mole/l.
[b]DSV (INHERENT VISCOSITY); GEL = 0.
[c]TRICYCLOHEXYL-18-CROWN-6.
[d]BICYCLOHEXYL-18-CROWN-6.

Such complexation may stabilize the live end and preclude
the termination via disproportionation of allylic anion in the
polymer chain.

The relative rate of polymerization of 1,3-butadiene
with n-butylsodium-crown ether initiator has also been studied[27]
(Figure 1). As expected, the rate of polymerization at 70°C is
faster than at 50°C, which in turn is faster than 30°C. However,
the polymerization rates are all faster than the system without
crown ether as a modifier as shown in Fig. 3. The difference in
ratio at 30°C is believed to be associated with the solubility of
n-butylsodium with crown ether in hexane.

The microstructure of the butadiene polymer in this system
has been determined by infrared spectroscopy. The vinyl content
shown in Table IV was 18 to 20 percent higher in the presence of
crown ether than that observed with the n-butylsodium system
alone. Thus, a steady increase in the crown ether concentration
showed a moderate increase in the vinyl content (68 to 80 percent)
at 30°C polymerization temperature. However, we did not observe
the temperature dependence of vinyl content as is typical in the
polymerization of 1,3-butadiene initiated with organolithium
compounds modified by polar modifiers.[36] Instead, we only
observed a moderate change in the vinyl content. As can be seen
in Table V, the vinyl content of polybutadiene prepared by
n-butylsodium/tricyclohexyl-18-crown-6 initiator decreases only
from 83 percent to 70 percent as the polymerization temperature
increased from 30°C to 70°C, respectively. The relatively small
change of vinyl content, or 1,2 content, of polybutadiene can be
interpreted as the existence of a stable complex between the
allylic anion and crown ether. This would lead to less dissociation than observed with simple ethers. Thus, the microstructure
of the polymer would be less sensitive to temperature.

(ii) Copolymerization of Butadiene and Styrene. It is well
known that the polymerization of butadiene with alkyllithium
imitator has a slower polymerization rate than styrene in a
homopolymerization system, but a faster rate than styrene in a
copolymerization system.[37] This phenomenon is generally known
as the reversal of activity. As a result, the initial portion of
the copolymer chain is rich in butadiene. This is then followed
by a block of polystyrene due to the exhaustion of the butadiene.

It is also known that the butadiene-styrene copolymerization with alkylsodium in the presence or absence of polar
modifiers give only randomized copolymers.[31] Furthermore,
the molecular weight of the copolymer is too low. As a result,
the reaction does not produce a high quality elastomer. The
situation is totally changed when n-butylsodium-crown ether
system is used as a catalyst for the copolymerization of

PLOT OF CONVERSION VS. TIME AT THREE DIFFERENT
TEMPERATURES AT Na/CROWN = 11.0; CATALYST
CONCENTRATION, 0.006 mole/l; MONOMER
CONCENTRATION, 2.10 mole/l

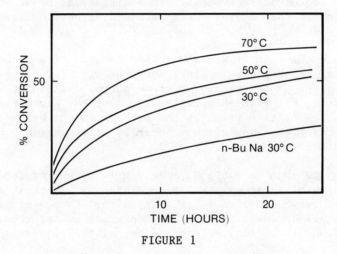

FIGURE 1

TABLE V

EFFECT OF TEMPERATURE ON MICROSTRUCTURE OF
POLYBUTADIENE POLYMERIZATION WITH
n-BUTYLSODIUM CROWN ETHER[a]

TEMPERATURE, °C	INFRARED ANALYSIS			CONVERSION, %
	1,2 %	trans-1, 4, %	cis-1, 4, %	
30	83.0	17.0	0.0	96.5
50	75.0	22.3	0.0	95.0
70	70.0	27.5	2.6	94.5

[a]Na/CROWN = 11.0; CATALYST CONCENTRATION, 0.006 mole/l.; MONOMER
CONCENTRATION, 2.10 mole/l.

butadiene and styrene in hexane. The resulting polymer is high
molecular weight with high conversion.

As can be seen in Table VI and Figure 2, the polymer
chain contains 35 percent (by IR) of the styrene initially,
then the styrene content tends to level off and remain constant
throughout the rest of the copolymerization. This result is
totally different from what we have been observing in either
lithium or sodium based systems. In addition, no block styrene
can be found in the polymer chain based on either chemical or NMR
analysis.

At this time, it is appropriate to consider the microstructure of the copolymers since the presence of polar modifiers
not only increases the rate of styrene polymerization in alkylsodium system, but also changes the ratio of the geometrical
isomers in the diene portion of the polymer. It is believed that
a similar phenomenon may also be observed in crown ether modified
n-butylsodium systems.

As can be seen in Table VI, the vinyl content of the butadiene portion of the copolymer is consistently higher in the
presence of crown ether modifier than in the absence of modifier
(68 percent). In addition, the vinyl content of the copolymer
was also found to be constant regardless of the conversion.

The reason for the increase in vinyl content of the butadiene
portion of the copolymer in the presence of crown ether modifier
is believed to be associated with complex formation between
allylic anion and sodium crown ether, as shown below.

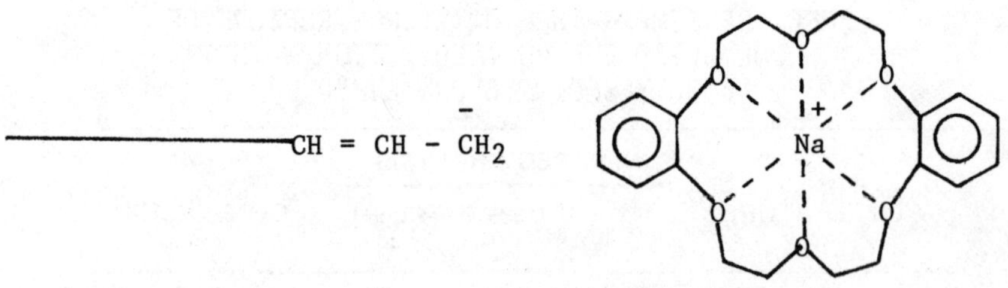

$$\text{———————CH} = \text{CH} - \text{CH}_2^{-}$$

The result of electron donation by the six oxygen atoms from
crown ether is an enhancement of the ionic character of the carbon-
sodium bond. An increase in ionic character in the copolymerization
results in an increased amount of vinyl structure. Such an explanation has been suggested by Antkowiak[36] in the n-butyllithium
initiated polymerization.

TABLE VI

COPOLYMERIZATION OF 1,3-BUTADIENE — STYRENE (80/20) WITH
n-BUTYLSODIUM — CROWN ETHER AT 30°C[a]

TIME, HR	CONVER-SION, %	STYRENE CONTENT, %		INFRARED ANALYSIS		
		IR	NMR	cis-1,4%	trans-1,4%	1,2, %
1.0	3.7	35.0	33.0	0	19.9	80.1
1.7	13.0	34.4	33.0	0	20.4	79.9
17.0	28.0	23.0	22.0	0	20.3	79.7
25.0	32.0	22.0	21.0	0	20.0	80.0
40.0	39.0	21.0	19.8	0	19.6	79.4
49.0	44.0	21.0	19.0	0	20.0	80.0
113.0	97.0	21.0	19.0	0	20.0	80.0

[a]CATALYST CONCENTRATION, 0.006-0.009 mole/l.; MONOMER CONCENTRATION, 2.10 mole/l.

COPOLYMERIZATION OF BD/St (80/20) WITH
n-BUTYLSODIUM-CROWN ETHER AT 30°C IN HEXANE
(Na/CROWN = 11)

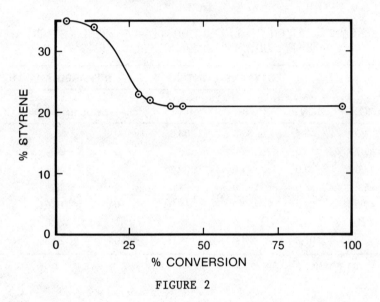

FIGURE 2

The Gel Permeation Chromatographic (GPC) analysis of this co-polymer showed a very broad molecular weight distribution (Figure 3). The broadness of the molecular weight distribution may be due to the uncomplexed initiator metalating the performed polymer chain, since the n-butylsodium to crown ether (tricyclohexyl-18-crown-6) ratio showed no effect on the molecular weight, conversion and styrene distribution over the range of 11.3 to 113.0 (Table VII).

The relative lack of dependence of molecular weight on butyl-sodium to crown ether ratios within the range of 11.3 to 113 suggests that the allylic sodium is chelating with the crown ether, forming a tight ion-paired species. These ion-paired species in hexane may form slowly and could be the rate-determining step of the polymerization. Such tight complexation of cation by crown ethers promotes ion separation and enhanced anion activity. It thus increases the rate of polymerization. The resulting polymer will have little change in molecular weight.

CONCLUSION

The nature of the activity of crown ethers and the advantages of the use of crown ethers as catalysts in chemical reactions make it applicable for polymer synthesis.

For the group I metals, it was found that stable complexes beween dicyclohexyl-18-crown-6 and sodium, potassium, rubidium and cesium metals have been obtained in benzene. These new complexes demonstrated the ability to act as active catalysts for the polymerization of butadiene and isoprene. Such catalysts increased the yield and rate of polymerization as compared with conventional alkali metal anionic systems. In addition, the microstructure of the polymer is different from that of the same polymer prepared by metals alone.

For the organosodium system, it was found that the polymeriza-tion of butadiene with crown ether-n-butylsodium catalysts also showed a rate enhancement. The presence of crown ethers as modifiers in the n-butylsodium system produced a polymer with a high molecular weight, a high vinyl content, and a high conversion as compared with an unmodified system.

GPC MOLECULAR WEIGHT DISTRIBUTION OF
BD/St COPOLYMER

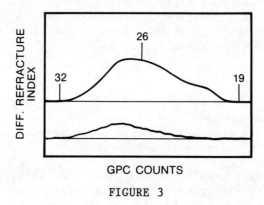

FIGURE 3

The copolymerization of butadiene and styrene with n-butyl-sodium-crown ether catalyst also produces polymers at high conversion and high molecular weight. The resulting copolymer prepared from this catalyst showed no block styrene at Na/crown ether ratios of 11.3 to 113. The data also showed a higher initial rate of styrene incorporation in the polymer chains at low conversion, but leveled off at about 40 percent conversion.

COPOLYMERIZATION OF 1,3-BUTADIENE-SYTENE (80/20) WITH
n-BUTYLSODIUM AND CROWN ETHER AT 30°C[a]

Na/ CROWN	INFRARED ANALYSIS				BLOCK STYRENE %	DSV, g/dl[b]	CON-VER-SION, %
	cis-1, 4, %	trans-1, 4, %	1,2, %	STYRENE, %			
11.3	0.0	26.3	73.7	22.7	0	2.43	80
15.8	0.9	24.3	74.8	23.5	0	2.20	80
14.1	1.0	26.0	72.9	21.7	0	2.40	87
63.5	4.8	24.7	70.5	18.8	0	2.97	94
113.0	0.0	20.2	79.0	20.1	0	2.92	97

[a]CATALYST CONCENTRATION, 0.006-0.009 mole/l.; MONOMER CONCENTRATION, 2.10 mole/l.
[b]DSV = INHERENT VISCOSITY; GEL = 0.

TABLE VII

REFERENCES

1. A. A. Morton and E. Grovenstein, Jr., J. Am.Chem. Soc., 74, 5434 (1952).
2. T. C. Cheng, A. F. Halasa, D. P. Tate, J. Polymer Sci., A-1, 9, 2493 (1971).
3. C. J. Petersen, J. Am. Chem. Soc., 89, 7017 (1967).
4. Chem. Eng. News, 48 (9), 26 (1970).
5. Ruggli, Ann. 392, 92 (1912).
6. A. Luttringhaus and K. Ziegler (1937).
7. A. Luttringhaus, Ann. 528, 181 (1937).
8. A. Luttringhaus, Ann. 528, 211 (1937).
9. A. Luttringhaus, Ann. 528, 223 (1937).

10. A. Luttringhaus and I. Sichert-Modrow, Makromol Chem., 18-19, 511 (1956).

11. R. G. Ackman, W. H. Brown and G. F. Wright, J. Org. Chem 20, 1147 (1955).

12. D. G. Stewart, D. Y. Waddan and E. T. Borrows, British Patent, 785, 229 (1957).

13. J. L. Down, J. Lewis, B. Moore and G. W. Wilkinson, Proc. Chem. Soc., 209 (1957).

14. J. L. Down, J. Lewis, B. Moore and G. W. Wilkinson, J. Chem. Soc., 3767 (1959).

15. C. J. Petersen, J. Am. Chem. Soc. 89, 7017 (1967).

16. C. J. Petersen, Fed. Proc. Fed. Am. Soc. Exp. Biol., 27, 1305 (1968).

17. C. J. Petersen, J. Am. Chem. Soc., 92, 386 (1970).

18. C. J. Petersen and H. K. Frensdorff, Angew. Chem. Internat. Edit., 11 (1), 16 (1972).

19. R. N. Greene, Tetrahedron Letters, 18, 1793 (1972).

20. F. Wada and T. Matsuda, Bull. Chem. Soc. Jpn., 53, 421 (1980).

21. M. Tomoi, O. Abe, M. Ikeda, K. Kihara and H. Kabinchi, Tetrahedron Letters, 33, 3031 (1978).

22. A. Warshawsky, R. Kaliv, A. Deske, H. Berkovitz and A. Patchovonik, J. Am. Chem. Soc., 101, 4249 (1979).

23. D. N. Reinhoudt, R. T. Gray, C. J. Smit and I. Veenstra, Tetrahedron 32, 1161 (1976).

24. K. H. Wang, G. Konizer and J. Smid, J. Am. Chem. Soc., 92, 666 (1970).

25. J. Almy, D. C. Garwood and D. J. Cram, J. Am. Chem. Soc., 92, 4321 (1970).

26. S. Boilean, B. Kaemf, J. M. Lehn and F. Schue, Polymer Letters, 13, 203 (1974).

27. T. C. Cheng and A. F. Halasa, J. Polymer Sci.; Polymer Chem. Ed., 14, 583 (1976).

28. S. Alev, F. Schire and B. Kaemf, Polymer Letters Ed., 13, 397 (1975).

29. J. P. Kennedy and E. G. M. Tornquist, Polymer Chemistry of Synthetic Elastomers, Interscience, New York, 1969, Part II, p. 561.

30. T. C. Cheng and A. F. Halasa, J. Polymer Sci., Polymer Chem. Ed., 14, 573 (1976).

31. T. C. Cheng, unpublished results.

32. R. V. Basova, et al, Proc Acad. Sci., USSR, 149, 312 (1963).

33. A. W. Meyer, et al, J. Am. Chem. Soc., 74, 2294 (1952), and F. C. Foster and J. L. Binder, Handling and Use of Alkali Metals, 19, of Advances in Chemistry series, Am. Chem. Soc., Washington D.C., 1957, p. 26.

34. A. Rembaum, F. R. Ells and R. C. Morrow and A. V. Tobolsky, J. Polymer Sci., 61, 155 (1962).

35. A. T. Tsatsas, R. W. Stearns, and W. M. Risen, J. Am. Chem. Soc., 94, 5247 (1972).

36. T. A. Antkowiak, A. E. Obesta, A. F. Halasa and P. P. Tate,
     J. Polymer Sci., A-1, 10, 1319 (1972).
37. S. N. Ushakov and P. A. Matuzov, Zh. Obshch. Khim., 17,
     435 (1944).

SYNTHESIS OF ULTRAHIGH MOLECULAR WEIGHT NYLON 4

WITH ONIUM SALT AND CROWN ETHER-CONTAINING CATALYSTS

R. Bacskai

Chevron Research Company
576 Standard Avenue
Richmond, California 94802

INTRODUCTION

The discovery of Ney, Nummy, and Barnes,[1] that 2-pyrrolidone can be polymerized to nylon 4 by basic catalysts, stimulated much subsequent work on the preparation, physico-chemical characterization, and utilization of the polymer. The great interest in this polymer is due to the fact that nylon 4 textile fibers combine the good strength properties of nylon 6 or 66 with the hydrophilicity of cotton.[2]

Early work in the nylon 4 area focused on increasing the rate of polymerization by adding N-acyl initiators (e.g., N-acetylpyrrolidone, N-benzoylpyrrolidone, etc.) or compounds which can form N-acylpyrrolidones (e.g., acid anhydrides, acid chlorides, esters, amides, isocyanates, etc.) to the 2-pyrrolidone/alkali catalyst system.[3] It was found, however, that nylon 4 produced with the aid of N-acyl initiators is relatively low in molecular weight and has a very broad molecular weight distribution (e.g., N-acetylpyrrolidone used in Experiment B of Reference 4 gave: η_{inh} = 2.1 dl/g, $\overline{Mw}/\overline{Mn}$ = 19.2, $\overline{Mw}$ ~ 65,000). Barnes[4] subsequently discovered that nylon 4 prepared with carbonated potassium pyrrolidonate (py-K/CO_2) is higher in molecular weight and has a narrower molecular weight distribution (Example 3: η_{inh} = 4.4 dl/g, $\overline{Mw}/\overline{Mn}$ = 2.5, $\overline{Mw}$ ~ 200,000) than polymer prepared without CO_2. Polymerization rates, however, are low with the

py-K/CO_2 catalyst; and the addition of an N-acyl initiator reduces
the molecular weight.

In the course of our work, directed towards finding reaction
conditions which accelerate the polymerization without reducing the
polymer's molecular weight, we examined the effect of onium salts
and crown ethers on the py-K/CO_2 catalyzed polymerization of
2-pyrrolidone. Some details of the onium salt[5] and crown ether[6]
polymerization work, discussed here, are described in patents.

EXPERIMENTAL

Polymerization

A pyrrolidone solution of py-K/CO_2 was prepared from
2-pyrrolidone, KOH, and CO_2 by the technique of Barnes.[4] Details
of the experimental procedure are described in Reference 7. Polymerizations were carried out in dry polyethylene bottles, under
nitrogen, by mixing and heating an appropriate portion of the reaction mixture with a preweighed amount of the onium salt, or crown
ether. Following polymerization and water wash, the crown ether,
if used, was removed by chloroform extraction from the polymer.

In experiments carried out from the same batch of pyrrolidone-
py-K/CO_2 solution the reproducibility with regard to conversion and
molecular weight was better than ±5%. The reproducibility between
runs prepared from different batches of py-K/CO_2 solutions was
somewhat poorer, probably because of the small, variable amounts of
inhibitor (e.g., 3-aminobutyric acid) formed during dehydration.
Therefore, each set of experiments, reported in the tables,
includes a control run (no onium salt or crown ether) from the same
batch of py-K/CO_2 solution.

Viscosity and Molecular Weight

The specific viscosity of nylon 4 was determined in m-cresol
at C = 0.1 g/100 ml, 25°C. Intrinsic viscosity[8] was calculated
from:

$$[\eta] = \sqrt{\frac{2(\eta sp - \ln \eta rel)}{c^2}}.$$

Weight average molecular weight[9] was calculated from: $[\eta] = 3.98$ x 10^{-4} $\overline{Mw}^{0.77}$.

RESULTS AND DISCUSSION

Onium Salt Polymerization

The reaction sequence leading to nylon 4 from 2-pyrrolidone, KOH, CO_2, and tetramethylammonium chloride (TMAC) is shown in Scheme 1.

Scheme 1.   Nylon 4 Synthesis with Onium Salts

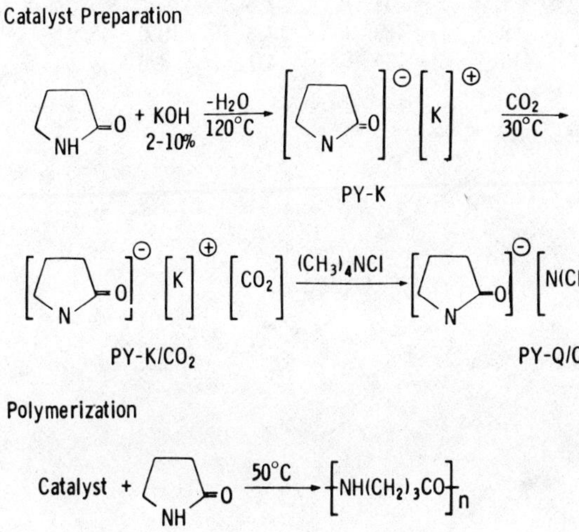

Catalyst Preparation

Polymerization

Since the carbonated tetramethylammonium pyrrolidonate (py-Q/CO_2) was not isolated, its formation can only be inferred from the polymerization results. No definite assignment for the

Table I. Polymerization of 2-Pyrrolidone with py-K/CO_2
and Tetramethylammonium Chloride (22-Hour Run)

Catalyst, Mol %		Conversion[2] %	$[\eta]$ dl/g	$\bar{M}_W \times 10^{-3}$
PY-K				
10		4	1.3	35
PY-K + TMAC				
5	5	35	4.1	165
PY-K/CO_2 [1]				
2		20	5.2	220
5		45	8.0	380
10		48	9.8	500
PY-K/CO_2 [1] + TMAC				
2	2	38	7.3	330
5	5	69	17.6	1050
10	10	59	14.7	820

[1] 30% CO_2 Based on K
[2] 50°C, 22 Hr

CO_2 attachment can be made at the present, and catalyst composition
is derived from the amount of CO_2 absorbed by the potassium pyrrolidonate (py-K) solution. In this work 30 mole % of CO_2 (based
on K) was always used.

Polymerizations carried out with py-K/CO_2 + TMAC are summarized in Tables I, II, and III.

Table II. Polymerization of 2-Pyrrolidone with py-K/CO$_2$
and Tetramethylammonium Chloride (8-Hour Run)

Catalyst, Mol %		Conversion,[2] %	$[\eta]$ dl/g	$\bar{M}_W \times 10^{-3}$
PY-K/CO$_2$ [1]				
2		9	3.1	117
5		16	6.3	285
10		14	6.7	305
PY-K/CO$_2$ [1] + TMAC				
2	2	15	5.0	210
5	5	40	11.1	575
10	10	54	16.8	980

[1] 30% CO$_2$ Based on K
[2] 50°C, 8 Hr

Table III. Effect of Tetramethylammonium Chloride
Concentration on the Polymerization of
2-Pyrrolidone[1]

(CH$_3$)$_4$NCl, Mole %	$\dfrac{[(CH_3)_4NCl]}{[K]}$	Conversion,[1] %	$\bar{M}_W \times 10^{-3}$
0	0	45	380
1.0	0.2	63	550
2.0	0.4	69	630
5.0	1.0	69	1050
7.7	1.5	69	610

[1] 5 Mole % PY-K/CO$_2$ (30% CO$_2$)
50°C, 22 Hr

For comparison, Table I also includes experiments run with py-K, py-K + TMAC, and py-K/CO_2. The concentrations reported are based on monomer. The data show that using py-K/CO_2 + TMAC catalyst, higher conversions and molecular weights are obtained than with py-K/CO_2 alone. In most reactions an increase in catalyst concentration results in higher conversion and molecular weight, but at very high catalyst levels a decrease in conversion and molecular weight was observed. This reversal is probably caused by the high viscosity of the reaction medium and by chloride ion termination (see below). It is significant that the molecular weight of nylon 4 obtained with onium salt catalysts is greater than that reported before. Estimates based on polymer solution viscosities indicate that with the appropriate catalyst combination ultrahigh molecular weight polymers, Mw > 1,000,000, can be readily prepared.

Interestingly, TMAC, although it is insoluble in 2-pyrrolidone, apparently reacts with py-K/CO_2 and influences the outcome of the polymerization. On the other hand, other insoluble onium salts, such as tetramethylammonium perchlorate, tetrafluoroborate, and hexafluorophosphate, are unreactive and give approximately the same conversion and molecular weight as the control run without onium salt. Among the soluble onium salts tested, tetraethylammonium chloride and tetramethylammonium acetate are quite effective to increase conversion and molecular weight but are not better than the insoluble TMAC (Table IV).

Table IV.   Polymerization of 2-Pyrrolidone with Soluble or Insoluble Onium Salts

Onium Salt	Solubility	$\dfrac{[K]}{[\text{Onium Salt}]}$	Conversion,[1] %	$\bar{M}_W \times 10^{-3}$
—	-	-	37	405
$(CH_3)_4N^{\oplus}Cl^{\ominus}$	Insoluble	1.0	69	1050
$(CH_3)_4N^{\oplus}ClO_4^{\ominus}$	Insoluble	1.0	31	400
$(CH_3)_4N^{\oplus}BF_4^{\ominus}$	Insoluble	1.1	35	415
$(CH_3)_4N^{\oplus}PF_6^{\ominus}$	Insoluble	1.0	33	405
$(C_2H_5)_4N^{\oplus}Cl^{\ominus}$	Soluble	1.0	73	1025
$(CH_3)_4N^{\oplus}OCOCH_3^{\ominus}$	Soluble	1.2	56	720

[1] 5% PY-K/CO_2 (30% CO_2)
50°C, 22 Hr

The comparative experiments of Table I show that both py-K/CO_2 and TMAC are essential for the formation of ultrahigh molecular weight polymers. This observation is consistent with the results of earlier investigators who obtained relatively low molecular weight ($[\eta]$ = 1.6 dl/g, $\overline{M}w$ = 48,000) nylon 4 with pure[10] or in-situ generated[11] quaternary ammonium pyrrolidonate catalysts.

The effect of onium salt anion on nylon 4 polymerization is shown in Table V.

Table V.   Effect of Onium Salt Anion on the
            Polymerization of 2-Pyrrolidone

Onium Salt	$\dfrac{[K]}{[\text{Onium Salt}]}$	Conversion,[1] %	$\overline{M}_W \times 10^{-3}$
-	-	33	390
$(CH_3)_4N^{\oplus}Cl^{\ominus}$	1.0	68	730
$(CH_3)_4N^{\oplus}Br^{\ominus}$	1.1	35	420
$(CH_3)_4N^{\oplus}I^{\ominus}$	1.3	33	380
$(C_2H_5)_4N^{\oplus}Cl^{\ominus}$	1.3	71	730
$(C_2H_5)_4N^{\oplus}Br^{\ominus}$	1.1	52	610
$(C_2H_5)_4N^{\oplus}I^{\ominus}$	1.1	33	390

[1] 5% PY-K/CO_2 (30% CO_2)
50°C, 22 Hr

Among the tetraalkylammonium salts tested, the order of polymerization activity is Cl > Br > I. Since the order of halide nucleophilicity is I > Br > Cl, and nucleophiles are chain terminators, this may explain why chlorides give the highest polymerization rates and molecular weights (Scheme 2).

Scheme 2.   Chain Termination by Halides in
            2-Pyrrolidone Polymerization

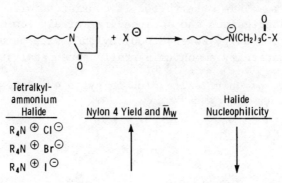

Tetralkyl- ammonium Halide	Nylon 4 Yield and $\bar{M}_W$	Halide Nucleophilicity
$R_4N^{\oplus} Cl^{\ominus}$		
$R_4N^{\oplus} Br^{\ominus}$	↑	↓
$R_4N^{\oplus} I^{\ominus}$		

Phosphonium salts also increase the conversion and molecular weight in nylon 4 polymerization.   This is shown in Table VI for tetrabutylphosphonium bromide.

Table VI.   Polymerization of 2-Pyrrolidone with
            py-K/CO_2 and Phosphonium Salt

Phosphonium Salt	$\dfrac{[K]}{[\text{Onium Salt}]}$	Conversion,[1] %	$\bar{M}_W \times 10^{-3}$
——	-	33	390
$(C_4H_9)_4P^{\oplus} Br^{\ominus}$	2.1	45	635

[1] 5% PY-K/CO_2 (30% CO_2)
50°C, 22 Hr

However, not every onium salt increases conversion and molecular weight in nylon 4 polymerization.   It was found that a large number of onium salts act as inhibitors for the polymerization reaction.   These results are summarized in Table VII.

Table VII. Onium Salts Which Inhibit
2-Pyrrolidone Polymerization

Onium Salt	$\dfrac{[K]}{[\text{Onium Salt}]}$	Conversion,[1] %
—	-	31
$(C_6H_5)_3C^{\oplus}$ $PF_6^{\ominus}$	~1	0
$(C_6H_5)_3C^{\oplus}$ $BF_4^{\ominus}$	~1	0
$C_7H_7^{\oplus}BF_4^{\ominus}$	~1	0
$(CH_3)_3O^{\oplus}$ $SbCl_6^{\ominus}$	~1	0
$(CH_3)_3O^{\oplus}$ $PF_6^{\ominus}$	1.4	9.4
$(CH_3)_3O^{\oplus}$ $BF_4^{\ominus}$	1.2	2.2
$NO_2^{\oplus}$ $PF_6^{\ominus}$	~1	0
$NO_2^{\oplus}$ $BF_4^{\ominus}$	~1	0
$NO^{\oplus}$ $PF_6^{\ominus}$	~1	0
$NO^{\oplus}$ $BF_4^{\ominus}$	~1	0
$C_6H_5N_2^{\oplus}$ $PF_6^{\ominus}$	~1	0
$(CH_3)_3HN^{\oplus}$ $Cl^{\ominus}$	1.1	0
$NH_4^{\oplus}$ $Cl^{\ominus}$	1	0

[1] 5% PY-K/CO$_2$ (30% CO$_2$)
50°C, 22 Hr

The inhibition by these compounds is probably caused by the
irreversible neutralization of the reacting pyrrolidone anion by
the strong electrophile, e.g.,

Role of Onium Salts in
Nylon 4 Polymerization

Since onium salts increase the rate and conversion in nylon 4
polymerization, one may think that they act as initiators for the
reaction. However, the rate increase is not achieved at the
expense of polymer molecular weight, indicating that onium salts do
not function as initiators. Further evidence that onium salts are
not initiators was obtained by determining the apparent number of
chains, n, in selected nylon 4 polymerizations. These results are
summarized in Table VIII.

Table VIII.   Evidence that Onium Salts are not Initiators

Catalyst	Conv.[1]	$\bar{M}_W \times 10^{-3}$	Apparent No. of Chains (n),[2] $\times 10^{-5}$
10% PY-K	4	35	11
10% PY-K/CO_2	48	500	10
10% PY-K/CO_2 +10%(CH_3)$_4$NCl	59	820	7
10% PY-K+0.8% N-Acetyl Pyrrolidone	55	28	196

[1] 50°C, 22 Hours
[2] Conversion/$\bar{M}_W$

The data show that n is approximately the same for polymerizations
carried out with py-K/CO_2 in the presence and absence of an onium
salt.   If TMAC were an initiator, a large increase in n would be
expected. The fact that N-acetylpyrrolidone, a typical initiator,
gives a much larger n is consistent with this analysis. It should
be pointed out that for TMAC and N-acetylpyrrolidone polymerizations the difference between the apparent number of chains would be
even greater if $\bar{M}_n$ were used to calculate n because of the narrower
molecular weight distribution of the onium salt prepared nylon 4.

In order to explain the polymerization results of the present
work, it is postulated that onium salts exert their influence in
the propagation step. The data are consistent with the formation
of py-Q/CO_2, where the anion-cation separation is greater than in
the corresponding py-K/CO_2. Considering the polymerization mechanism proposed by Hall,[12] the "free" pyrrolidone anion should add
faster in the rate-determining propagation step and, in the absence

of termination, should lead to higher molecular weight polymers. This is shown in Scheme 3.

Scheme 3.   Role of Onium Salts in
2-Pyrrolidone Polymerization

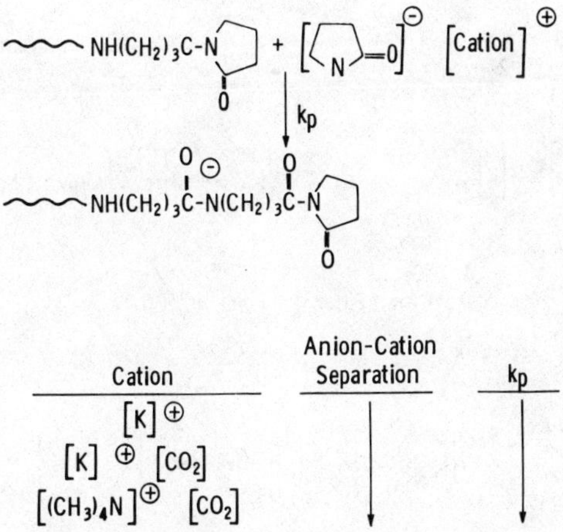

Cation	Anion-Cation Separation	k_p
$[K]^{\oplus}$		
$[K]^{\oplus} [CO_2]$		
$[(CH_3)_4N]^{\oplus} [CO_2]$		

Crown Ether Polymerization

Crown ethers form strong complexes with alkali metal ions, and the resulting "free" or "naked" anions cause large rate accelerations in many organic and polymerization[13] reactions. Since the polymerization of 2-pyrrolidone is catalyzed by organo alkali metal compounds, it was of interest to investigate the effect of crown ethers on the reaction.

The catalysts used in these reactions were prepared by either adding the crown ether to py-K/CO_2 or by mixing the crown ether with the reactants before dehydration (Scheme 4).

Scheme 4.   Nylon 4 Synthesis with Crown Ethers

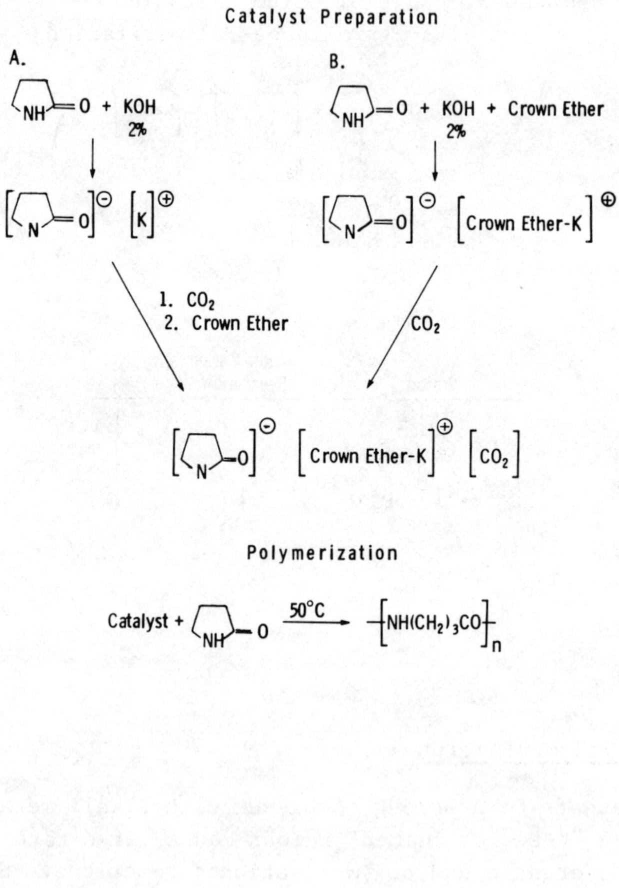

Catalyst Preparation

Polymerization

The polymerization of 2-pyrrolidone catalyzed by py-K/CO$_2$ and various crown ethers is summarized in Table IX.

Table IX. Polymerization of 2-Pyrrolidone with
py-K/CO$_2$ and Crown Ethers

Crown Ether[2]	$\dfrac{[K]}{[\text{Crown Ether}]}$	Conversion,[1] %	$\overline{M}_W \times 10^{-3}$
-	-	12	270
18-Crown-6	1.4	55	440
Dibenzo-18-Crown-6	1.3	31	320
Dicyclohexyl-18-Crown-6	1.5	17	145
15-Crown-5	1.3	25	175
Diethylene Glycol Dimethyl Ether	1.3	13	155

[1] 2% PY-K/CO$_2$ (30% CO$_2$)
50°C, 22 Hr
[2] Crown Ether Added to PY-K/CO$_2$

The results show that among the compounds tested, 18-crown-6 and
dibenzo-18-crown-6 are the most effective to increase the rate and
molecular weight in nylon 4 polymerization. The order of crown
ether addition is apparently important in these reactions. By adding the crown ether to the reaction mixture before dehydration,
higher rates and molecular weights were obtained (Table X).

Table X. Effect of Order of Crown Ether Addition

Crown Ether	Order of Crown Ether Addition	$\dfrac{[K]}{[\text{Crown Ether}]}$	Conv.,[1] %	$\overline{M}_W \times 10^{-3}$
-	-	-	12	270
18-Crown-6	Before Dehydration	1.4	63	610
18-Crown-6	After Dehydration	1.4	55	440
DBZ-18-Crown-6	Before Dehydration	1.4	52	470
DBZ-18-Crown-6	After Dehydration	1.3	31	320

[1] 2 Mole % PY-K/CO$_2$ (30% CO$_2$)
50°C, 22 Hr

One explanation of this result is that the crown ether promotes the dehydration reaction and reduces the formation of inhibiting impurities (e.g., 3-amino-butyric acid).

Among the bicyclic crown ethers tested, [2.2.1]-cryptate was the most effective, giving rates and molecular weights similar to that obtained with 18-crown-6 (Table XI).

Table XI.   Comparison of Mono- and Bicyclic Crown Ethers
in 2-Pyrrolidone Polymerization

Crown Ether	$\dfrac{[K]}{[\text{Crown Ether}]}$	Conversion,[1] %	$\bar{M}_w \times 10^{-3}$
-	-	2	175
18-Crown-6	1.3	29	365
[2.2.1]-Cryptate	2.0	21	395
[2.2.2]-Cryptate	1.7	14	175

[1]2 Mole % PY-K/CO_2 (30% CO_2)
50°C, 8 Hr

In order to produce ultrahigh molecular weight polymer, the presence of both py-K/CO_2 and crown ether are essential. For example, 2% py-K and 1.4% 18-crown-6 gave only 14% polymer with a molecular weight of 105,000 ($[\eta]$ = 2.9 dl/g). This is consistent with Sekiguchi's[14] observation. He obtained relatively low molecular weight nylon 4 ($[\eta]$ = 1.3 dl/g, $\bar{M}_w$ = 36,000) with py-K and [2.2.2]-cryptate.

Role of Crown Ethers in
Nylon 4 Polymerization

As it was shown, crown ethers increase significantly the rate and molecular weight in nylon 4 polymerization. In analogy to the onium salt polymerizations discussed above, it is postulated that these effects are caused by changes in the propagation rate of the polymerization. Thus, cation complexation by the crown ether breaks up the relatively tight py-K/CO_2 ion pair and produces a "free," more reactive pyrrolidone anion. This will propagate

faster, resulting in higher conversion and molecular weight
(Scheme 5).

Scheme 5.   Role of Crown Ethers in
                 2-Pyrrolidone Polymerization

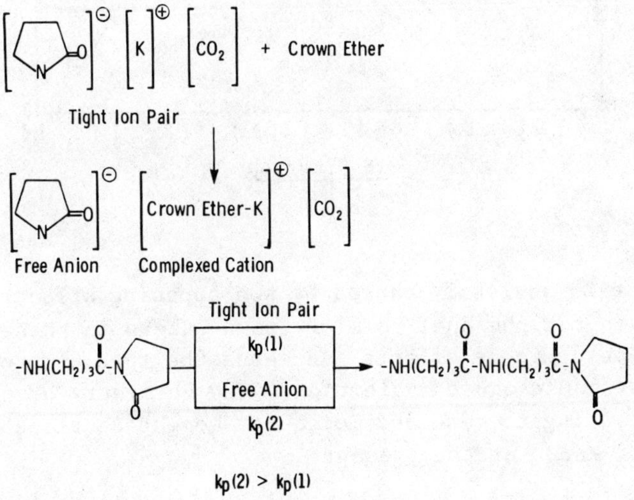

While this is an attractive explanation, it does not account for
all of the experimental observations. Thus, it was noticed that a
plot of conversion and molecular weight against crown ether concentration gives curves with maxima at a crown ether to py-K/CO_2 ratio
of approximately 0.7 (Figure 1).

Figure 1.   Effect of Crown Ether Concentration
on Nylon 4 Polymerization

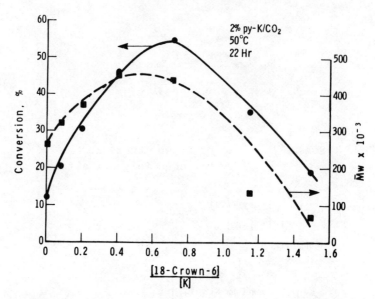

This behavior is obviously caused by two opposing effects and provides evidence for the dual role of crown ethers in these polymerizations. One of these effects increases the rate of propagation,
k_p; the other could cause inhibition. The chemistry of the
inhibition leading to rate and molecular weight decrease by crown
ethers is not known at the present.

The author thanks Dr. S. J. Lapporte and Professors J. P.
Collman and H. K. Hall, Jr. for stimulating discussions.

REFERENCES

1. W. O. Ney, Jr., W. R. Nummy, and C. E. Barnes, U.S. Patent
2,638,463 (1953).

2. E. M. Peters and J. A. Gervasi, Chemtech., p 16, January 1972.

3. L. Szego, La Chimica E L'Industria, 55, 280 (1973).

4. C. E. Barnes, U.S. Patent 3,721,652 (1973).

5. R. Bacskai, U.S. Patents 4,098,774 (1978), 4,100,146 (1978),
4,101,531 (1978), and 4,218,558 (1980).

6. J. P. Collman, U.S. Patent 4,073,778 (1978); R. Bacskai, U.S.
    Patent 4,169,936 (1979).

7. R. Bacskai and B. A. Fries, J. Polym. Sci. Chem. Ed., 20, 2341
    (1982).

8. J. B. Kinsinger, In Encyclopedia of Polymer Science and
    Technology, Vol 14; H. F. Mark, N. G. Gaylord, and N. M.
    Bikales, Eds., Interscience, New York, 1971, p 727.

9. Z. Tuzar, J. Kleteckova, and J. Kralicek, Coll. Czech. Chem.
    Comm., 39, 2206 (1974).

10. H. Sekiguchi, P. R. Tsourkas, and B. Coutin, J. Polymer Sci.,
    Symposium No. 42, 51 (1973).

11. N. Sakata, Japanese Patent 71-26195 (1972).

12. H. K. Hall, Jr., J. Am. Chem. Soc., 80, 6404 (1958).

13. L. J. Mathias, J. Macromol. Sci. Chem., A15 (5), 853 (1981).

14. H. Sekiguchi, P. Tsourkas, and B. Coutin, Makromol. Chem.,
    178, 2135 (1977).

SYNOPSIS

SYNTHESIS OF ULTRAHIGH MOLECULAR WEIGHT NYLON 4 WITH ONIUM SALT AND
CROWN ETHER-CONTAINING CATALYSTS. R. Bacskai, Chevron Research
Company, 576 Standard Avenue, Richmond, California 94802.

Catalysts based on carbonated potassium pyrrolidonate
(py-K/CO_2) plus an onium salt or crown ether produce high rates and
molecular weights in nylon 4 polymerization. Best results, leading
to polymers with a weight average molecular weight greater than one
million, are obtained by using tetramethylammonium chloride or
tetraethylammonium chloride as the catalyst additive. With tetraalkylammonium halides, the order of activity is Cl > Br > I. Among
the crown ethers tested, 18-crown-6 is the most effective, giving
maximum conversion and molecular weight at a crown ether-to-
py-K/CO_2 ratio of 0.7. The rate and molecular weight increasing
effect of crown ethers and onium salts can be explained by greater
ion pair separation which produces "freer," more reactive pyrrolidone anions.

MECHANISMS OF CATALYSIS BY POLYMER-SUPPORTED

QUATERNARY AMMONIUM AND PHOSPHONIUM IONS

Warren T. Ford

Department of Chemistry
Oklahoma State University
Stillwater, Oklahoma 74078

INTRODUCTION

Quaternary ammonium anion exchange resins have been used as catalysts for organic reactions for thirty years.[1-3] Until 1975 their uses had been limited to two phase reaction mixtures containing a solution and the insoluble polymeric catalyst. Catalysis in triphase reaction mixtures apparently was investigated first independently in three laboratories. Regen[4] reported effects of the structure of quaternary ammonium ions and the percent ring substitution on the triphase reaction of 1-bromooctane with aqueous sodium cyanide. Brown and Jenkins[5] reported effects of percent ring substitution and separation of the onium ion from the polymer backbone on the benzylation of 2-naphthoxide ion. Montanari, Tundo and co-workers[6] reported polymer-bound quaternary phosphonium ions, crown ethers and cryptands for catalysis of the reactions of 1-bromooctane with aqueous potassium iodide and potassium cyanide. Subsequently many more catalysts and applications of triphase catalysis have been reported, and reviews are available.[7-9]

Our entry into polymer-supported phase transfer catalysts was an investigation of several empirical factors that influence catalytic activity in these complex triphase mixtures. By approaching the problem from the standpoint of reaction mechanisms one can understand better the chemical and physical features of heterogeneous catalysis. Such understanding will aid in the design of new catalysts and enable rational choice of catalysts for triphase reactions, and will be particularly important in the development of industrial processes for manufacture of fine organic chemicals and pharmaceuticals.

201

This review focuses on mechanisms of catalysis and is limited to quaternary ammonium and phosphonium catalysts: more examples of triphase catalysis are available elsewhere.[7-9] Topics discussed by other contributors to this volume, such as chiral quaternary ammonium ions for asymmetric syntheses, have been omitted.

PHASE TRANSFER CATALYSIS

Catalysis of reactions between components of immiscible aqueous and organic solutions by quaternary ammonium salts was first reported in 1951[10] and was brought to the attention of synthetic chemists by Makosza[11] and by Starks.[12] Makosza investigated the alkylation of active methylene compounds such as phenylacetonitrile with benzyltriethylammonium chloride in the presence of concentrated aqueous sodium hydroxide (eq 1). Starks used more lipophilic onium

$$C_6H_5CH_2CN + C_2H_5Br + NaOH(aq) \rightarrow C_6H_5CH(C_2H_5)CN + NaBr \qquad (1)$$

salts such as methyltrioctylammonium chloride and hexadecyltributyl-phosphonium bromide as catalysts for nucleophilic displacement reactions of alkyl halides (eq 2). Hundreds of other examples of

$$\underline{n}\text{-}C_8H_{17}Cl + NaCN(aq) \rightarrow \underline{n}\text{-}C_8H_{17}CN + NaCl \qquad (2)$$

phase transfer catalysis now are known,[13-15] but the alkylation of active methylene compounds and nucleophilic displacements are by far the most widely used.

Nucleophilic displacement reactions are thought to proceed entirely in the organic phase according to Scheme 1. Alkylations of active methylene compounds are thought to proceed by a somewhat different mechanism because of the extremely low solubility of quaternary ammonium hydroxides in most organic solvents. The benzyl-triethylammonium ion is thought to attract hydroxide ion to the aqueous/organic interface, where it abstracts a proton from phenyl-acetonitrile. The carbanion then reacts with the alkylating agent in the organic phase.

Scheme 1.[13]

organic phase $C_8H_{17}Cl + R_4N^+CN^- \rightarrow C_8H_{17}CN + R_4N^+Cl^-$

aqueous phase $\quad NaCl + R_4N^+CN^- \rightleftharpoons NaCN + R_4N^+Cl^-$

Quaternary ammonium and phosphonium ions bound to insoluble polystyrene present an even more complicated mechanistic problem. Polystyrene beads lacking onium ions (or crown ethers, cryptands, or other polar functional groups) have no catalytic activity. The onium ions are distributed throughout the polymer matrix in most catalysts. The reactive anion must be transferred from the aqueous phase to the polymer, where it exists as the counter ion in an anion exchange resin, and the organic reactant must be transferred from the external organic phase into the polymer to meet the anion. In principle, catalysis could occur only at the surface of the polymer beads, but kinetic evidence supports catalysis within the beads for most nucleophilic displacement reactions and for alkylation of phenylacetonitrile.

Advantages and Disadvantages of Polymer-Supported Catalysts

The major advantages of polymeric catalysts are: 1) The insoluble beads can be separated from reaction mixtures by simple filtration. Soluble quaternary ammonium and phosphonium salts are difficult to recover, and sometimes are surfactants which form emulsions. Insoluble catalysts cannot contaminate the desired reaction products. 2) The insoluble catalysts can be reused. With soluble catalysts, new catalyst must be used for each reaction. Moreover, disposal of soluble catalysts could cause trouble in sewage systems because most lipophilic quaternary ammonium salts have biocidal activity. 3) Insoluble catalysts can be used in continuous flow reactors, which are advantageous for large scale industrial processes. Although these advantages are highly attractive, many of the catalysts investigated so far are unsatisfactory for large scale use because of physical attrition, difficulty in filtration, and chemical decomposition of the active sites. These problems and their likely solutions are addressed at the end of the review.

The major disadvantages of the polymer-supported quaternary ammonium and phosphonium ion catalysts are: 1) They have higher initial cost. Unless they can be used in a flow system or recovered from batch reactors and reused many times, they will be more expensive to use than soluble catalysts. 2) In most cases the insoluble catalysts are less active than their soluble analogues. Their lesser activity is an intrinsic property of heterogeneous catalysts. If activity is the sole criterion for choice of a catalyst, one should use a soluble catalyst. The reasons for use of supported catalysts are ease of separation and reuse.

FACTORS THAT CONTROL ACTIVITY OF POLYMER-SUPPORTED CATALYSTS

Many empirical parameters affect the activity of polymer-supported quaternary ammonium and phosphonium catalysts:
   1) Active site structure.

2) Percent ring substitution.
3) Separation of the active site from the polymer backbone.
4) Structure of the polymer.
5) Mixing of the triphase system.
6) Particle size of the catalyst.
7) Cross-linking of the polymer.
8) Organic solvent.
9) Concentration of salt in the aqueous phase.
10) Molecular size of the organic reactant.

Active Site Structure

Most catalysts investigated have been tetraalkylammonium or
tetraalkylphosphonium ions. The major factor controlling activity
in nucleophilic displacement reactions is not the charged atom, but
the sizes of the alkyl groups bound to the charged atom. Tri-$\underline{n}$-butyl
onium ions are much more active than trimethyl onium ions, as shown
in Table I. The same effect is well known with soluble phase transfer catalysts,[13-15] where tetrabutylammonium, methyltrioctylammonium
and hexadecyltributylphosphonium ions are highly active, but surfactants such as dodecyltrimethylammonium ion are much less active in
nucleophilic displacement reactions. The best explanation for this
behavior is that the trimethylammonium ions with small ionic radii
are more hydrophilic than the tributylammonium ions with larger ionic
radii. Water hydrogen bonds to the reactive anions, reducing their
activity. The water concentration at the active site is higher when
the onium ion radius is small.

Table I. Effect of Onium Ion Structure on Activity[16]

$$\underline{n}\text{-}C_8H_{17}Br \quad + \quad KI \quad \xrightarrow[90°C]{toluene} \quad \underline{n}\text{-}C_8H_{17}I \quad + \quad KBr$$

catalyst, 2% cross-linked	mequiv/g	$\underline{k} \times 10^6$ s^{-1}
P—⟨⟩—CH$_2$NHCO(CH$_2$)$_{10}$$\overset{+}{N}$(CH$_3$)$_3$	1.2	20.4
P—⟨⟩—CH$_2$NHCO(CH$_2$)$_{10}$$\overset{+}{N}$($\underline{n}$-C$_4H_9$)$_3$	1.1	330
P—⟨⟩—CH$_2$NHCO(CH$_2$)$_{10}$$\overset{+}{P}$($\underline{n}$-C$_4H_9$)$_3$	1.4	272

Percent Ring Substitution

The degree of substitution of onium ions in the cross-linked polystyrene affects dramatically the rates of nucleophilic displacement reactions. In the first kinetic study Regen[17] found that catalysts 1 and 2 with 21% or less of rings substituted were active for reaction of cyanide ion with 1-bromooctane while catalysts with 46% or more of rings substituted, as in commercial ion exchange resins, were inactive as shown in Table II. Reeves[18] also found higher activity of 18% than of 65% ring substituted phosphonium ion catalysts 3 for reaction of 1-bromopentane with aqueous potassium cyanide. This fundamental discovery of greater activity of lightly substituted polystyrene-supported onium ions has enabled a large number of synthetic applications of polymer-bound catalysts that would not work with highly substituted commercial anion exchange resins.

$$P\text{---}\bigcirc\text{---} CH_2\overset{+}{N}(CH_3)_2(\underline{n}\text{-}C_4H_9)\ Cl^-\quad \underline{1}$$

$$P\text{---}\bigcirc\text{---} CH_2\overset{+}{N}(CH_3)_3\ Cl^-\qquad\qquad \underline{2}$$

$$P\text{---}\bigcirc\text{---} (CH_2)_n\overset{+}{P}(\underline{n}\text{-}C_4H_9)_3\ X^-\quad \begin{array}{l}\underline{3a},\ n=1,\ X=Cl\\ \underline{3b},\ n=2,\ X=Br\\ \underline{3c},\ n=3,\ X=Br\end{array}$$

For reaction of $\underline{n}$-decyl methanesulfonate with aqueous sodium chloride (eq 3), five percent ring substituted phosphonium ion catalysts 3a have activities 0.6-0.7 times that of the highly active soluble benzyltributylphosphonium bromide.[19] The activities decrease rapidly with increasing ring substitution to <0.3 times that of the soluble catalyst at ≥20% ring-substitution.[19]

Table II. Effect of % Ring Substitution on Rate[17]

$$\underline{n}\text{-}C_8H_{17}Br + NaCN \xrightarrow[90°C]{toluene} \underline{n}\text{-}C_8H_{17}CN + NaBr$$

catalyst	% ring subst.	k_{rel}
1	10	0.7
1	21	0.7
2	10	1.0
2	46	0.005
2	76	< 0.005

$$\underline{n}\text{-}C_{10}H_{21}OSO_2CH_3 + NaCl(aq) \rightarrow \underline{n}\text{-}C_{10}H_{21}Cl \qquad (3)$$

Low percent ring substitution is not critical for activity in all nucleophilic displacements. Brown and Jenkins[5] found 40% to 100% ring substituted catalysts 4 highly active for O-alkylation of 2-naphthoxide ion with benzyl bromide.

$$\text{CH}_2\text{O}_2\text{C}(\text{CH}_2)_n \overset{+}{\text{N}}(\text{CH}_3)_3 \quad \text{Cl}^- \quad \begin{array}{l} \underline{4a}, \, n = 5 \\ \underline{4b}, \, n = 11 \end{array}$$

Separation of the Active Site from the Polymer Backbone

Several groups have studied catalysts with varied lengths of aliphatic chains separating the onium ion catalyst from the polystyrene backbone. However, some of them have used more than one degree of ring substitution or more than one kind of polystyrene, which prevents meaningful evaluation of the effect of chain length. Spacer chains of 13 and of 25 atoms give about equally active catalysts (Table III), both of which are more active than the polymer-bound benzyltributylphosphonium ion for reaction of 1-bromooctane with aqueous potassium iodide.[16] Tomoi[20] has observed increased activity by lengthening the chain between the aromatic ring and the phosphonium ion from one to seven carbon atoms for the reaction of 1-bromooctane with aqueous sodium cyanide (Table IV). Tomoi's results were obtained with a single mesh fraction of polymer beads and nearly identical percent ring substitution throughout the series of catalysts. Reeves[21] obtained similar results with

Table III. Effect of Long Spacer Chains on Activity[16]

$$\underline{n}\text{-}C_8H_{17}Br \quad + \quad KI \quad \xrightarrow[90°C]{toluene} \quad \underline{n}\text{-}C_8H_{17}I \quad + \quad KBr$$

catalyst, 2% cross-linked	mequiv/g	$\underline{k} \times 10^6$ s^{-1}
P—⟨benzene⟩—CH$_2\overset{+}{\text{P}}(\underline{n}\text{-}C_4H_9)_3$	2.0	118
P—⟨benzene⟩—CH$_2$NHCO(CH$_2$)$_{10}\overset{+}{\text{P}}(\underline{n}\text{-}C_4H_9)_3$	1.1	272
P—⟨benzene⟩—CH$_2$[NHCO(CH$_2$)$_{10}]_2\overset{+}{\text{P}}(\underline{n}\text{-}C_4H_9)_3$	1.4	308

Table IV. Effect of Short Spacer Chains on Activity.[20]

$$\underline{n}\text{-}C_8H_{17}Br \quad + \quad NaCN \quad \xrightarrow[90°C]{toluene} \quad \underline{n}\text{-}C_8H_{17}CN \quad + \quad NaBr$$

catalyst, 2% cross-linked 100/200 mesh	mequiv/g	$\underline{k} \times 10^5$ s^{-1}
P—◯—$CH_2\overset{+}{P}(\underline{n}\text{-}C_4H_9)_3$	1.08	20
P—◯—$CH_2O(CH_2)_3\overset{+}{P}(\underline{n}\text{-}C_4H_9)_3$	1.01	31
P—◯—$(CH_2)_4\overset{+}{P}(\underline{n}\text{-}C_4H_9)_3$	1.09	43
P—◯—$(CH_2)_7\overset{+}{P}(\underline{n}\text{-}C_4H_9)_3$	0.99	50

catalysts 3a, 3b, and 3c in reaction of 1-bromopentane with aqueous potassium cyanide.

Structure of the Polymer

Most investigations of polymer-supported onium ion phase transfer catalysts have used cross-linked polystyrenes. Not all of them have the same structure, even when they have the same formal degree of cross-linking with divinylbenzene. (The effect of percent cross-linking is considered in a later section). Two principal methods have been used to functionalize polystyrene for phase transfer catalysts, chloromethylation of pre-formed beads and copolymerization of chloromethylstyrene monomer with styrene and divinylbenzene. The chloromethylation route employs chloromethyl methyl ether (a cancer suspect agent), and a Lewis acid, usually stannic chloride.[22] Substitution proceeds >90% para and is accompanied by some intrapolymer alkylation, which puts additional cross-links into the polymer

$$P\text{—}◯ \xrightarrow{ClCH_2OCH_3, \ SnCl_4} P\text{—}◯\text{—}CH_2Cl$$

$$+ \quad P\text{—}◯\text{—}CH_2\text{—}◯\text{—}P \qquad \qquad (4)$$

(eq 4).[22] Thus one can never be sure of the exact degree of cross-
linking of the polymer when the catalyst is prepared by the chloromethylation route. Most commercial samples of chloromethylstyrene
monomer consist of a 60/40 mixture of meta/para isomers, although
the composition may change from batch to batch. The copolymer
reactivity ratios for chloromethylstyrenes (M_1) and styrene (M_2) are
r_1 = 1.08 and r_2 = 0.72,[23] indicating that the functional monomers
are incorporated almost randomly throughout the polymer. The degree
of cross-linking of the copolymers is known directly from the monomer
composition when the polymerization is carried to total conversion.
However, the catalytic activity of the m-substituted onium ions may
not be the same as that of the p-substituted onium ions. The meta
positions are closer to the polymer backbone and may be less accessible due to steric hindrance. No direct comparison of the activities of catalysts prepared by the two routes. chloromethylation and
copolymerization with chloromethylstyrenes, exists in the literature.
Most research groups have used commercial samples of "chloromethylated polystyrene" for which the method of preparation is unknown
to the customer. It is possible to distinguish the two kinds of
"chloromethylated polystyrenes" qualitatively from the ^{13}C NMR
linewidths of $CDCl_3$-swollen samples cross-linked nominally with
1% or 2% divinylbenzene.[24]

   Tundo has compared activities of onium ions supported on conventional and isoporous[25] cross-linked polystyrenes and silica
gel.[26-28] (Silica gel has excellent strength and porosity for use
as a catalyst support, and it is easy to functionalize with onium
ions. It dissolves in alkaline media and thus is unsatisfactory for
a large number of organic reactions.) Typical data are in Table V,
but one should interpret the results only qualitatively because of
the large variations in catalyst particle size, surface area and
percent ring substitution. The most intriguing result is the
high activity of the isoporous cross-linked catalyst, which was
prepared by Friedel-Crafts alkylation of soluble polystyrene with
1,6-dibromohexane in a mixture containing aluminum chloride and a
small amount of toluene in nitrobenzene solvent. (See Scheme 2).
The term isoporous has been used by Davankov[29] to describe polystyrenes cross-linked by Friedel-Crafts alkylation because they are
claimed to swell extensively even in nonsolvents. Tundo's isoporous
catalyst was highly swellable.[25] Its activity probably should be
compared with those of 0.5% or 1% divinylbenzene-cross-linked catalysts. Such lightly cross-linked polystyrenes generally are not
physically stable enough under conditions of use for large scale
manufacturing processes. The physical stability of isoporous catalysts has not been reported.

   Macroporous[30,31] or macroreticular[32] polystyrenes have also been
used as catalyst supports. (The terms macroporous and macroreticular
are synonomous. They originated in different laboratories.) Macroporous polymers are prepared by suspension polymerization of

Table V. Effect of Support Structure on Catalyst Activity [16,28]

$$RBr + KI \rightarrow RI + KBr$$

catalyst	$\dfrac{k(\text{polymer catalyst})}{k(\underline{n}\text{-}C_{16}H_{33}\overset{+}{P}(\underline{n}\text{-}C_4H_9)_3Br^-)}$
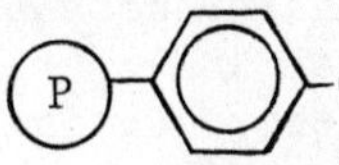 (P)—⬡—$CH_2[NHCO(CH_2)_{10}]_2\overset{+}{P}(\underline{n}\text{-}C_4H_9)_3$ 2% cross-linked, 1.4 mequiv/g	0.39
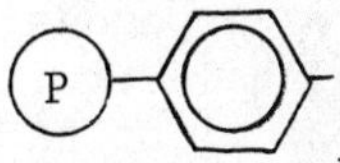 (P)—⬡—$(C_6H_{12})\overset{+}{P}(\underline{n}\text{-}C_4H_9)_3$, 0.83 mequiv/g isoporous cross-linked	0.96
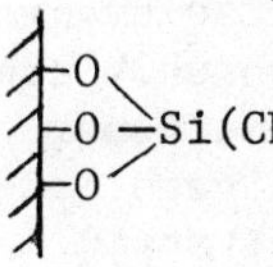 $-O-Si(CH_2)_3\overset{+}{P}(\underline{n}\text{-}C_4H_9)_3$ 1.24 mequiv/g, 100/140 mesh, 500 m²/g	0.14

Scheme 2

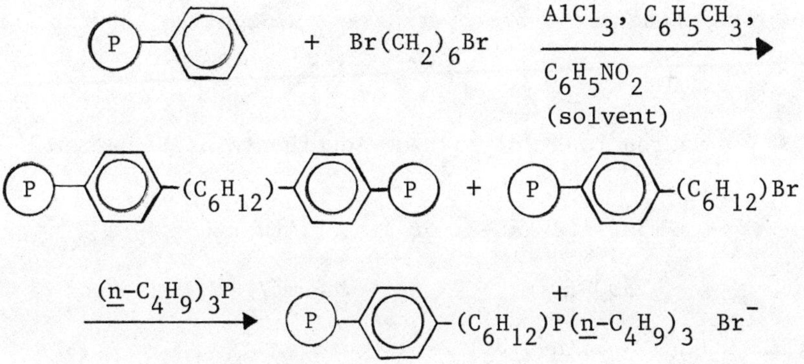

monomer containing an organic diluent. The polymer precipitates
within each droplet during the polymerization. A wide variety of
polymer morphologies may be produced by variation of the degree of
cross-linking and the amount and chemical nature of the diluent.[31]
Solvents and nonsolvents for polystyrene have been used as diluents.
Removal of the solvent at the end of the polymerization leaves
permanent macropores when the polymer is dried from a nonsolvent.
Conventional polystyrene beads prepared without diluent contain no
macropores. They are microporous, meaning they become porous when

swollen by a solvent. The micropores are the solvent-filled spaces
between polymer chains of the matrix. A macroporous solvent-swollen
polystyrene has both macropores and micropores. The strictly microporous polymers have often been called "gel polymers." Such use of the
term "gel polymer" is misleading because the macroporous polymers
also have a gel phase in solvent-swollen form. The term gel refers
to a substance which is liquid-like at the microsopic level but does
not exhibit macrosopic flow. Two broad classes of macroporous
polystyrenes have been tested as supports for phase transfer catalysts: highly cross-linked adsorbents such as Amberlite® XAD-2
and XAD-4, and lightly cross-linked copolymers designed for use in
ion exchange resins such as Amberlite® XE-305. (Amberlite is a
trade name of the Rohm and Haas Co.) In principle the macroporous
supports should provide more active catalysts. Their high internal
surface areas insure that most of the catalytic sites are close to
the surface, so that reaction rates should be limited less by intraparticle diffusion of the reactants. In practice both highly cross-
linked macroporous adsorbents and lightly cross-linked macroporous
chloromethylstyrene/styrene copolymers have led to catalysts less
active than the non-macroporous lightly cross-linked polystyrenes.[19,33] (See Scheme 3 and Figure 1). Nothing is wrong with
the principle of high activity of high surface area catalysts, but
there must be diffusional problems with the macroporous catalysts
that are not yet understood.

Modified dextran ion exchangers also have served as phase
transfer catalysts for iodide and cyanide ion displacement reactions
with 1-bromobutane and 1-bromooctane.[34] Lipophilic substituents were

Scheme 3. Macroporous Polystyrene Adsorbents As Supports.[19]

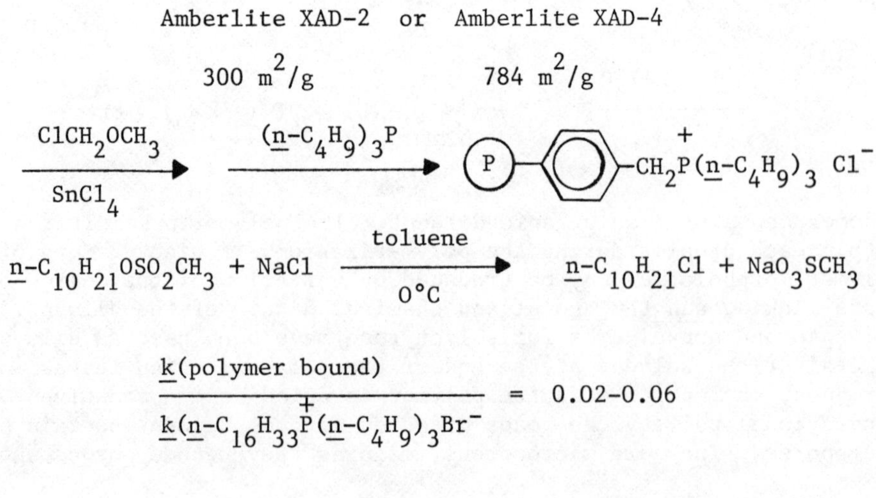

Amberlite XAD-2   or   Amberlite XAD-4

$300 \ m^2/g$ $784 \ m^2/g$

$$\xrightarrow[SnCl_4]{ClCH_2OCH_3} \xrightarrow{(n-C_4H_9)_3P} \text{(P)} - \langle \bigcirc \rangle - CH_2 \overset{+}{P}(n-C_4H_9)_3 \ Cl^-$$

$$n-C_{10}H_{21}OSO_2CH_3 + NaCl \xrightarrow[0°C]{toluene} n-C_{10}H_{21}Cl + NaO_3SCH_3$$

$$\frac{\underline{k}(polymer \ bound)}{\underline{k}(n-C_{16}H_{33}\overset{+}{P}(n-C_4H_9)_3Br^-)} = 0.02-0.06$$

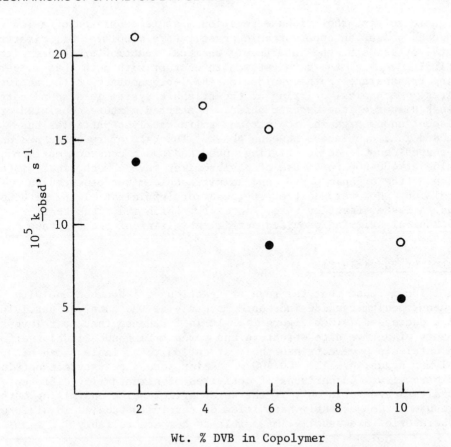

Wt. % DVB in Copolymer

Figure 1. Comparison of activities of microporous 0 and macroporous
           ● catalysts 3a, 100/200 mesh, 11-18% ring substitution,
           for reaction of 1-bromooctane in toluene with aqueous
           sodium cyanide at 90°C.

necessary for high activity of dextran catalysts, just as were the
lipophilic environments attained by low percent ring substitution
in cross-linked polystyrenes.

Mixing of the Triphase System

Reaction rates of phase transfer catalyzed nucleophilic displacement reactions depend upon stirring speed.[16,35] Tomoi studied
the effect of stirring speed on the rate of reaction of 1-bromo-
octane with aqueous sodium cyanide using 15-18% ring substituted
benzyltributylphosphonium ion catalysts.[36] With toluene as the

organic solvent the triphase reaction mixture consisted of a lower
aqueous phase, an upper organic phase, and a monolayer of polystyrene
catalyst beads at the interface. Chemical reaction proceeded without
stirring because the catalyst was in contact with both phases, permitting transport of both cyanide ion and 1-bromooctane to the active
sites. Mechanical stirring of the triphase system in a round bottom
flask dispersed the organic phase. A stirred mixture consisted of
discontinuous organic droplets and solid catalyst particles suspended in a continuous aqueous phase. The rate of reaction had an
S-shaped dependence on stirring speed and had a constant maximum
value at 400-600 rpm. Use of a vibromixer gave a much finer dispersion of the organic phase, and an ultrasonic mixer produced an
emulsion. However, neither of these more efficient mixing methods
gave a faster reaction rate (Table VI), which indicates that
mechanical stirring provided sufficient mixing of the triphase
system.

Catalyst Particle Size

    Tomoi found that the rate of reaction of 1-bromooctane with
aqueous sodium cyanide increased markedly as the catalyst particle
size decreased. (See Figure 2). It is fortunate that most investigators of active site structure and percent ring substitution effects
on catalysis have confined their attention to a single batch of
polymer beads, usually 200/400 mesh microporous polystyrene obtained
commercially. Comparisons of activites of structurally different
catalysts are valid only if the same particle size and mixing method
are used. Consequently activities of catalysts studied in different
laboratories are usually impossible to compare reliably.

Cross-linking of the Polymer

    Systematic studies of the effect of polymer cross-linking on
activity of polystyrene-bound benzyltributylphosphonium ion catalysts

Table VI. Dependence of Rate Constants on Mixing Method[36]

$$\underline{n}\text{-}C_8H_{17}Br + NaCN(aq) \xrightarrow[90°C]{\underset{toluene}{3a}} \underline{n}\text{-}C_8H_{17}CN + NaBr$$

Method	Particle Size, μm	$10^5 \underline{k}, s^{-1}$
Mechanical, 600 rpm	150 - 300	15
Vibromixer		15
Ultrasound		11
Mechanical, 600 rpm	100 - 200	21
Vibromixer		19

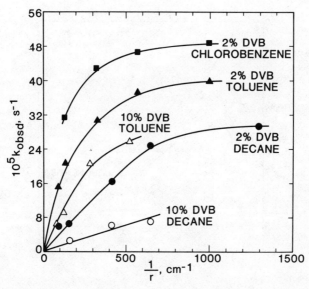

Figure 2.   Dependence of Reaction Rates
on Catalyst 3a Particle Size.

(Reprinted with permission from ref. 36.   Copyright 1981 American
Chemical Society).

have been reported by Tomoi[36] for the reaction of 1-bromooctane with
aqueous sodium cyanide and by Regen[19] for the reaction of decyl
methanesulfonate with aqueous sodium chloride.   Both found that
activity of microporous catalysts decreased as the percent divinylbenzene in the copolymer increased.   (See Figure 3).

Organic Solvent

The rate of reaction of 1-bromooctane with aqueous potassium
iodide using phosphonium ion catalysts decreased with decreasing
polarity of the organic solvent in the order o-dichlorobenzene >
toluene > decane.[16]   Tomoi[36] obtained similar results for the
reaction of 1-bromooctane with aqueous sodium cyanide and catalyst
3a, shown in Figure 3.   In both cases the reactivity decreased in
the same order as the degree of swelling of the catalyst by the
solvent.   Swelling of the catalyst should promote reactant transport
to the active sites.   More polar solvents also should increase the
intrinsic reactivity between nucleophile and 1-bromooctane.

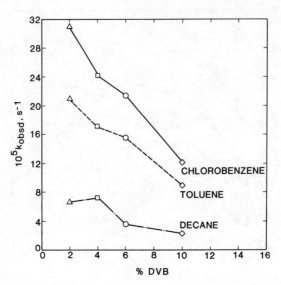

Figure 3.   Effect of Polystyrene Cross-linking
on Catalyst <u>3a</u> Activity.

(Reprinted with permission from ref. 36.   Copyright 1981 American
Chemical Society).

Concentration of Salt in the Aqueous Phase

Regen[37] has reported rates of reaction of decyl methanesulfonate
with chloride ion under both normal triphase catalytic conditions and
stoichiometric conditions, in which there was no chloride ion external to the catalyst.   The catalysts were 17% and 52% ring substituted,
1% cross-linked benzyltributylphosphonium ions.   When the catalysts
were suspended in pure water, they swelled much more than in concentrated aqueous sodium chloride because of the high osmotic pressure.
As shown in Table VII the osmotic effect was greater with the 17%
ring substituted polymer.   The increased hydration of the chloride
ion in the 17% ring substituted catalyst suspended in salt-free water
decreased its activity by a factor of 73.   These results confirm that
a hydrophobic environment speeds nucleophilic displacement reactions.

Molecular Size of the Organic Reactant

In any reaction whose rate is limited by diffusivity of the
reactant through the polymer matrix to the active site, larger

Table VII. Influence of Imbibed Water on Activity.[37]

$$\underline{n}\text{-}C_{10}H_{21}OSO_2CH_3 + NaCl \xrightarrow[70°C]{\text{toluene}} \underline{n}\text{-}C_{10}H_{21}Cl + NaO_3SCH_3$$

catalyst = 3a, 1% cross-linked

% ring subst.	conditions	mol H_2O/mol P^+	$\underline{k} \times 10^5$, s^{-1}
17	stoichiometric	297	0.06
17	catalytic	11.0	4.4
52	stoichoimetric	96	4.2
52	catalytic	18.9	12

molecules may be expected to react slower than analogous smaller molecules. 1-Bromoalkanes of more than four carbon atoms have identical reactivities with nucleophiles in solution.[38] However, using 2% cross-linked onium ion catalysts 1-bromooctane reacts 2.4 to 6.7 times faster than 1-bromohexadecane with aqueous sodium cyanide as shown in Table VIII.[39]

A MECHANISM OF CATALYSIS

All of the empirical parameters that affect the activity of polystyrene-supported onium ion catalysts in nucleophilic displacement reactions fit into a general mechanism. The reaction rates may be limited by 1) mass transfer of reactants from the bulk liquid phases to the surface of the catalyst, 2) diffusion of the reactants from the catalyst surface to the active site, and 3) intrinsic reactivity at the active site, as diagrammed in Figure 4. Mass transfer refers to transport of molecules to the catalyst surface first by agitation and diffusion in bulk liquid and then by film

Table VIII. Effect of Reactant Size on Catalyst Activity.[39]

catalyst	particle size, mesh	10^5 $\underline{k}$, s^{-1}	
		$C_8H_{17}Br$	$C_{16}H_{33}Br$
3a	325/400	38	16
2	60/100	1.4	0.21
2	325/400	1.4	0.52

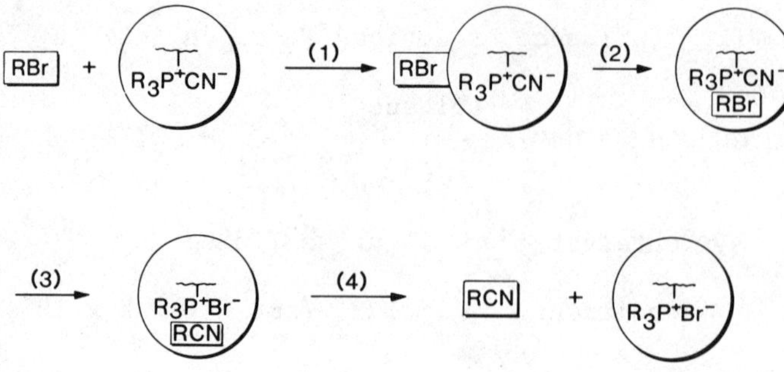

Figure 4. Kinetic Steps in Triphase Catalysis.

(Reproduced with permission from ref. 36.   Copyright 1981 American
Chemical Society).

diffusion to the particle surface. Mass transfer rates increase
with more efficient mixing, with higher concentration of the
reactant in the liquid phase, and with increased external surface
area of the catalyst. The mixing experiments of Tomoi[36] are the
clearest indication of the importance of thorough agitation of
reaction mixtures in polymer-supported phase transfer catalysis.
In fact, unless the mixing methods used in other laboratories are
highly reproducible, many reports of synthetic yields after a certain time at a certain temperature may not be reproducible either.
If the mixing is not done reproducibly, all other comparisons of
the factors affecting catalytic activity may be invalid.

     After the reactants reach the catalyst surface, they diffuse
to an active site, where they react. Several empirical factors
prove limitation of reaction rates by intraparticle diffusion:
particle size, degree of cross-linking, and size of the reactant.
If mass transfer limitations have been overcome, rate dependence on
particle size of uniformly functionalized catalysts means that the
reaction rate is limited by intraparticle diffusion as well as by
the intrinsic reactivity at the active site. In all of Tomoi's[36,39]
experiments on the reaction of 1-bromooctane with aqueous sodium
cyanide, there is only one example in which intraparticle diffusion
did not limit the rate: 60/100 and 325/400 mesh benzyltrimethylammonium catalysts had the same activity (Table VIII). In that case
the observed rate of reaction was miserably slow, with a half life
of 13.7 hours. We conclude that the particle size of any reasonably
active polystyrene-supported phase transfer catalyst will affect the
reaction rate. Increased cross-linking decreases reaction rates
because reactant diffusion to the active sites is slower through

more highly cross-linked polymer matrices. The lesser reactivity
of 1-bromohexadecane than of 1-bromooctane with polystyrene-supported catalysts also shows that the reaction rate of at least the
1-bromohexadecane is retarded by intraparticle diffusion.

    The other empirical factors that influence reaction rates with
polystyrene-supported onium ion catalysts probably affect the intrinsic reactivity of the anion and the substrate. Nucleophilic displacement reactions of anions and alkyl halides proceed much faster
in dipolar aprotic solvents such as dimethylsulfoxide, N,N-dimethyl-
formamide, and hexamethylphosphoramide than in aqueous or alcoholic
solvents.[40] Nucleophilic anions have much higher activity in the
absence of hydrogen bond donors. Similarly in polymer-supported
phase transfer catalysts, the less hydration of the onium/anion pair
or aggregate within the polymer matrix, the faster the intrinsic
reaction rate. Regen's[37] hydration numbers of the ionic sites of
benzyltributylphosphonium catalysts 3 as a function of sodium
chloride concentration in the aqueous phase, and the corresponding
rates of reaction of the chloride ion in the polymer with decyl
methanesulfonate, prove that hydration decreases chloride ion
reactivity. Similar effects should be expected on the reactivity
of other anions such as cyanide, hydroxide, and phenoxide which also
are strong hydrogen bond acceptors. Much smaller hydration effects
on reaction rates should be expected with anions which are weaker
hydrogen bond acceptors, such as iodide or thiophenoxide.

    The ionic radius of the onium ion is the principal feature of
active site structure that affects activity. Alkyltrimethylammonium
ions are more hydrated than alkyltributylammonium and alkyltributyl-
phosphonium ions. With alkyl groups as large as n-butyl, it makes
little difference whether the central atom is nitrogen or phosphorus.
(The central atom does influence the stability, the ease of synthesis,
and the cost of the catalyst however.)

    The percent of rings substituted with onium ions strongly
influences the overall degree of hydration of an anion exchange
resin. For most nucleophilic displacement reactions catalysts which
swell more in aromatic and halogenated solvents than in water are
much more active than the highly substituted commercial anion exchange resins. If the water content of a catalyst is too low,
however, transport of the reactive anion from the particle surface
to the active site could become a rate-limiting step.[33]

    The separation of the active site from the polymer backbone
probably affects mainly the time required to transport the reactants
to the active site. Diffusivity of a small molecule in a gel should
be faster several molecular dimensions away from the polymer chain
than right next to the slow-moving polymer-chain. The degree of
hydration of the active site might also be affected by its distance
from the polymer backbone. However, Montanari, Tundo and co-

workers[16] found that chloride, bromide, and iodide ion hydration
numbers of the catalysts in Table III in toluene/water were independent of the length of the spacer chain and were equal within
experimental error to the hydration numbers of the halides in the
corresponding hexadecyltributylphosphonium salts in toluene/water.

The chemical nature and morphology of the polymer might affect
both the intraparticle diffusion of reactants and the intrinsic
reactivity at the active site. The isoporous polystyrene support
of Tundo[25] may be highly active because of faster reactant diffusion
in an isoporous matrix than in a conventional divinylbenzene-cross-
linked matrix. The isoporous resin retained 9.4 g of dichloromethane
per g of resin compared with 4.4 g of dichloromethane per g of a
conventional 2% cross-linked benzyltributylphosphonium catalyst. No
information is available on the degree of hydration of the isoporous
catalyst.

ALKYLATION OF PHENYLACETONITRILE

Nucleophilic displacement reactions and alkylations of active
nitriles and carbonyl compounds comprise the two most important
classes of phase transfer catalyzed reactions. Balakrishnan[41,42]
has carried out a detailed kinetic study of alkylation reactions
using polystyrene-supported benzyltrimethylammonium ion catalysts.
A triphase mixture of 1-bromobutane, phenylacetonitrile, 50% aqueous
sodium hydroxide and catalyst at 80°C produced 2-phenylhexanenitrile
as shown in Scheme 4. The same qualitative dependences on stirring
rate, particle size, and cross-linking of the catalyst were observed
as in reaction of 1-bromooctane with aqueous sodium cyanide. Reaction rates increased with stirring rate up to about 500 rpm and were
constant from 500 to 700 rpm. Vibromixing did not increase the rate.
Rates increased as particle sizes decreased using 500 rpm stirring.
Rates decreased markedly as polymer cross-linking increased in the
order 2%, 4%, 6% and 10%. Mass transfer and intraparticle diffusion
are important rate limiting factors in the alkylation of phenylacetonitrile.

Scheme 4

$$C_6H_5CH_2CN \ + \ NaOH(aq) \ \xrightarrow{80°C} \ C_6H_5\bar{C}HCN$$

$$C_6H_5\bar{C}HCN \ + \ \underline{n}\text{-}C_4H_9Br \ \rightarrow \ C_6H_5\overset{\overset{\textstyle C_4H_9}{|}}{C}HCN$$

An especially dramatic illustration of the importance of polymer
swelling in catalysis of phenylacetonitrile butylation is shown in
Figure 5. When the catalyst was conditioned in phenylacetonitrile
and 1-bromobutane was added at time zero, alkylation rates were up
to eight times faster than when the catalyst was conditioned in 1-
bromobutane and phenylacetonitrile was added at time zero. At 80°C
phenylacetonitrile swelled the catalyst to 3.0 times its dry volume
in less than 5 minutes, while 1-bromobutane swelled the catalyst to
only 1.4 times its dry volume over a period of 12 hours. The activity of slowly swelling catalysts depends on the time of conditioning
the catalyst before use. Montanari, Tundo and co-workers[16] observed
such a dependence of rates of reaction of 1-bromooctane with aqueous
potassium iodide on conditioning of the catalyst using conventional
divinylbenzene-cross-linked catalysts but not using more swellable
isoporous catalyst.

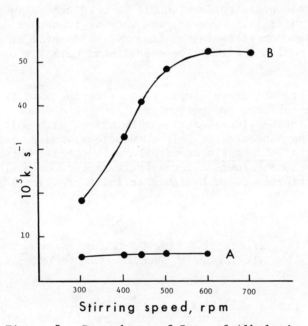

Figure 5.   Dependence of Rate of Alkylation of
            Phenylacetonitrile on Stirring Speed.

A: 1-Bromobutane added before phenylacetonitrile.   B: Phenylacetonitrile added before 1-bromobutane. (Reprinted with permission from
Tetrahedron Letters, 22, 4377 (1981), Pergamon Press, Ltd.)

The major difference in the properties of effective catalysts
for phenylacetonitrile alkylation and for nucleophilic displacement
reactions is that the alkylations proceed rapidly even with high
degrees of ring substitution of onium ions. With 2% cross-linked,
microporous catalysts the 50% ring substituted catalyst was more
active than the 17% ring substituted catalyst. Commercial ion
exchange resins also were active,[42,43] as shown in Table IX. A
hydrophobic environment is not required for the rate-limiting carbanion generation from phenylacetonitrile and hydroxide ion. The
same reaction using soluble quaternary ammonium ion catalysts is
thought to proceed at the aqueous/organic phase boundary because the
hydroxide ion concentration in the organic phase of such reaction
mixtures is too low to account for the overall rate of alkylation.[44-46] Hydroxide ion is such a strong hydrogen bond acceptor
that it likely never loses all of its hydration in any phase transfer
reaction with either soluble or insoluble catalysts. In the 16% ring
substituted catalyst in Table IX the hydroxide ion would be expected
to be less hydrated than in catalysts having 40% or more substituted
rings. The less hydrated form should be more active in deprotonation
of phenylacetonitrile. However, the rate of transport of hydroxide
to the ion exchange sites in the less hydrated catalyst may be much
slower, thus making the 16% ring substituted catalyst overall less
active.

Active site structure, separation of active site from the backbone, organic solvent, and molecular size of the reactant have not
been studied in phenylacetonitrile alkylations with polystyrene-
supported onium ion catalyst. In conventional phase transfer catalyzed ethylation of phenylacetonitrile with tetra-n-butylammonium
bromide, Chiellini[46] found a formal rate dependence of $[NaOH]^{5.3}$,
which was attributed to much higher activity of hydroxide ion in

Table IX. Effect of % Ring Substitution on Activity
for Alkylation of Phenylacetonitrile.[42]

catalyst	% ring subst.	av. diameter μm	$k \times 10^5$, $s^{-1}mequiv^{-1}$
2% cross-linked	16	223	57.8
2% cross-linked	50	229	171
Amberlyst A-27	40	466	82.2
Amberlyst A-27	40	411	105
Amberlyst A-26	90	502	49.5

Amberlyst catalysts are macroporous from Rohm and Haas Co.

a 19 $\underline{M}$ solution than in an 11 $\underline{M}$ solution. The same effect of hydroxide ion concentration on alkylation rate would be expected with a
polymer-supported catalyst.

CHEMICAL REACTIONS AS PROBES OF ACTIVE SITE ENVIRONMENT

     Alkylation of phenoxide and naphthoxide ions can occur at either
oxygen or carbon, depending on solvent, counter ion and alkylating
agent. Protic solvents which hydrogen bond to the oxygen anion
favor C-alkylation, while aprotic solvents favor O-alkylation.[47]
Brown and Jenkins[5] found that 98% ring substituted catalyst $\underline{4b}$ gave
98% O-benzylation of 2-naphthoxide ion, and that catalyst $\underline{2}$ (2%
cross-linked, >90% substituted) gave about 65% O-benzyl product (eq
5). The long aliphatic spacer chain creates a more hydrophobic

$$(5)$$

environment in $\underline{4b}$ than in $\underline{2}$ Regen[48] investigated the same reaction
with 17% and 52% ring substituted phosphonium ion catalysts $\underline{3a}$ and
found 94% and 19% O-alkylation. The regioselectivities of the reaction correlate qualitatively with the degrees of hydration of the
catalysts in chloride form: In water/toluene in 17% ring substituted
catalyst imbibed 0.81 g of water per g of resin while the 52% ring
substituted catalyst imbibed 11.91 g of water.[48]

     Montanari and Tundo[16] found 95-98% O-alkylation of sodium
phenoxide in dichloromethane/water using phosphonium catalysts $\underline{5}$
bound to polystyrene but only 74% O-alkylation with a similar phos-

phonium ion bound to silica gel. They concluded that in the polystyrene catalysts the phenoxide was solvated only by dichloromethane, while in silica gel the water at the silica surface helped
solvate the phenoxide ion, causing more C-alkylation. It seems
unlikely to this reviewer that an ionic catalyst would contain $\underline{no}$
water, but the degrees of hydration of $\underline{5}$ in phenoxide form have not
been determined.

   The rate of decarboxylation of 6-nitrobenzisoxazole-3-carboxylate
ion (eq 6)[49] is highly sensitive to solvent. Polystyrene-supported
catalysts 6 having 2-15% divinylbenzene cross-linking and 22-92%

$$ (6) $$

ring substitution were more active than the corresponding soluble
polystyrene-bound catalysts and more active than micellar cetyltri-
ethylammonium bromide in aqueous solution at pH 9.0.[50] Since the
decarboxylation rate depends on the extent of hydrogen-bonding
stabilization of the starting carboxylate, the results indicate that
the active sites of the cross-linked polymers have much less water
available for hydrogen bonding than the active sites of the soluble
polymer, which has a lower water content than the micelles of a
cationic surfactant.

CATALYST ACTIVITY IN FLOW SYSTEMS

   There is one report of triphase liquid/liquid/solid catalysis
in a continuous flow reactor. Ragaini and Saed[51] passed a mixture
of 1-bromooctane in o-dichlorobenzene and aqueous potassium iodide
upward through a bed of polystyrene-bound phosphonium ion catalyst.
A magnetic stirring bar at the entrance to the packed tube mixed the
liquid phases. At a flow rate of 3.0 L/h up to 90% conversion was
attained in 8 h at 80°C by continuous recycling of about 100 g of
liquid phases through 0.5 g of catalyst. The overall rate of conversion was slower than in a stirred batch reactor using the same
ammounts of reactants and catalyst.

   Tundo has studied gas phase flow of alkyl halides through packed
beds of solid salts and phase transfer catalysts and obtained the
products of nucleophilic displacement.[52] His most recent results
appear elsewhere in this volume.

CATALYST STABILITY

   The economic promise of polystyrene-supported phase transfer
catalysts depends on their reuse in industrial processes. Under some
of the reaction conditions described for nucleophilic displacement
reactions and for alkylation of active methylene compounds, the

catalysts slowly lose activity. Benzyltributylphosphonium ions
decompose under triphase conditions in the presence of concentrated
aqueous sodium cyanide.[33,36] Benzyltrimethylammonium catalysts
decompose in the presence of 50% aqueous sodium hydroxide.[42,53]
Both are stable in the presence of more weakly basic anions such as
halides and phenoxide. Decomposition of quaternary ammonium and
phosphonium ions has been studied under liquid/liquid phase transfer
conditions and in single phase systems.[53,54] The major decomposition pathways for tetraalkylammonium ions are Hoffman elimination
and nucleophilic attack at methyl and benzyl carbon atoms.
Phosphonium ions also can react by ylide formation and by attack
of hydroxide ion on phosphorus to give a phosphine oxide and
the hydrocarbon derived from the most stable carbanionic leaving
group, as in eq. 7. Hexadecyltributylphosphonium bromide was reported to be stable in 50% aqueous sodium hydroxide for 16 h at
100°C.[45] However, no tetraalkylphosphonium ion is expected to be

$$C_6H_5CH_2PR_3^+ \ + \ OH^- \ \rightarrow \ C_6H_5CH_3 \ + \ R_3PO \tag{7}$$

stable indefinitely in polymer-bound catalysts under phase transfer
conditions that require the presence of hydroxide ion at the ion
exchange site. Benzyltrialkylammonium and phosphonium ions are much
less stable in base than non-benzylic tetraalkylammonium and phosphonium ions.[54] Industrial applications of polystyrene-supported
onium ion catalysts under strongly basic conditions will require
catalysts such as 7,[21] rather than the usual commercially available

anion exchange resins. Polymer-bound crown ethers and oligomeric
ethylene oxides, described elsewhere in this symposium, are more
stable in base than onium ions. They may, however, be more expensive (crown ethers) or less active (ethylene oxide oligomers).

Physical instability of polystyrene beads could also prevent
recycling of the catalyst. On a laboratory scale one can minimize
catalyst breakdown by use of a shaker or overhead stirrer to agitate
the reaction mixture. Magnetic stirring bars trap beads against the
wall of the flask and grind them to a powder. The very lightly
cross-linked Merrifield resins (≤ 1% divinylbenzene) often break
under vigorous agitation, but as little as 2% cross-linking usually
prevents such breakage. On a large scale highly swollen polymers
may prove impossible to filter because they are too gelatinous.
Higher degrees of cross-linking will reduce the swelling problem,
but the catalysts may also be less active because of intraparticle
diffusional limitations. Macroporous polystyrenes are a possible
solution to the diffusional limitations of more highly cross-linked
polymers, but their potential for high activity has not yet been

achieved. Polystyrene-based ion exchange resins, both microporous and microporous, have been used continuously in industrial water treatment systems for years. Problems of physical instability of polystyrene-supported catalysts should be surmounted when the financial incentive of an industrial process is great.

CONCLUSION

Rational design of polymer-supported phase transfer catalysis processes is now possible. We have a qualitative understanding of the importance of rate limiting mass transfer and intraparticle diffusion processes in triphase catalysis. Extensive data is available on the effects of active site structure, degree of ring substitution, separation of the active site from the polymer backbone, and organic solvent. For industrial processes in which ion exchange resins currently available are inadequate, new catalysts must be synthesized. Vigorous mixing of batch reactions or flow systems should be used. The catalyst particle size should be as small as the conditions of the flow system or filtration permit. The polymer should be more highly cross-linked than a Merrifield resin to limit swelling in a packed column and aid filtration of a batch reaction mixture. When aggressive anions are required, the onium ion must be separated from the polymer backbone by a chain of a least three carbon atoms or a polyether must be used to minimize active site decomposition.

REFERENCES

1. R. Kunin, "Ion Exchange Resins," 2nd edn., Wiley, New York, 1958, pp. 258-259.
2. F. Helfferich, "Ion Exchange", McGraw-Hill, New York, 1962, Chpt. 11.
3. R. Kunin, "Amber-hi-lites," Rohm and Haas Co., Philadephia, PA, No. 128 (May 1972) and No. 135 (July 1973).
4. S. L. Regen, J. Am. Chem. Soc., 97, 5965 (1975).
5. J. M. Brown and J. A. Jenkins, J. Chem. Soc., Chem. Commun., 458 (1976).
6. M. Cinquini, S. Colonna, H. Molinari, F. Montanari, and P. Tundo, J. Chem. Soc., Chem. Commun., 394 (1976).
7. S. L. Regen, Angew. Chem., Int. Ed. Engl., 18, 421 (1979).
8. D. C. Sherrington, "Polymer-supported Reactions in Organic Synthesis", (P. Hodge and D. C. Sherrington, Eds.), Wiley, New York, 1980, pp. 180-194.
9. A. Akelah and D. C. Sherrington, Chem. Rev., 81, 557 (1981).
10. J. Jarrouse, C. R. Hebd. Seances Acad. Sci., Ser. C, 232, 1424 (1951).
11. M. Makosza and B. Serafinowa, Rocz. Chem., 39, 1223 (1965).

12.   C. M. Starks, J. Am. Chem. Soc., 93, 195 (1971).
13.   C. M. Starks and C. Liotta, "Phase Transfer Catalysis",
Academic Press, New York, 1978.
14.   W. Weber and G. Gokel, "Phase Transfer Catalysis in Organic
Synthesis", Springer Verlag, Berlin, 1977.
15.   E. Dehmlow and S. Dehmlow, "Phase Transfer Catalysis",
Verlag Chemie, Weinheim, 1980.
16.   H. Molinari, F. Montanari, S. Quici, and P. Tundo, J. Am.
Chem. Soc., 101, 3920 (1979).
17.   S. L. Regen, J. Am. Chem. Soc., 98, 6270 (1976).
18.   M. S. Chiles and P. C. Reeves, Tetrahedron Lett., 3367
(1979).
19.   S. L. Regen, D. Bolikal, and C. Barcelon, J. Org. Chem.,
46, 2511 (1981).
20.   M. Tomoi, Yokohama National University, personal communication.
21.   M. S. Chiles, D. D. Jackson, and P. C. Reeves, J. Org.
Chem., 45, 2915 (1980).
22.   K. W. Pepper, H. M. Paisley, and M. A. Young, J. Chem.
Soc., 4097 (1953).
23.   J. Brandrup and E. H. Immergut, Eds., "Polymer Handbook",
2nd edn., Wiley, New York, 1975, p. II-350.
24.   W. T. Ford and S. A. Yacoub, J. Org. Chem., 46, 819 (1981).
25.   P. Tundo, Synthesis, 315 (1978).
26.   P. Tundo, J. Chem. Soc., Chem. Commun., 641 (1977).
27.   P. Tundo and P. Venturello, J. Am. Chem. Soc., 101, 6606
(1979).
28.   P. Tundo and P. Venturello, J. Am. Chem. Soc., 103, 856
(1981).
29.   V. A. Davankov and M. P. Tsyurupa, Angew. Makromol. Chem.,
91, 127 (1980) and references therein.
30.   J. R. Millar, D. G. Smith, W. E. Marr, and T. R. E.
Kressman, J. Chem. Soc., 218 (1963).
31.   J. Seidl, J. Malinský, K. Dusek, and W. Heitz, Adv. Polym.
Sci., 5, 113 (1967).
32.   R. Kunin, E. Meitzner, and N. Bortnick, J. Am. Chem. Soc.,
84, 305 (1962).
33.   W. T. Ford, J. Lee, and M. Tomoi, Macromolecules, 14, in
press (1982).
34.   H. Kise, K. Araki, and M. Seno, Tetrahedron Lett., 22,
1017 (1981).
35.   S. L. Regen and J. J. Besse, J. Am. Chem. Soc., 101, 4059
(1979).
36.   M. Tomoi and W. T. Ford, J. Am. Chem. Soc., 102, 7140
(1980); 103, 3821 (1981).
37.   N. Ohtani and S. L. Regen, Macromolecules, 14, 1594 (1981).
38.   A. Streitwieser, Jr., "Solvolytic Displacement Reactions",
McGraw-Hill, New York, 1962, pp. 13-30.
39.   M. Tomoi and W. T. Ford, J. Am. Chem. Soc., 103, 3828
(1981).

40. A. J. Parker, Chem. Rev., $\underline{69}$, 1 (1969).

41. T. Balakrishnan and W. T. Ford, Tetrahedron Lett., $\underline{22}$, 4377 (1981).

42. T. Balakrishnan and W. T. Ford, submitted for publication.

43. H. Komeili-Zadeh, H. J.-M. Dou, and J. Metzger, J. Org. Chem., $\underline{43}$, 156 (1978).

44. M. Makosza and E. Bialecka, Tetrahedron Lett., 183 (1977).

45. E. V. Dehmlow, M. Slopianka, and J. Heider, Tetrahedron Lett., 2361 (1977).

46. R. Solaro, S. D'Antone, and E. Chiellini, J. Org. Chem., $\underline{45}$, 4179 (1980).

47. N. Kornblum, R. Seltzer, and P. Haberfield, J. Am. Chem. Soc., $\underline{85}$, 1148 (1963).

48. N. Ohtani, C. A. Wilkie, A. Nigam, and S. L. Regen, Macromolecules, $\underline{14}$, 516 (1981).

49. D. S. Kemp and K. G. Paul, J. Am. Chem. Soc., $\underline{97}$, 7305 (1975).

50. N. Yamazaki, S. Nakahama, A. Hirao, and J. Kawabata, Polymer J., $\underline{12}$, 231 (1980).

51. V. Ragaini and G. Saed, Z. Phys. Chem., $\underline{119}$, 117 (1980).

52. P. Tundo, J. Org. Chem., $\underline{44}$, 2048 (1979).

53. H. J.-M. Dou, R. Gallo, R. Hassanaly, and J. Metzger, J. Org. Chem., $\underline{42}$, 4275 (1977).

54. Ref. 13, pp 62-65.

CHIRAL POLYMER SUPPORTED CATALYSTS IN PHASE TRANSFER REACTIONS

Emo Chiellini, Roberto Solaro, and Salvatore D'Antone

Istituto di Chimica Generale, Facoltà di Ingegneria
Istituto di Chimica Organica Industriale, Università
di Pisa, Pisa, Italy

INTRODUCTION

Over the past few years a great deal of interest has been focused on reactions carried out under phase transfer conditions in the presence of low and high molecular weight catalysts[1-3].

Among the rather large number of reports aimed at stressing the powerfulness of this new synthetic procedure both for the practical and more speculative implications involved, particular attention has been reserved to reactions performed on prochiral substrates or racemates in the presence of chiral catalysts based on onium salts and crown ethers either soluble in organic solvents or supported on insoluble polymeric matrices.

In the present paper we wish to report on the most significant examples of chiral polymer-supported onium salts and polymeric amines which have been used in heterophase reactions. In the first part attention will be paid to describe the Michael-type reactions that have been rather extensively studied by several research groups. Even though they constitute an example of application placed right at the borderline of base catalyzed reactions performed under conventional and phase transfer conditions, they appear worthy of comment by virtue of the achieved valuable and reproducible optical yields in the chemical transformation of several prochiral substrates.

In the case of either liquid-liquid biphase systems or liquid--liquid-solid triphase systems, to the best of our knowledge no significant enantiomeric discrimination efficiency has been clearly reported until now both in the kinetic resolution of racemates and in the transformation of prochiral substrates.

1. MICHAEL-TYPE REACTIONS

The base-catalyzed Michael-type addition of active hydrogen compounds to activated double bonds is generally performed under homogeneous conditions[4-10] and lays therefore at the borderline of the scope of this paper, even if it has been sometimes carried out under typical phase transfer conditions[11-13]. However, considering that the catalysts promoting phase transfer reactions, i.e. ammonium salts, amines and crown ethers, are generally active in the Michael addition, the reported reactions will be discussed here in some de-tails also for the reasons mentioned in the introduction.

The polymeric catalysts used can be divided into two main clas-ses: polymer bound alkaloids[14-21] and polymeric amines[22-33], even if it is fair to mention that the first known asymmetric Michael addition was carried out in the presence of alcoholates supported on optically active natural quartz[34].

In the first class of catalysts there are copolymers of acrylo-nitrile with cinchona alkaloids[17,19,20] (**1a–1d**),

		R^1	C_3	C_4	C_8	C_9
1a	Quinine	OMe	R	S	S	R
1b	Quinidine	OMe	R	S	R	S
1c	Cinchonidine	H	R	S	S	R
1d	Cinchonine	H	R	S	R	S

homo' and copolymers of O-acryloylquinine[14,15] and O-acryloylcin-chonine[17] (**2a,2b**),

		R^1	C_4	C_8	C_9
2a	OMe		S	S	R
2b	H		S	R	S

and cinchona alkaloids bound to crosslinked poly(styrene)[16](**3-5**).

3   **4**   **5**

The addition of active hydrogen compounds (alcohols, thiols, β-ketoesters and nitroalkanes) to activated double bonds performed in the presence of the **1-5** catalysts occurs with a maximum enantiomeric excess of about 60%.

Whereas in the case of low molecular weight analogs the stereochemical pathway of the addition reaction is generally determined by the C_8, C_9 absolute configurations[8,12] and by the C_3 absolute configuration only in the case of bulky ·substituted alkaloids[18], in the polymers the control is played only by the C_3 absolute configuration[19,20], thus indicating a key role of the macromolecular structure. This last effect is also recognizable in the generally higher efficiency of the polymeric systems in affecting the enantiomeric excess of the reaction products.

No apparent difference was detectable in the results obtained in Michael reactions performed in the presence of poly(O-acryloyl-quinine) (**2a**) and poly(2-quinuclinidinylmethyl acrylate) (**6**), thus indicating that the quinoline moiety is not essential for the reaction stereochemistry[18]. On the contrary, the presence of a free hydroxyl group, both in low molecular weight compounds and in the corresponding polymers, is vital for the control of the enantiomeric excess during the addition step[12]. These results seem coherent with the existence of a multicenter transition state involving the activated double bond, the nitrogen atom which concurs to the extraction and successive transfer of the active proton, and the hydroxyl group which interacts by hydrogen bonding with the electron rich atom present on the prochiral substrate[20].

6

In Scheme 1 the mentioned steps have been sketched for the addition of a thiol to an α,β-unsatured carbonyl compound.

Scheme 1. Concerted reaction mechanism in Michael type addition in the presence of cinchona alkaloids.

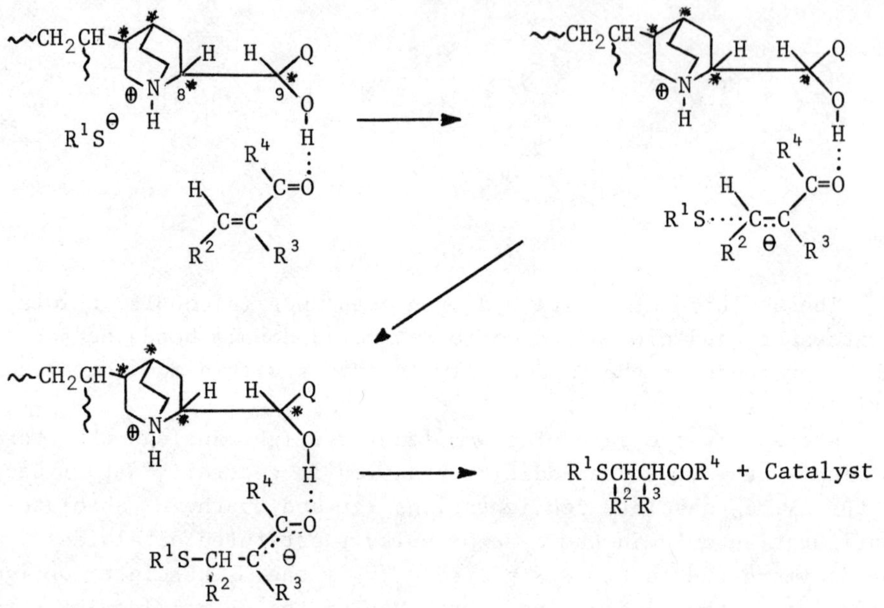

The same type of mechanism has been invoked for reactions carried out in the presence of ephedra alkaloid derivatives[12].

The other systems utilized for the same type of reactions can be divided in two groups characterized respectively by the presence of amine type nitrogen atoms in the main chain[22-25,32], obtained by ring opening polymerization of aziridine derivatives (7-9) and poly(α-aminoacid)s capped with a primary and/or tertiary amine function (10)[26-30] or containing an amine group in the side chain[22] (11-12). Poly(acrylate)s containing a N-benzyl-2-pyrrolidinylmethyl group (13) have been also used[31]

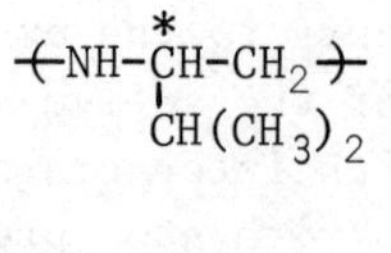

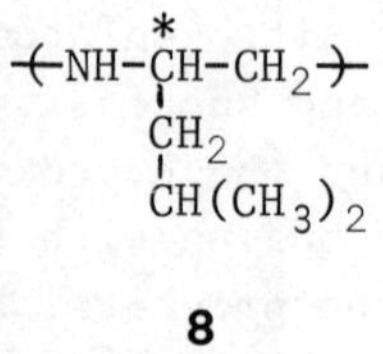

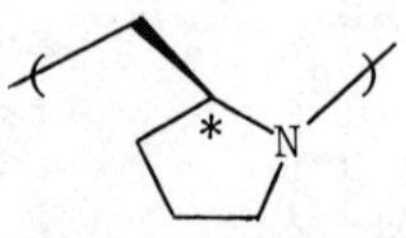

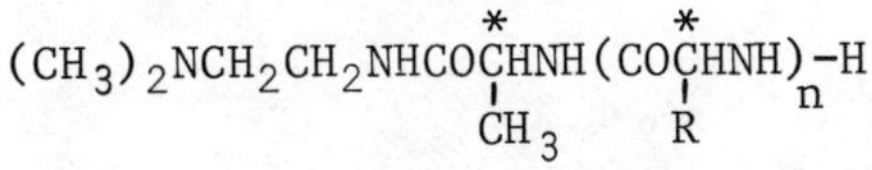

$$(CH_3)_2NCH_2CH_2NHCO\overset{*}{C}HNH(CO\overset{*}{C}HNH)_n-H$$
$$\underset{CH_3 \quad\quad R}{}$$

10

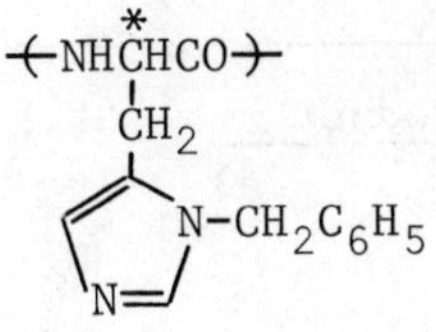

11

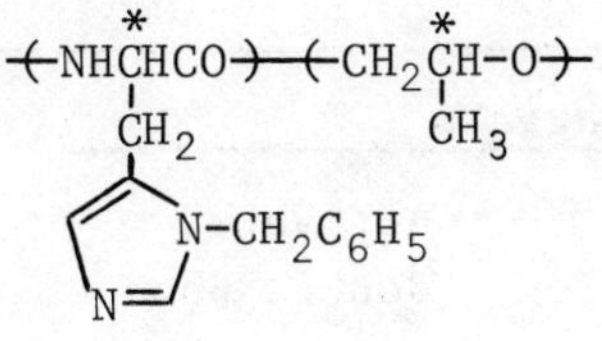

12

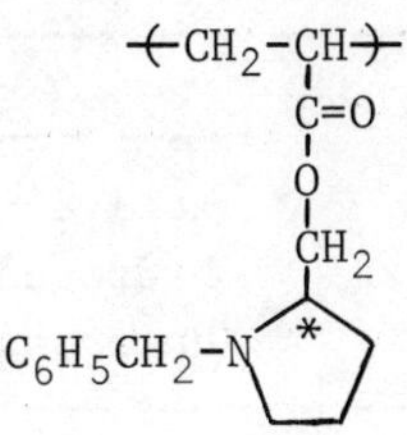

13

As reported in Table 1 the enantiomeric excesses observed in the thiol addition to activated double bonds are in the range of the values obtained in the presence of polymeric cinchona alkaloids and in most cases the stereochemical discrimination of the polymeric systems is higher than that observed for the corresponding low molecular weight structural models, thus clearly evidencing a polymer effect in affecting the reaction pathway.

For the reaction catalyzed by oligo(α-aminoacid)s[26-30] a correlation between the main chain conformation and asymmetric induction has been reported. In particular the α-conformation appears as the most suitable to guarantee fairly good asymmetric inductions, whereas a random coil conformational situation does not lead to any significant enantiomeric excess.

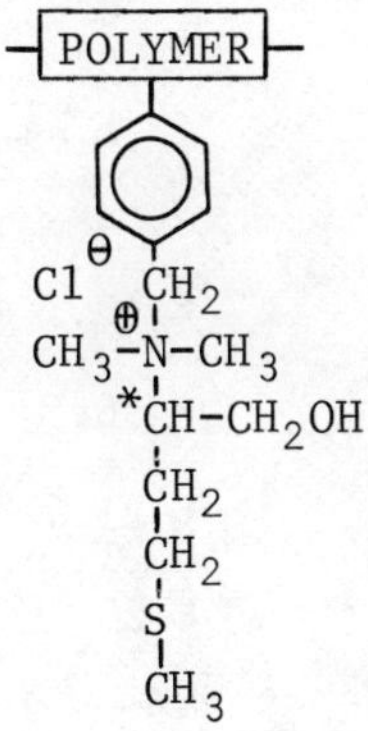

14

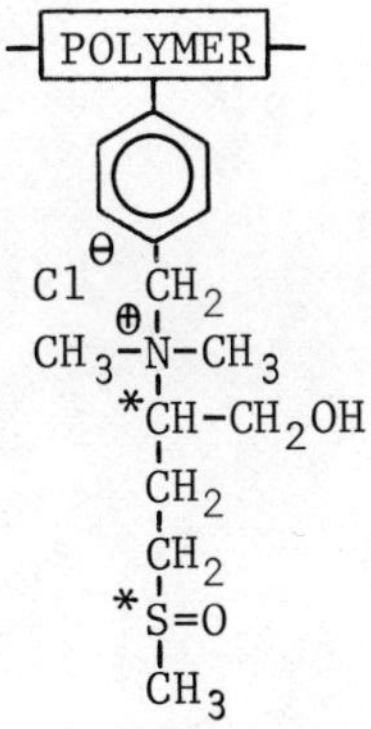

15

Recently in the search for new and more versatile chiral phase transfer catalysts, onium salts **14** and **15** derived from L-methionine

Table 1

Michael type reactions carried out in the presence of optically active polymeric catalysts.

Catalyst	Active hydrogen compound	Activated double bond compound	$\|\alpha\|^{25}_D$	e.e.[a] (%)	Ref.
2a			− 16.2	14.9	15
2b			+ 38.1	35.0	15
8	CH_3OH	$CH_3C=C=O$ C_6H_5	− 2.0	1.8	32
13			− 5.5	5.0	31
9			+ 11.0	10.0	32
6			− 10.0	9.0	14
1a			− 24.1	30	17
1b			+ 33.7	42	17
3	–COOCH$_3$ (indanone)		− 6.5	8	16
4			− 8.5	11	16
5			− 1.9	2	16
14			0	0	33
1a			+ 18.7	9.3	17,19
1b			+ 36.3	18.0	17,19
1c	$C_6H_5CH_2SH$	$C_6H_5CH=CHNO_2$	+ 25.7	12.7	19
1d			+ 25.1	12.4	19
14			0	0	33

a) enantiomeric excess.

Catalyst	Reagents		Product		Ref.
	Active hydrogen compound	Activated double bond compound	$\lvert\alpha\rvert_D^{25}$	e.e.[a]	
8		$CH_2{=}\underset{CH_3}{C}{-}COOCH_3$	+ 0.6		22
11			~ 0		22
12			~ 0		22
8		$CH_3CH{=}CH{-}COOCH_3$	+ 0.19		23
8		$CH_2{=}\underset{CH_3}{C}{-}COCH_3$	+ 0.37	2	24
10			+ 1.84	11	29
8		$CH_3CH{=}CH{-}COCH_3$	0		24
8	$n{-}C_{12}H_{25}SH$	$C_6H_5CH{=}CHCOCH_3$	− 0.04		24
8		$C_6H_5CH{=}CHCOC_6H_5$	− 0.08		24
8		$CH_2{=}C{<}^{CN}_{CH_3}$	+ 0.08		24
7		$CH_3OOC{>}C{=}C{<}^{COOCH_3}_{H}$ (H)	− 1.55		25
7		$CH_3OOC{>}C{=}C{<}^{H}_{COOCH_3}$ (H)	+ 1.49		25
1a		benzene-OCH₃ $-CH{=}CHCOC_6H_5$	+16.0	20	20
1b			+ 6.1	8	20
1c			+ 2.9	4	20
1d			+ 9.3	12	20

a) enantiomeric excess.

have been used in Michael type reactions carried out on prochiral substrates[33]. In spite of the presence of multiple binding sites and of a chiral sulphur atom, no asymmetric induction has been obtained for both the reported polymeric systems and the corresponding low molecular weight model compounds.

2. PHASE TRANSFER REACTIONS

It has been known for several years that tertiary amines are active in nucleophilic substitution reactions carried out under phase transfer conditions[35]. The reaction mechanism is however still disputable[35-39]. Recently optically active poly(amine)s (**7**) and aminated poly(ether)s (**16**)

$$\text{(-O-CH}_2\text{-CH-)}$$
$$\underset{\underset{\underset{\overset{|}{\text{CH}_3}}{\text{NH-CH}_2\text{-}\overset{*}{\text{CH}}\text{-N(CH}_3\text{)}_2}}{|}}{\overset{|}{\text{CH}_2}}$$

16

have been used in the esterification reaction of achiral salts, chlorides and anhydrides of organic acids with racemic alkyl bromides and alcohols[40,41]. Similarly chiral esters have been prepared starting from racemic derivatives of carboxylic acids and achiral alkylating reagents.

$$\text{RCOOK + R'Br} \xrightarrow[\text{benzene}]{\textbf{7 or 16}} \text{RCOOR' + KBr}$$

R = CH_3, C_2H_5, $i\text{-}C_3H_7$ R' = $s\text{-}C_4H_9$, 1-phenylethyl

R = $s\text{-}C_4H_9$ R' = $n\text{-}C_4H_9$

$$\text{RCOX + R'OH} \xrightarrow[\text{benzene}]{\textbf{7}} \text{RCOOR' + HX}$$

X = Cl, $\underset{\overset{\|}{O}}{O\text{-}C\text{-}R}$

R = CH_3 R' = $s\text{-}C_4H_9$

R = $s\text{-}C_4H_9$ R' = C_2H_5

In all cases esters with an enantiomeric purity never exceeding 5% were obtained, thus indicating that enantiomeric discrimination is very poor under the adopted conditions. In spite of that, indications have been achieved on the coexistence of the two proposed mechanisms in the esterification reaction catalyzed by tertiary amines and implying either the activation of the carboxylate anion or the formation of quaternary ammonium salt by reaction of the alkylbromide with the tertiary amine.

No more encouraging enantiomeric excess values were obtained in the dynamic resolution of racemic 1-phenylethanol performed by acetylation in the presence of optically active polymers [poly(O--acryloylquinine) (**2a**), poly(N-benzyl-2-pyrrolidinylmethyl acrylate) (**13**) and poly(O-acryloyl-N-benzylephedrine) (**17**)]

$$
\begin{array}{l}
\text{+CH}_2\text{-CH+} \\
\quad | \\
\quad \text{C=O} \\
\quad | \\
\quad \text{O} \\
\quad | \\
\quad \overset{*}{\text{CH}}\text{-C}_6\text{H}_5 \\
\text{CH}_3\text{-}\overset{*}{\text{CH}} \\
\quad | \\
\quad \text{N-CH}_3 \\
\quad | \\
\quad \text{CH}_2\text{C}_6\text{H}_5
\end{array}
$$

17

containing tertiary amine groups [42]:

$$C_6H_5\underset{\underset{CH_3}{|}}{C}H\text{-OH} + (CH_3CO)_2O \xrightarrow[CCl_4]{\text{catalyst}} C_6H_5\underset{\underset{CH_3}{|}}{C}HOCOCH_3$$

As a general comment on all the reported esterification reactions we have to mention that no real phase transfer of the nucleophile from the aqueous to the organic phase occurs because water is not used as inorganic phase.

18a   n = 1   X = Cl

18b   n = 12   X = Br

By using polymeric onium salts chiral epoxides have been prepared by either epoxidation of olefinic substrates[43] or carbenation of prochiral and chiral carbonyl compounds[44].

In the former case quininium chloride or bromide salts, either directly bound (**18a**) or spaced apart (**18b**) from a 2% DVB crosslinked poly(styrene) have been used and enantiomeric excesses of 0-4% have been obtained in the epoxidation of substituted chalcones.

$$R-\langle\bigcirc\rangle-CH=CH-COC_6H_5 \xrightarrow[\text{NaOH/toluene}]{\textbf{6}} R-\langle\bigcirc\rangle-\underset{\underset{O}{\diagdown\diagup}}{CH-CH}-COC_6H_5$$

R = H, Cl

Rather surprisingly, the mentioned polymeric systems show an enantiomeric discrimination efficiency that is one order of magnitude lower than that shown by the corresponding low molecular weight analogs[45] and accordingly the most spaced system displays a discrimination efficiency. However the stereochemical efficiency drops down in recycling experiments, whereas the catalytic activity increases.

An analogous trend is observed in the Darzens reaction between acetaldehyde or benzaldehyde and α-chlorophenylacetonitrile.[44]

$$RCHO + C_6H_5\underset{Cl}{\overset{|}{C}H}-CN \xrightarrow[\text{NaOH/CH}_2\text{Cl}_2]{\textbf{19}} RCH-C\diagdown^{CN}_{\diagdown C_6H_5}$$

catalyzed by ephedrinium salts supported on 2% DVB crosslinked poly(styrene) (**19**):

$$\text{(P)}-\langle\bigcirc\rangle-CH_2-\underset{\underset{Cl^\ominus}{}}{\overset{\oplus}{N}}-\underset{CH_3}{\overset{CH_3 \quad CH_3}{C}H}-\underset{OH}{\overset{*}{C}H}-C_6H_5$$

19

Higher discrimination efficiencies are claimed on the contrary in the carbenation of prochiral and chiral carbonyl compounds performed with chloromethyl-*p*-tolyl sulphone in the presence of the same polymeric catalysts.[44]

$$\underset{R}{\overset{R'}{\diagdown}}C=O + p-CH_3C_6H_4SO_2CH_2Cl \xrightarrow[\text{NaOH/CH}_2\text{Cl}_2]{\textbf{19}} \underset{R}{\overset{R'}{\diagdown}}\underset{\underset{O}{\diagdown\diagup}}{C}-CHSO_2C_6H_4CH_3$$

R = H   ;   CH_3   ; C_2H_5 ; C_6H_5

R' = CH_3, C_6H_5 ;   CH_3, C_2H_5, *i*-C_3H_7 ; C_2H_5 ; C_6H_5

The chemical and stereochemical catalytic efficiency does not seem substantially affected by recycling experiments.
By starting from the confusion that still remains on the effecti-

veness of some chiral low and high molecular weight quaternary ammonium salts in catalyzing the chemical transformation of prochiral
substrates with a certain extent of asymmetric induction, Sherrington and coworkers have reported[46] the preparation of a series of
rather complex polymer-supported optically active catalysts that
for convenience have been reduced here to four groups **(20 - 23)**

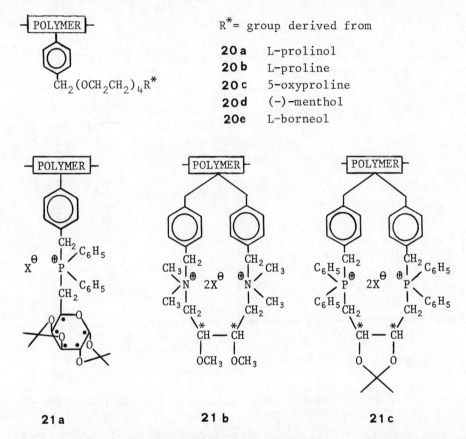

$R^* =$ group derived from

20a	L-prolinol
20b	L-proline
20c	5-oxyproline
20d	(-)-menthol
20e	L-borneol

21a   **21b**   **21c**

The formulation of the reported groups arises from the general
idea of assembling the active sites in a chiral environment generated directly by rather chemically stable (at least under the
adopted phase transfer reaction conditions) asymmetric groups
bound to either commercially available or especially prepared sty-
rene/DVB resins.

In Table 2 are reported the loads of the chiral group and of
the catalytically active moiety as evaluated by elemental analysis
and IR spectroscopy. No data have been published up to now on the

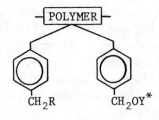

2 2 – 2 3

R	Y*	
	1,2:3,4-di-O-*iso*-propylidene-α-D-galactopyranose	1,2:4,5-di-O-*iso*-propylidene-β-D-fructopyranose
$-\overset{\oplus}{P}(C_4H_9)_3 \; Cl^{\ominus}$	**2 2 a**	**2 3 a**
$-\overset{\oplus}{N}(C_4H_9)_3 \; Cl^{\ominus}$	**2 2 b**	**2 3 b**
$-\overset{\oplus}{N}(n-C_8H_{17})_3 \; Cl^{\ominus}$	**2 2 c**	**2 3 c**
$-(OCH_2CH_2)_4OCH_3$	**2 2 d**	**2 3 d**
$-(OCH_2CH_2)_4OCH_2-$ (pyridyl)	**2 2 e**	**2 3 e**
$-(OCH_2CH_2)_4O-$ (quinolyl)	**2 2 f**	**2 3 f**

use of the reported systems in the transformation of prochiral sub-strates, however as far as we know, by a private communication[47] they have been tested in the borohydride reduction of prochiral ketones and apparently alcohols with very low enantiomeric excess have been obtained.

At the end of the general overview of what has been reported by other authors on asymmetric syntheses performed under phase transfer conditions and in the presence of polymer supported cata-lysts it is fair to mention that most of the criticism [48] on the reliability of the asymmetric inductions observed by several authors

Table 2

Functional group content in polymer-supported optically active phase transfer catalysts

Type	Content of		Type	Content of	
	chiral groups (%-mol)	active groups (%-mol)		chiral groups (%-mol)	active groups (%-mol)
20a – 20f	14	14	23a	70	25
			23b	70	20
22a	80	10	23c	70	low
22a	40	45	23d	70	30
22b	40	30	23e	20	20
22c	40	40	23f	80	low
22d	85	15			
22e	40	40	21a	40	40
22f	40	25	21b	50	50
			21c	30	30

in the presence of low molecular weight onium salts can be applied as well to high molecular weight systems.

In fact under rather strong alkaline conditions even polymer supported onium salts suffer the Hoffman-type decomposition reaction (Scheme 2) that gives rise to a loss of specific activity accompanied by an undesiderable drop of chiral requirements.
Moreover the possible formation of significant amounts of optically active low molecular weight compounds, not easily separable from the reaction products leads to a lack of reliability in quantitative aspects of the enantiomeric discrimination efficiency of the catalysts employed.

Systems able to overcome the drawbacks connected with the low chemical and stereochemical stability of the catalysts used, are in our opinion those having chiral groups tied up with a chemically and stereochemically stable macromolecular backbone.

In this respect linear copolymers of optically active α-olefins, alkyl acrylate and alkyl methacrylate with vinylaromatic co-monomers, for which it has been unequivocally proven that the aromatic units are imbedded in a highly dissymmetric environment[49,50]

Scheme 2.   Degradation of quaternary ammonium salts under alkaline
            conditions.

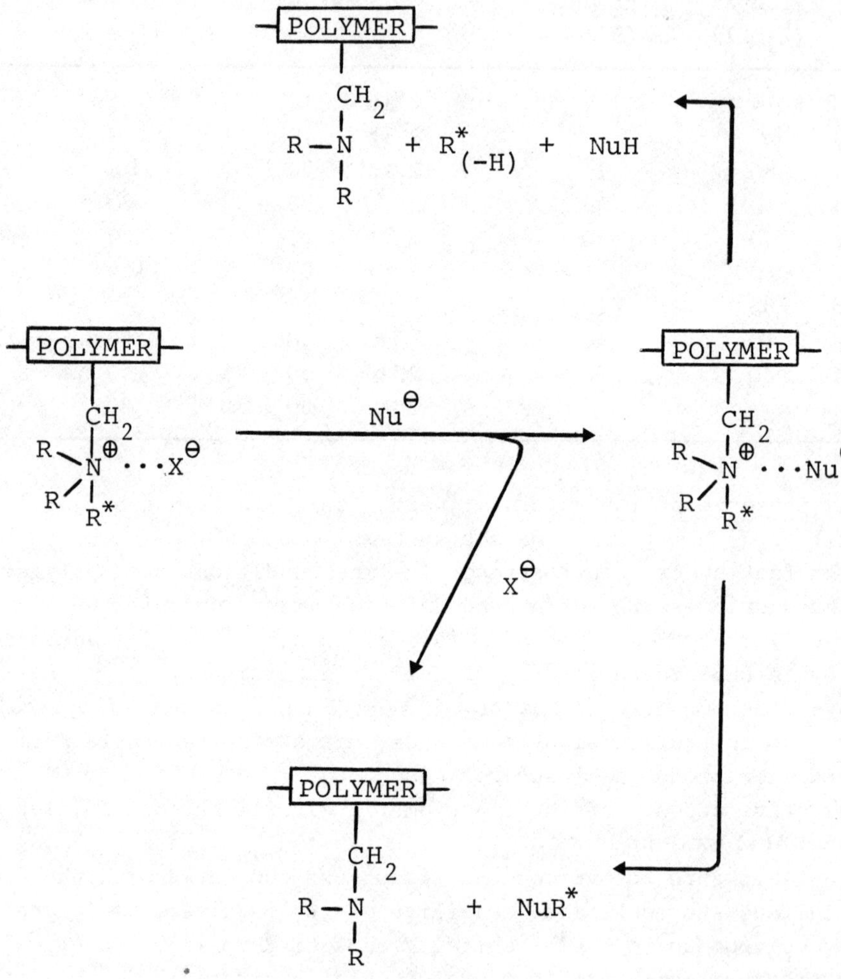

appeared to be very promising. The functionalization of these systems at the level of the aromatic moiety via the conventional route
for the insertion of onium groups in the para position of aromatic
nuclei (Scheme 3) occurs in all cases without any appreciable variation of the stereochemical requirements of the chiral centers present
in the main chain as well as in the lateral ones[51].

Scheme 3. Polymeric onium salts by chemical transformation of pre-
          formed polymers.

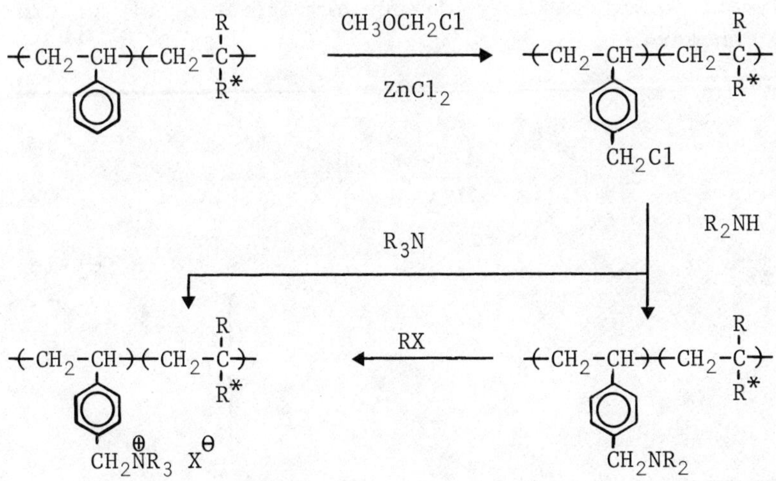

R = H   , R* = -COOMent*
R = CH$_3$, R* = -COOMent*
R = H   , R* = -CH$_2$CH(CH$_3$)C$_2$H$_5$
R = H   , R* = -CH(CH$_3$)CH$_2$CH$_2$CH$_2$CH(CH$_3$)$_2$

It is worth mentioning that depending on the primary structure
of the chiral comonomer, chemical composition, distribution of monomeric units and degree of substitution of aromatic nuclei a large
series of polymeric onium salts with a different amphiphilic character can be prepared starting from suitable comonomer mixtures.
     Alternatively chiral copolymers based on acrylic and metha-
crylic monomers containing onium groups were prepared by radical

copolymerization of (-)-menthyl acrylate (MnA) and (-)-menthyl methacrylate (MnMA) with 2-(N,N,N-trimethylammonium)ethyl acrylate
bromide (TMAB) and 2-(N,N,N-triethylammonium)ethyl methacrylate
bromide (TEMAB) or by quaternization of the corresponding polymeric
tertiary amines (Scheme 4). In all cases optically active polymeric
products containing 10 - 20 % mol of functional groups can be obtained
and apparently no substantial difference is evidenced between copolymers of comparable composition prepared by the two different
routes[52].

Scheme 4. Polymeric onium salts by direct polymerization of functional monomers.

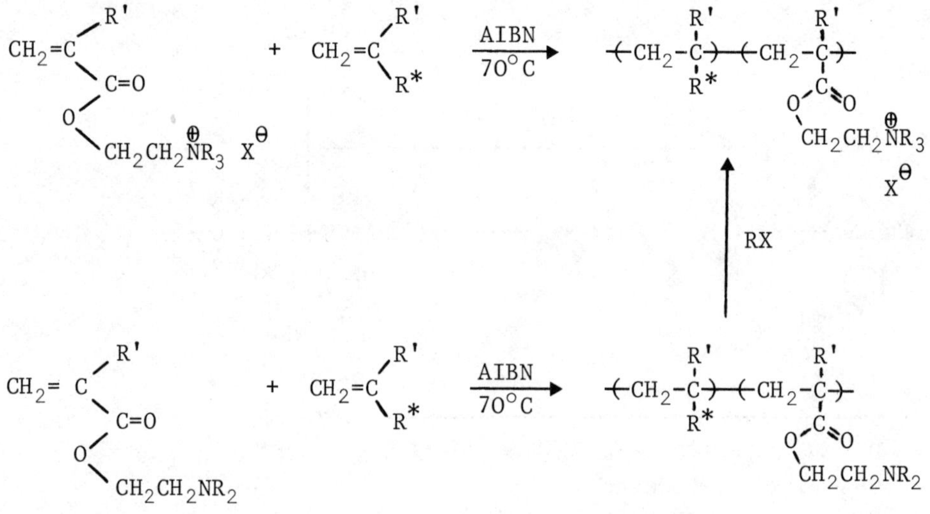

R' = H ,   R = CH_3 ,   X = Br
R' = CH_3,   R = C_2H_5 ,   X = Br
R' = CH_3,   R = C_2H_5 ,   X = I
R* = -COOMenthyl

Starting from polymeric matrices prepared by suspension polymerization in the presence of divinylbenzene (DVB) or ethylenglycol
dimethacrylate as crosslinking agent and by following the above reported procedures an analogous series of crosslinked chiral polymers
containing onium groups were obtained[52]

The activity of the polymeric salts as catalysts in phase transfer reactions was tested in alkylation and dichlorocarbenation reactions carried out on chiral or prochiral substrates under poly-phase conditions in the presence of concentrated aqueous sodium hydroxide. They were tested also in oxidation reaction of racemic alcohols under acidic conditions[52].

Both linear and crosslinked catalysts show a comparable reactivity, about one order of magnitude lower than that observed for the corresponding low molecular weight model compounds[53-55]. The very low asymmetric induction observed indicates that under the adopted conditions the asymmetric perturbation around the onium salt is not strong enough to appreciably affect the stereochemical reaction pathway. As preliminarly indicated by C.D. studies[56] an appreciable chirality on the reactive anion is observable only in tight ion-pairs. It seems therefore very likely that under the usual phase transfer conditions the possible asymmetric induction arising from the tight ion-pairs is hidden by the overwhelming reactivity of the non-chirally perturbed free-anions.

During recycle experiments the catalytic activity is progressively reduced probably due to Hoffman-type degradation reactions occurring at the level of the onium group, thus likely preventing any large scale application of the polymeric ammonium salts as catalysts under conventional phase transfer reaction conditions.

No loss of the stereochemical requirements of the polymers takes place, indicating the preservation of chiral environment of the catalytic sites regenerated either in situ or after the recover of the spent polymeric systems.

To overcome the drawbacks of the regeneration of onium sites the investigation has addressed the synthesis of poly-glyme and crown ether containing polymers (**24-26**) both under conditions leading to linear and crosslinked structures[57].

24 $\quad -(CH_2-CH)_{76}(CH_2-CH)_{24}-$
$\qquad\qquad\qquad$ OMent $\qquad\qquad$ O$(CH_2CH_2O)_4CH_3$

25 $\quad -(CH_2-\underset{\underset{COOMent}{|}}{\overset{\overset{CH_3}{|}}{C}})_{67}(CH_2-\underset{\underset{CO(OCH_2CH_2)_4OCH_3}{|}}{\overset{\overset{CH_3}{|}}{C}})_{33}-$

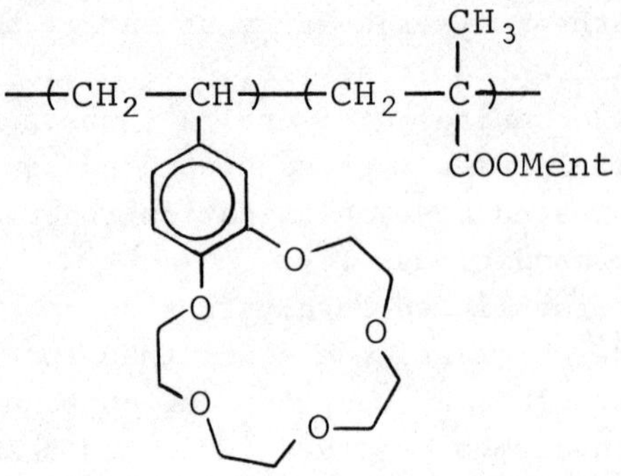

26

They have been used in the $NaBH_4$ reduction of several alkyl aryl ketones under phase transfer conditions and very poor asymmetric inductions were achieved in all cases (Table 3).

Table 3

$NaBH_4$ reduction of alkyl phenyl ketones in the presence of chiral copolymers[a].

Catalyst	C_6H_5COR R =	Conv. after 3.5hr (%)	$|\alpha|_D^{25}$
	CH_3	78.4	- 0.03
25	C_2H_5	36.1	- 0.05
	i-C_3H_7	8.8	- 0.07
	CH_3	83.9	- 0.23
	C_2H_5	55.1	- 0.05
24	i-C_3H_7	30.4	- 0.07
	t-C_4H_9	50.1	- 0.04

a) 10 mmol of ketone in 10 ml of benzene, 6 mmol of $NaBH_4$ in 10 ml of water, 0.5 mmol of catalyst under magnetic stirring at 25°C.

By considering that when crosslinked polymers are used the reaction pathway most likely implies the simultaneous interaction of substrate and reagent within the core of the swollen polymer particles (Scheme 5), it is expected that systems able to discriminate between two enantiofaces or two chiral centers on the basis of a built-in preferential pathway within chiral cavities with a permanent physico-mechanical stability, should be the most valuable ones[58,59].

Scheme 5. Mechanism of nucleophilic substitution reaction carried out under phase transfer conditions.

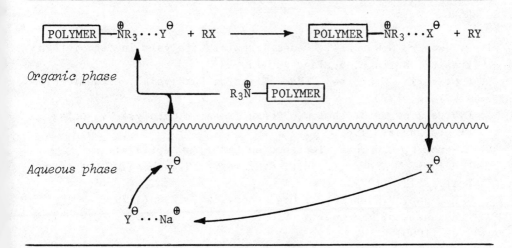

In this contest we have been investigating the realization of macroporous systems with a suitable degree of crosslinking and prepared in the presence of low and high molecular weight chiral porogen molecules that can be removed after the polymerization without collapsing the chiral architecture of the built-in-cavities[60].

CONCLUDING REMARKS

The rather exciting and stimulating "phase transfer" synthetic procedure based on the use of polymer supported active functional moieties, such as onium groups, crown ethers, cryptands and polyglymes, becomes even more naive when the active sites are bound to chiral matrices, whose prevalent chirality can be either intrinsically built in or determined by anchoring low molecular weight optically active molecules.

The results reached until now in the transformation of chiral or prochiral substrates under the usual phase transfer conditions are rather frustrating as far as the optical yield is concerned.

In our opinion the more encouraging results obtained in Michael type reactions are to be more accurately considered as they indicate that multifunctional systems, chemically stable under reaction conditions and able to selectively bind the substrate and the reacting anion, are nowadays the most promising and reliable route to undertake.

REFERENCES

1. W.P.Weber, G.W.Gokel, "Phase Transfer Catalysis in Organic Synthesis", Springer-Verlag, Berlin, 1977.
2. C.M.Starks, C.Liotta, "Phase Transfer Catalysis", Academic Press, New York, 1978.
3. E.V.Dehmlow, S.S.Dehmlow, "Phase Transfer Catalysis", Verlag Chemie, Basel, 1980.
4. H.Wynberg, R.Helder, Tetrahedron Lett., 4057 (1975).
5. R.Helder, R.Arends, W.Bolt, H.Hiemstra, H.Wynberg, Tetrahedron Lett., 2181 (1977).
6. K.Hermann, H.Wynberg, J. Org. Chem., 44, 2238 (1979).
7. P.Annunziata, M.Cinquini, S.Colonna, J. Chem. Soc., Perkin I, 2422 (1980).
8. S.I.Bhole, V.N.Gogte, Indian J. Chem., 20 B, 218 (1981).
9. N.Kobayashi, K.Iwai, J. Org. Chem., 46, 1823 (1981).
10. H.Pracejus, F.W.Wilcke, K.Hanemann, J. Prakt. Chem., 319, 219 (1977); Chem. Abstr. 87, 136342 (1977).
11. S.Colonna, H.Hiemstra, H.Wynberg, J. Chem. Soc., Chem. Comm., 238 (1978).
12. S.Colonna, A.Re, H.Wynberg, J. Chem. Soc., Perkin I, 547 (1981).
13. D.J.Cram, G.D.Y.Sogah, J. Chem. Soc., Chem. Comm., 625 (1981).
14. T.Yamashita, H.Yasueda, N.Nakamura, Chem. Lett., 585 (1974); Chem. Abstr., 81, 49390 (1974).
15. T.Yamashita, H.Yasueda, Y.Miyauchi, N.Nakamura, Bull. Chem. Soc. Japan, 50, 1532 (1977).
16. K.Hermann, H.Wynberg, Helv. Chim. Acta, 60, 2208 (1977).
17. N.Kobayashi, K.Iwai, J. Am. Chem. Soc., 100, 7071 (1978).
18. T.Yamashita, H.Yasueda, N.Nakamura, Bull. Chem. Soc. Japan, 52, 2165 (1979).

19. N.Kobayashi, K.Iwai, Tetrahedron Lett., 2167 (1980).

20. N.Kobayashi, K.Iwai, Polymer J., 13, 263 (1981).

21. N.Kobayashi, K.Iwai, J. Polymer Sci., Polym. Lett. Ed., 20, 85, (1982).

22. S.Inoue, S.Ohashi, A.Tabata, T.Tsuruta, Makromol. Chem., 112, 66 (1968).

23. S.Ohashi, S.Inoue, Makromol. Chem., 150, 105 (1971).

24. S.Ohashi, S.Inoue, Makromol. Chem., 160, 69 (1972).

25. S.Ohashi, S.Inoue, Y.Unno, Polymer J., 3, 611 (1972).

26. K.Uenayagi, S.Inoue, Makromol. Chem., 177, 2807 (1976).

27. S.Inoue, Adv. Polym. Sci., 21, 77 (1976).

28. K.Uenayagi, S.Inoue, Makromol. Chem., 178, 375 (1977).

29. K.Uenayagi, S.Inoue, Makromol. Chem., 179, 887 (1978).

30. S.Inoue, Y.Kawano, Makromol. Chem., 180, 1405 (1979).

31. T.Yamashita, H.Yasueda, N.Nakatami, N.Nakamura, Bull. Chem. Soc. Japan, 51, 1183 (1978).

32. T.Yamashita, H.M.Tsui, H.Watanabe, N.Nakamura, Polymer J., 13, 179 (1981).

33. S.Banfi, M.Cinquini, S.Colonna, Bull. Chem. Soc. Japan, 54, 1841 (1981).

34. A.P.Terentev, E.I.Klabunovskii, Sbornik Statei Obshchei Khim., 2, 612 (1953); Chem. Abstr., 49, 5263b (1955).

35. H.Normant, T.Cuvigny, P.Savignac, Synthesis, 805 (1975).

36. H.E.Hennis, J.P.Rasterly, L.R.Collins, L.R.Thompson, Ind. Ing. Chem. Prod. Res. Dev., 6, 193 (1967).

37. H.E.Hennis, L.R.Thompson, J.P.Long, Ind. Ing. Chem. Prod. Res. Dev., 7, 96 (1968).

38. M.Makosza, A.Kacprowicz, M.Fedorinski, Tetrahedron Lett., 219 (1975).

39. W.P.Weber, G.W.Gokel, "Phase Transfer Catalysis in Organic Synthesis", Springer-Verlag, New York, 1977, chap. 6.

40. Y.Kawakami, T.Sugiura, Y.Mizutani, Y.Yamashita, J. Polymer Sci., Polym. Chem. Ed., 18, 3009 (1980).

41. Y.Kawakami, Y.Yamashita, ACS Polymer Prep., 23, 181 (1982).

42. T.Yamashita, H.Yasueda, N.Nakamura, Bull. Chem. Soc. Japan, 51, 1247 (1978).

43. N.Kobayashi, K.Iwai, Makromol. Chem. Rapid Commun., 2, 105 (1981).

44. S.Colonna, R.Fornasier, U.Pfeiffer, J. Chem. Soc., Perkin I, 8 (1978).

45. B.Marshaw, H.Wynberg, J. Org. Chem., 44, 2312 (1979).

46. D.C.Sherrington, J.Kelly, ACS Polymer Prep., 23, 177 (1982).

47. D.C.Sherrington, private communication.

48. E.V.Dehmlow, P.Sing, J.Heider, J. Chem. Res., 5, 292 (1981).

49. F.Ciardelli, P.Salvadori, C.Carlini, E.Chiellini, J. Am. Chem. Soc., 94, 6536 (1972).

50. R.N.Majumdar, C.Carlini, R.Nocci, F.Ciardelli, R.C.Schulz, Makromol. Chem., 177, 3619 (1976).

51. E.Chiellini, R.Solaro, S.D'Antone, Makromol. Chem., Suppl. 5, 82 (1981).

52. E.Chiellini, S.D'Antone, R.Solaro, ACS Polymer Prep., 23, 179 (1982).

53. E.Chiellini, R.Solaro, J. Chem. Soc., Chem. Comm., 231 (1977).

54. E.Chiellini, R.Solaro, S.D'Antone, Makromol. Chem., 178, 3165 (1977).

55. E.Chiellini, R.Solaro, Chim. Ind. (Milan), 60, 1006 (1978).

56. D.C.Sherrington, E.Chiellini, R.Solaro, J. Chem. Soc., Chem. Comm., 1103 (1982).

57. R.Solaro, S.D'Antone, E.Chiellini, Proc. Int. Symp. on Polymer Supported Reagents in Organic Chemistry, Lyon, 1982, p. 162.

58. G.Wulff, A.Sarhan, K.Zabrocki, Tetrahedron Lett., 4329 (1973).

59. G.Wulff, A.Sarhan, "Chemical Approaches to Understanding Enzyme Catalysis: Biomimetic Chemistry and Transition-State Analogs", B.S.Green, Y.Ashani, D.Chipman Eds., Elsevier, Amsterdam, 1982, p. 106.

60. E.Chiellini, S.D'Antone, M.Penco, R.Solaro, to be published.

POLYMER-SUPPORTED OPTICALLY ACTIVE PHASE TRANSFER CATALYSTS

D.C. Sherrington and J. Kelly

Department of Pure and Applied Chemistry
University of Strathclyde
Glasgow, Scotland, U.K.

INTRODUCTION

Over the last few years there has been increasing pressure on organic chemists to devise reactions which are highly selective in order to improve efficiency and reduce costs. In this context enantioselective reactions are particularly important because of the growing demand for optically pure compounds in the pharmaceutical and agro-chemical industries. A number of groups have been examining the possibility of achieving either asymmetric induction or kinetic racemate resolution in phase transfer catalysed and related reactions[1-13]. One or two analogous resin-supported systems have also been reported[14-21]. In principle the latter have the advantage, in the long term, of efficiently retaining rare or expensive catalysts for re-use or re-cycling, or indeed for allowing reactions to be conducted continuously in columns.

NON-SUPPORTED SYSTEMS

The present situation with regard to non-supported catalysts has been admirably reviewed recently[1], with Wynberg's excellent article[13] being particularly informative. To date virtually all the work has employed quarternary ammonium salts based either on ephedrine [I] and its relatives or the cinchona alkaloids, notably quinine [II], as the active chiral catalyst. Some confusion still remains in the literature concerning the effectiveness of these catalysts[1]. However, the results of a number of Michael additions[4,6,10] seem to be genuine and have been reproduced in more than one laboratory. For example, a chemical yield of ∼100% in the addition of the cyclohexanone derivative, [III], to methyl

vinyl ketone [reaction 1] catalysed by N-methylquinium hydroxide
is accompanied by optical yields of up to ∿ 25%[10].

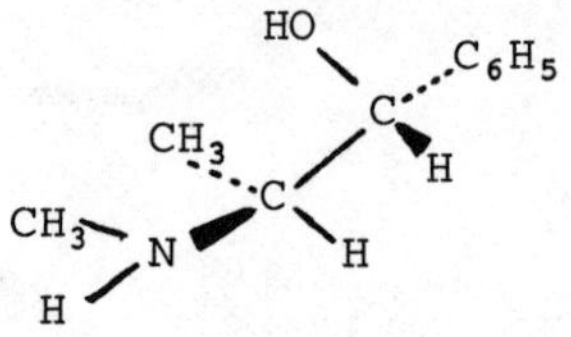

[I]

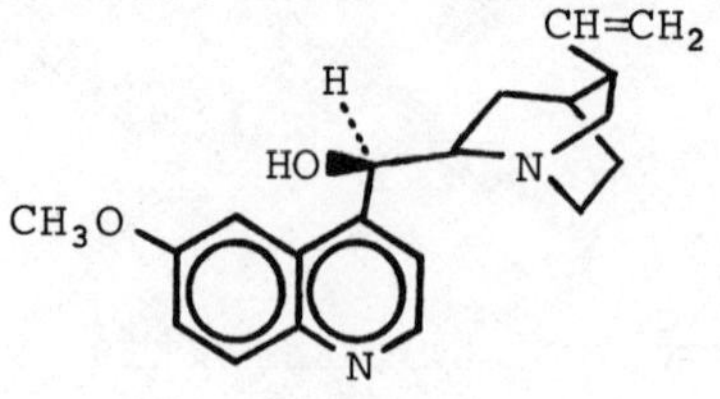

[II]

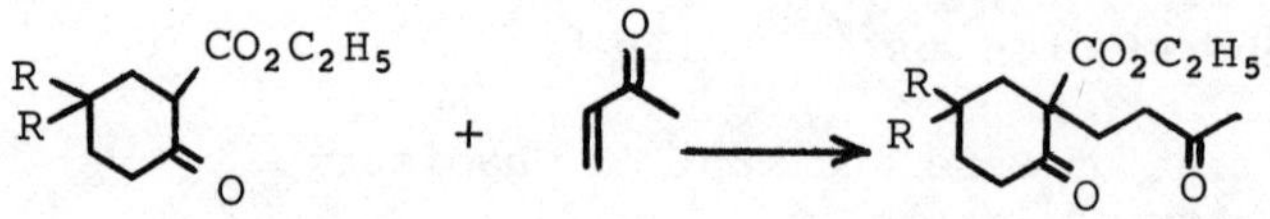

[III]

Similarly Michael additions of thiols and nitroalkanes to α,β-
unsaturated ketones [reactions 2 and 3] catalysed by anhydrous
potassium fluoride in the presence of the above ammonium salts[3]
produces optical yields of up to ∿ 35%.   Another unambiguous

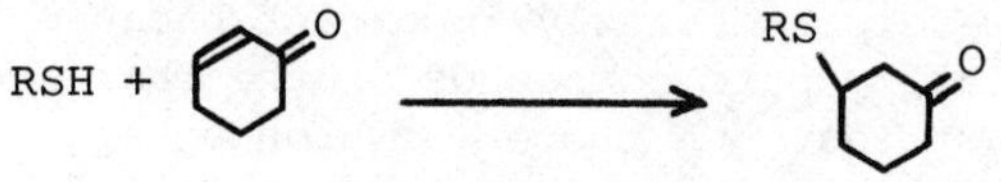

2.

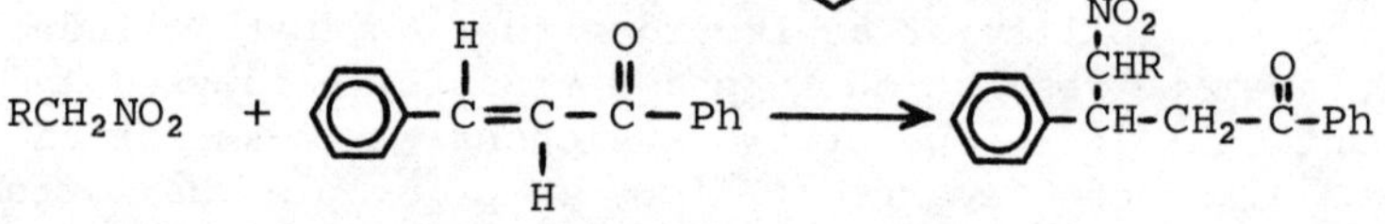

3.

case appears to be phase transfer catalysed epoxidations of
chalcone [IV] using alkaline hydrogen peroxide and benzylated
cinchona alkaloids as the catalyst[22,23] [reaction 4].

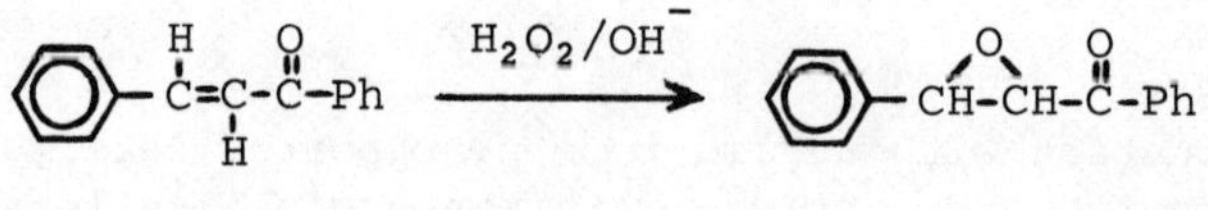

4.

[IV]

     A number of successful kinetic racemate resolutions have been
claimed in displacement reactions on (±) ethyl-2-bromopropionate[7,
9,11] [reaction 5].   Again the catalysts used were based on amines
[I] and [II].   These results and those of asymmetric sodium borohydride reductions of ketones under phase transfer catalysed
conditions[8,11] probably require further investigation before a
clear picture emerges.

$$(\underline{+})CH_3CHBrCO_2Et \; + \; X^- \longrightarrow CH_3CHXCO_2Et \; + \; Br^-$$

5.

$X^- \; = \; PhO^-, \; PhCO_2^-, \; PhS^-,$

Although there now seems to be good evidence that an hydroxy substituent β to the quarternary ammonium centre of the chiral catalyst is generally required for significant enantioselectivity to result, it is unfortunate that these structures are also responsible for some of the confusion which exists in the literature, since under basic conditions an optically active epoxide can be formed readily from quarternary derivatives of [I] and [II]. Hence reaction mixtures must be carefully purified before any observed optical rotations can be clearly attributed to an induction reaction.

Another important generalisation which seems to be emerging concerns the nature of the organic solvent employed. More often than not the highest inductions are observed with non-polar solvents. Media capable of hydrogen bonding or solvating chiral ammonium ion pairs generally inhibit induction and this would be perfectly in keeping with a weakening of the ion pair interaction.

The recent results reported by Cram[2] are extremely important because his catalysts are chiral crown ethers, and represent one of the first attempts to develop novel structures without the inherent disadvantages of the ammonium salts described above. Such crowns yield products from Michael additions of β-ketoesters to methyl vinyl ketone with optical purities up to ∿99%. A series of novel chiral onium salts derived from methionine have also been reported by Colonna and his coworkers[19]. Although so far these have displayed no asymmetric induction in their reactions, they are important as an imaginative new series of catalysts.

POLYMER-SUPPORTED SYSTEMS

With the above background of selective catalysis a number of attempts have been made to develop analogous polymer-supported systems. Chiellini and his coworkers[16] have used linear poly(4-vinylpyridine) partially quaternised with a chiral alkylbromide as a phase transfer catalyst in the ethylation of phenylacetonitrile. Unfortunately, the high chemical yield achieved was accompanied by a negligible enantiometic excess. However, the same group have prepared chiral catalysts by reaction of linear poly(4-chloromethyl-

styrene) with optically active amines[17].   In this instance application as a catalyst in the dichlorocarbenation of benzaldehyde to
yield mandelic acid [V][reaction 6] produced optical yields of up
to 10%.   In their most recent report[20] these workers have attached

$$CHCl_3/PhCHO \quad + \quad NaOH(aq) \longrightarrow PhCH(OH)CO_2H \qquad\qquad 6.$$

$$[V]$$

achiral crown ethers to a chiral polymer support in a novel approach
to this problem, and, although no induction has yet been seen,
exciting developments may come from this avenue of investigation.

Also of considerable encouragement were the results reported
in the Darzens reaction some time ago[15].   Carbonyl compounds when
reacted with chloromethyl-p-tolyl sulphone or with chlorophenyl-
acetonitrile in the presence of aqueous sodium hydroxide and chiral
onium salts derived from [I] [reaction 7 and 8] yielded product
epoxides with up to 2.5% enantiomeric excesses.   More importantly,

$$R^1R^2C{=}O \quad + \quad CH_3\!-\!\langle\bigcirc\rangle\!-\!SO_2CH_2Cl \longrightarrow R^1R^2\overset{O}{\overset{\|}{C}}\!-\!CHSO_2\langle\bigcirc\rangle CH_3 \qquad 7.$$

$$RCHO \quad + \quad PhCH(CN)Cl \longrightarrow R{-}\overset{O}{\overset{\triangle}{CH{-}C}}(CN)Ph \qquad\qquad 8.$$

however, when the same catalyst was supported on a 1% crosslinked
styrene-divinylbenzene copolymer resin, optical yields rose to
∿ 20-25%.   So far the novel onium salts derived from methionine,
though displaying higher catalytic efficiency when supported on a
resin, continue to show no induction properties[19].

In contrast a number of chalcone epoxidations catalysed by a
polystyrene supported quininium salt do parallel the analogous non-
supported reactions[14].   So far the enantiomeric excesses achieved
are small but nevertheless the experiments confirm that the solution
reactions can be translated successfully into the supported system.
One possible explanation for the poorer performance of the resin
material in this instance may be the method of attachment to the
chloromethylated polystyrene resin.   The authors report the
formation of the monobenzyl salt of the alicyclic nitrogen as the
effective linkage.   From our own experiments using linear poly-
(chloromethylstyrene), and corresponding copolymers of styrene and
chloromethylstyrene, we find it very difficult to avoid some reaction
with both nitrogen centes in quinine[36].   In our case this is clearly
manifest in the formation of crosslinked insoluble polymers.
Starting with a crosslinked resin such effects would not be easily
detected, and under the conditions reported for the attachment of
quinine[14] some bridging dibenzylated salt structures would almost
certainly result.   These would, of course, be conformationally

less mobile, and indeed approach to them may be stereochemically highly impaired.

The enantioselective ester syntheses from acid salts, chlorides and anhydrides with racemic alkyl halides, catalysed by optically active polyamines[18] almost certainly proceed via in situ formation of chiral ammonium salts, and therefore fall within the scope of phase transfer catalysts. Though the optical yields obtained are low, the work is important because it explores the use of polyamine species with a potential chirality derived from the polymerization of optically active oxazolines, and as such is again a novel approach.

At the outset of our own work there seemed to be considerable scope for synthesising alternative polymer-supported catalysts in order to explore the effectiveness of a wider range of chiral structures and also to provide a wider context in which the earlier results might be viewed. Our approach, however, has remained essentially a trial and error one, although more recently we have embarked upon a more rational, though longer term, examination of the problem using induced circular dichroism observed in achiral anions bound to chiral cations[24].

CATALYST SYNTHESES

We have prepared four groups of resin-bound catalysts. Group 1 employed Bio-Beads SX2, chloromethylated in our own laboratory using standard procedures[25], while groups 2-4 used a ∿ 2% crosslinked resin prepared inhouse using chloromethyl styrene as the main comonomer component. The suspension polymerization technique has been reported previously[26] and resin modifications were carried out using normal synthetic methods, although a simple rotating device[27] was employed to achieve efficient but gentle agitation of reaction mixtures.

Group 1

The first group of catalysts (VI a-e) represent an extension of some achiral systems which we have previously developed and characterized for use, in particular in solid/liquid phase separated reactions[28]. The catalytic entity is a linear oligo-ethylene oxide chain (an 'open' crown ether) and the chiral component is the optically active endgroup or headgroup. On complexing the alkali metal cation of a reaction ion pair it was hoped to bring the chiral structure to the site of reaction. The synthetic route employed was the same as previously stated[28]. It involved the displacement of the tosylate leaving group from species [VII] using L-prolinol (L-2-pyrrolidinemethanol), the tri-ethylammonium salts of L-proline and L-pyroglutamic acid (L-2-pyrrolidone-5-carboxylic acid) and the sodium salts of L-menthol and

L-borneol to produce catalysts [VI a-e] respectively (Group 1).

[VII]   [VI]

$$P—⬡—CH_2(OCH_2CH_2)_4OTs \rightarrow P—⬡—CH_2(OCH_2CH_2)_4R$$

R = [VIa]   [VIb]   [VIc]

[VId]   [VIe]

Group 1

In each case the starting resin was ∿40% chloromethylated and as before[28] tetraethylene glycol could be attached by a single end to only ∿28% of resin aromatic groups.   The remaining chloromethyl groups were consumed presumably in the formation of doubly attached or bridging oligoether residues.   The final loading of the species with the specific structures, [VIa-e] was ∿14% of aromatic groups. (All loadings were estimated from micro-analyses, and infra-red spectra confirmed qualitatively each modification).

Group 2

The second group of catalysts [VIIIa-f] consists of resins carrying a protected galactose residue and a number of different achiral phase transfer catalysts.   In these the stereochemical relationship of the optically active species and the catalyst itself is not well defined and the aim has been simply to load macromolecular chains as heavily as possible with both species. 1,2:3,4-Di-O-isopropylidene-α-D-galactopyranose was prepared by reaction of acetone with galactose in the presence of a mixture of phosphoric acid, phosphorus pentoxide and zinc chloride[29].   The protected sugar was attached to the resin first via in situ gener- ation of the alkoxide of the residual primary hydroxyl group using sodium hydride in toluene[30].   To aid penetration of the resin a phase transfer catalyst [Bu₃PCl⁻ or triglyme] was employed. Residual chloromethyl groups were then reacted with tributylphosphine, tri- butylamine and trioctylamine in refluxing DMF to yield the catalysts VIIIa, b and c respectively, while the sodium salts of the correspond- ing monoderivatised tetraethylene glycols[31] were generated using sodium hydride in refluxing toluene in order to attach the species

VIIId,e and f. The characteristic infra-red absorption doublet of the protected sugar at 1385 and 1395 cm^{-1} remained unchanged during catalyst attachment confirming the stability of the bound sugar residues under these conditions. Orders of magnitude estimates of the loadings of each functional group were calculated from the sequential elemental microanalysis and the results are shown in Table 1.

<u>Group 3</u>

The third group of catalysts [IXa-f] was prepared in a similar way to those in group 2 using 1,2:4,5-di-O-isopropylidene-β-D-fructopyranose as the sugar residue. Again the protected sugar was prepared according to a literature method using acetone and concentrated sulphuric acid[32]. The functional group content in these species is also summarised in Table 1.

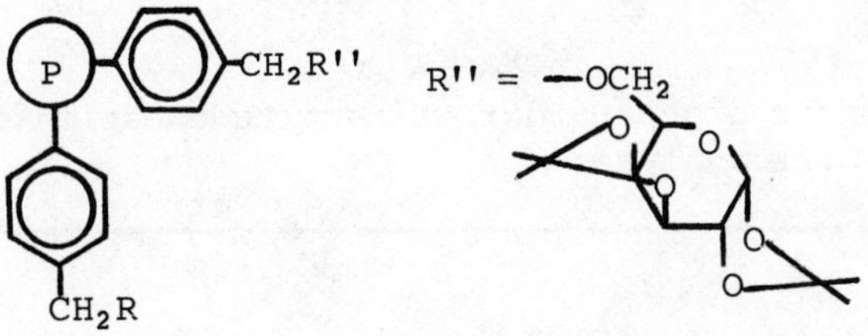

[VIII][Group 2]

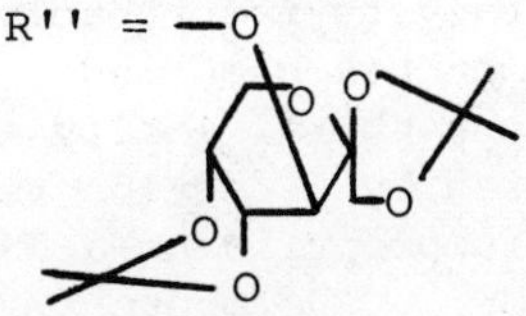

[IX][Group 3]

$R = -\overset{+}{P}Bu_3Cl^-$, $-\overset{+}{N}Bu_3Cl^-$, $-\overset{+}{N}Oct_3Cl^-$

a b c

$R = -(OCH_2CH_2)_4OR'$, where $R' = -CH_3$, $-CH_2-$ 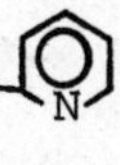 , 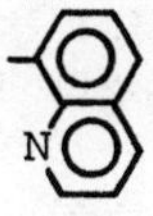

d e f

Groups 2 and 3

Table 1.   Functional Group Content[x] of Catalysts VIII and IX

Resin Catalyst	Sugar Residues	Phosphonium Salt	Ammonium Salt	Oligoether
a	85	10	–	–
a'	40	45	–	–
b	40	–	30	–
VIII c	40	–	40	–
d	85	–	–	15
e	40	–	–	40
f	40	–	–	25
a	70	25	–	–
b	70	–	20	–
c	70	–	low	–
d	70	–	–	30
e	20	–	–	20
f	80	–	–	low

[x] % of resin aromatic groups–order of magnitude estimation from elemental microanalyses

Group 4

This group of catalysts is probably the most sophisticated of all in that some attempt has been made to define the stereochemical relationship between the chiral entity and the onium ion catalyst. Species [X] was prepared by direction reaction of 6-diphenylphosphino-6-deoxo-1,2-3,4-di-O-isopropylidene-α-D-galactopyranose with chloromethylated resin in DMF.   The phosphinated galactose itself was prepared as described in the literature[33].   The resulting resin was ∿40% loaded.   Species [XI] was prepared in a similar manner from commercially available (S,S)-(+)-2,3-dimethoxy-1,4-bis(dimethylamino) butane.   Changes in elemental microanalyses indicated substitution of ∿50% of polymer aromatic groups and the N/Cl ratio suggested the bridging structure, [XI].   Finally, species [XII] was synthesised in an analogous fashion from 2,3-O-isopropylidene-2,3-dihydroxyl-1,4-bis(diphenylphosphino)butane (DIOP), itself prepared by a literature method[34] from D-tartaric acid.   The resin loading was ∿30% and again analysis suggested a predominance of the bridging structure [XII].

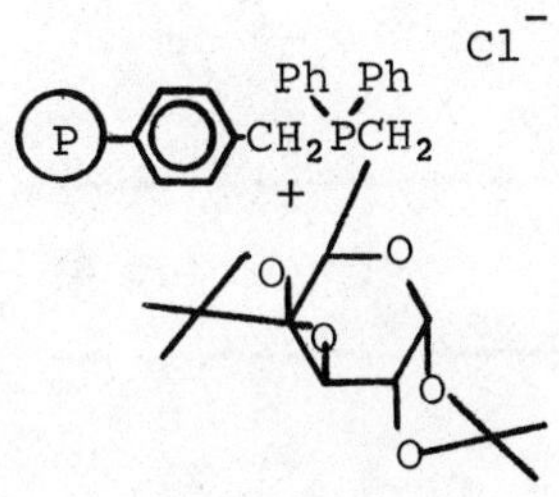

[X]

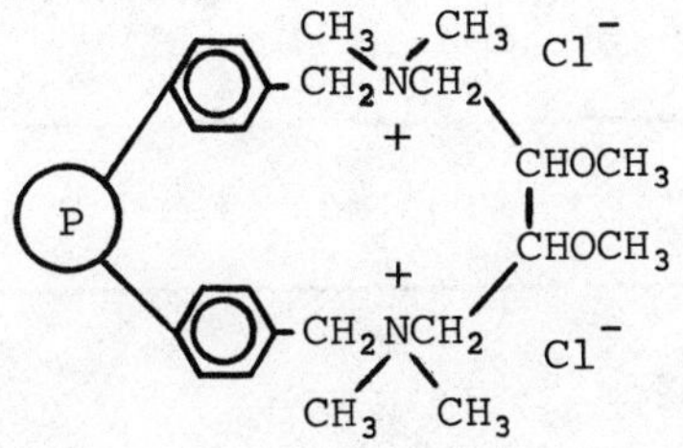

[XI]

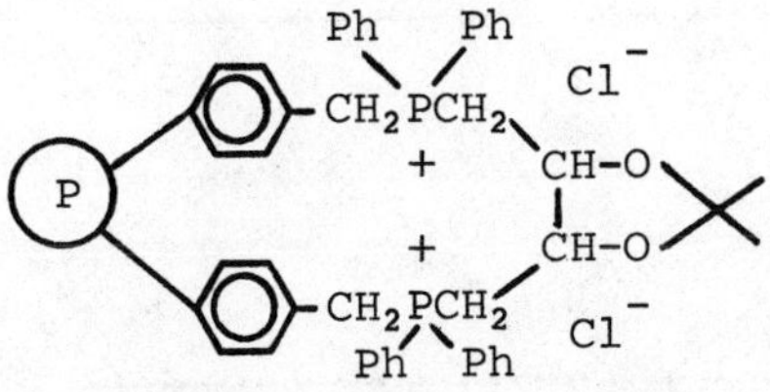

[XII]

Group 4

CATALYSES

A number of reactions have been examined using these phase transfer catalytic species. These include displacement reactions of solid potassium phenoxide, solid sodium azide and aqueous sodium azide on racemic ethyl-2-bromopropionate in toluene (similar to reaction 5), sodium borohydride (solid) reductions of acetophenone and octan-2-one, chalcone epoxidations using aqueous H_2O_2/OH^- (reaction 4) and Michael additions of nitromethane to chalcone in the presence of anhydrous potassium fluoride (reaction 3). Details of these results will be published in due course and at the moment it is possible to record only a few representative examples of the displacements and reductions.

Table 2 shows that many of the catalysts are chemically effective in both azide and phenoxide displacements and although in general their activity is lower than conventional non-supported onium salts and crown ethers, nevertheless their performance is respectable. In this particular instance an attempt was made to keep conversions below 50% since a kinetic racemate resolution was being sought. In the event, no significant optical activity was found in the products, indicating the absence of enantioselection. However, some consolation was taken in the fact that the resin catalysts remained intact, and in particular there was no evidence for the cleavage and leaching of the various optically active species from polymer matrix.

Table 2. Nucleophilic Displacements on (±)Ethyl-2-bromo-
         propionate (Note 1)).

Catalyst	Nucleophile	Time(hr)	Conversion(%)
None		24	8
XII	Solid K^+OPh^-	24	28
VIIId	(Note 2))	24	30
IXf		24	20
None		6	3
IXf		3	58
VIIIa	Solid $Na^+N_3^-$	7	36
IXe	(Note 3))	7	14
IXb		5	50
None		8	6
IXf	Aqueous $Na^+N_3^-$	6	60
IXa	(Note 4))	7	25
XII		7	36
VIIIa		7	17

1) 0.5 mmol RBr; 2.0mmol nucleophile; 0.075 mmol catalyst
2) 30°C; 2ml toluene.
3) 65°C; 2ml tetrahydrofuran
4) 30°C; 2ml toluene; 2ml H_2O

A large number of sodium borohydride reductions have been
performed and some of the results appear in Table 3.   After careful
evaluation, the solvent of choice was 1,2 dichloroethane to which
was added a controlled amount of water.   The latter is required to
complete the stoichiometry of the reaction and reactions performed
in this way were very reproducible.   Solvents such as tetrahydrofuran appear to contain sufficient adventitious water for this
addition not to be necessary, but as a result we found such reactions
difficult to reproduce and control.   Some reports in the literature
make no mention of this, nor report control reactions without
catalysts, and we feel these results require careful scrutiny and
interpretation.

Again many of our catalysts show high chemical activity with
virtually quantitative yields being achievable with some species.
Once again, however, in spite of careful analysis no significant

Table 3. Sodium Borohydride Reductions of Ketones in
1,2-dichloroethane (Note 1)

Ketone	Catalyst	Time(hr)	Conversion(%)
Acetophenone (Note 2))	None	6	～ 0
	VIa-e	24	85-98
	IXf	7	81
	XII	7	45
	VIIId	24	60
	VIIIa	7	99
Octan-2-One (Note 2))	None	7	～0
	VId	7	33
	IXf	7	87
Octan-2-One (Note 3))	None	24	～ 3
	IXf	24	62
	IXd	24	41
	IXe	24	21
	VIIIa	24	47

1) solvent/H_2O, 20/1, solvent 2 ml; 0.5mmol ketone;
0.075 mmol catalyst; 4.0 mmole $NaBH_4$
2) 75°C.   3) 30°C

optical activity was found in the products.   At the moment,
therefore, it seems that our results confirm the reservations
expressed in the literature[1] about certain reactions.   In the case
of sodium borohydride reductions employing a reagent modified by a
stoichiometric quantity of optically active co-reagent, high
inductions do appear to be possible[3,5] and at least some of our
resin species might be amenable to exploitation in this manner.

REFERENCES

1.   E.V. Dehmlow, P. Singh and J. Heider, J. Chem. Res. (S), 292
       (1981).
2.   D.F..Cram and G.D.Y. Sogah, Chem. Comm., 625 (1981).
3.   S. Colonna, A. Re and H. Wynberg, J.Chem. Soc., Perkin I, 54
       (1981).
4.   S. Julia, A. Ginebreda, J. Guixer, J. Masana, A. Tomas and
       S. Colonna, J. Chem. Soc., Perkin I, 574 (1981).
5.   H. Wynberg and B. Marsman, J. Org. Chem., 45, 158 (1980).

6.   R. Annunziata, M. Cinqunini and S. Colonna, J. Chem. Soc.,
      Perkin I, 2422 (1980).
7.   S. Julia, A. Ginebreda, J. Guixer and A. Tomas, Tet. Lett.,
      3709 (1980).
8.   S. Colonna and R. Fornasier, J. Chem. Soc., Perkin I, 372 (1978).
9.   S. Julia, A. Ginebreda and J. Guixer, Chem. Communications,
      742 (1978).
10.   K. Hermann and H. Wynberg, J. Org. Chem., 44, 2238 (1979).
11.   S. Julia, A. Ginebreda, J. Guixer, J. Masana, A. Tomas and
      S. Colonna, J. Chem. Soc. Perkin I, 574 (1981).
12.   R. Annunziata, M. Cinquini, S. Colonna and F. Cozzi, J. Chem.
      Soc. Perkin I, 3118 (1981).
13.   H. Wynberg, Recl. Trav. Chim. Pays-Pas, 100, 393 (1981).
14.   N. Kobayashi and K. Iwai, Makromol. Chem. Rapid Comm., 2, 105
      (1981).
15.   S. Colonna, R. Fornasier and U. Pfeiffer, J. Chem. Soc. Perkin
      I, 8 (1978).
16.   E. Chiellini, R. Solaro and S. D'Antone, Makromol. Chem., 178,
      3165 (1977),
17.   E. Chiellini and R. Solaro, Chem. Comm., 231 (1977).
18.   Y. Kawakami, T. Sogiura, Y. Mizutani and Y. Yamashita, J. Pol.
      Sci. Pol. Chem. Ed., 18, 3009 (1980).
19.   S. Banfi, M. Cinquini and S. Colonna, Bull. Chem. Soc. Jap.,
      54, 1841 (1981).
20.   E. Chiellini, S. D'Antone and R. Solaro, A.C.S. Pol. Preprints,
      23, 179 (1982).
21.   D.C. Sherrington and J. Kelly, A.C.S. Pol. Preprints, 23, 177
      (1982).
22.   R. Helder, J.C. Hummelen, R.W.P.M. Laane, J.S. Wiering and H.
      Wynberg, Tetrahedron Lett., 1831 (1976).
23.   H. Wynberg, Chimia, 30, 445 (1976).
24.   D.C. Sherrington, E. Chiellini and R. Solaro, J. Chem. Soc.
      Chem. Comm., submitted for publication.
25.   J.A. Greig and D.C. Sherrington, Eur. Polymer J., 15, 867 (1979).
26.   See Appendix in 'Polymer-supported Reactions in Organic
      Synthesis', P. Hodge and D.C. Sherrington, Eds. J. Wiley and
      Sons, London, 1980.
27.   J.A. Greig, W.M. MacKenzie and D.C. Sherrington, Polymer, 18,
      1291 (1977).
28.   J.G. Heffernan, W.M. MacKenzie and D.C. Sherrington, J. Chem.
      Soc., Perkin II, 514 (1981).
29.   H. Van Grunenberg, C. Bredt and W. Freudenberg, J. Amer. Chem.
      Soc., 60, 1507 (1938).
30.   K. Kobayashi and H. Sumitomo, Macromol. 13, 234 (1980).
31.   For preparation see K. Hiratani, P. Reuter and G. Manecke,
      Israel J. Chem., 18, 208 (1979).
32.   R.F. Brady, Carbohydrate Res., 15, 35 (1970).
33.   J. Benes and J. Hetflejs, Coll. Czech. Chem. Comm., 41, 2550
      (1976).
34.   H.B. Kagan and T.P. Daig, J. Amer. Chem. Soc., 94, 6429 (1972).

35. A. Hirao, S. Itsuno, M. Owa, S. Nagami, H. Mochizuki, H.H.A.
     Zoorov, S. Niakahama and N. Yamazaki, J. Chem. Soc., Perkin
     I, 900 (1981).
36. D.C. Sherrington, Unpublished Results.

ASYMMETRIC SELECTION BY OPTICALLY ACTIVE POLYMERS IN

PHASE-TRANSFER SYSTEM

Yuhsuke Kawakami and Yuya Yamashita

Department of Synthetic Chemistry
Faculty of Engineering, Nagoya University
Chikusa, Nagoya 464 Japan

INTRODUCTION

Phase-transfer catalyst systems have been deeply concerned
with the various aspects of polymer chemistry in recent years.
Imai et al showed that phase-transfer catalysts like crown ethers
or ammonium salts were very effective in condensation polymerizations.

$$ClSO_2\text{-}\langle O \rangle\text{-}O\text{-}\langle O \rangle\text{-}SO_2Cl \ + \ HO\text{-}\langle O \rangle\text{-}\langle O \rangle\text{-}OH \quad \underline{\text{TBAC, 18-CR-6, DC18-CR-6}}$$

$$\longrightarrow \text{Polymer with } [\eta]=1.1$$

Frechet[2] and Boileau[3] have studied the use of phase-transfer catalysts to modify the functional groups of polymers.

$$-(OCH_2CH)_n \ \underset{\substack{CH_2 \\ Cl}}{} \quad \xrightarrow[\text{hydrogensulphate}]{\text{Tetrabutylammonium}} \quad -(OCH_2CH)_{nx}(OCH_2CH)_{n(1-x)} \ \underset{\substack{CH_2 \\ Cl}}{} \ \underset{\substack{CH_2 \\ N}}{}$$

Tundo[4], Regen[5], and Ford[6] have been exploring polymeric phase-
transfer catalysts.

$$\langle O \rangle\text{-}CH_2CN \ + \ RBr \ \xrightarrow{\ \textcircled{P}\text{-}\langle O \rangle\text{-}CH_2\overset{+}{N}Me_3Cl^- \ } \ \langle O \rangle\text{-}\underset{R}{\overset{|}{C}HCN}$$

order of addition is important

So far, most efforts concerned activating reactions by phase-
transfer systems. We have been interested in the catalysts from
the viewpoints of not only the activation of the reactions but

also the asymmetric selection by optically active polymers in the
phase-transfer system.

We found that macrocyclic polyether acetals or 18-crown-6
were effective catalysts for esterification[7] or fluorination[8] in
the solid-liquid phase-transfer system. The efficiency of cyclic
acetals in activating the esterification reaction of equation (1)
is shown in Fig. 1.

$$n\text{-}BuBr + CH_3CO_2K \xrightarrow{\text{Cat.}} CH_3CO_2Bu^n + KBr \qquad \bullet\bullet\bullet \qquad (1)$$

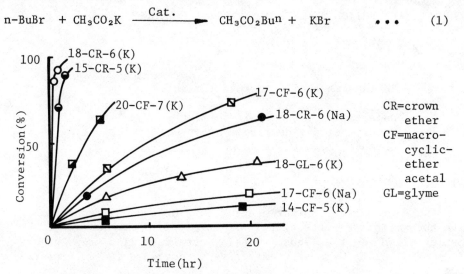

Fig. 1. Rate Enhancement by Addition of Macro-
        cyclic Ether Acetals.
        [BuBr]=0.01 M, [Alkali Metal Acetate]=
        0.01 M, [Activating Agent]= 0.01 M, Bz=
        10 ml, 90°C.

Although macrocyclic ether acetals are not as good activating
agents as crown ethers of similar ring size, they are much
better activating agents than linear glymes in the reaction (no
other glymes than 18-GL-6 showed any significant activation).

Potassium halide-18-crown-6 systems were shown effective to
exchange halogens of epihalohydrins. Recently, we have also shown
that tetrabutylammonium fluoride is a very effective reagent to
replace mesylate of glycerol 1,2-acetonide[9].

Meanwhile, polyethylene glycol is reported to act as a phase-
transfer catalyst[10,11], and it is also shown to form a double
helix in complexing with an alkaline metal cation[12]. If we use an
isotactic optically active polymer back-bone, there may be a
chance to see asymmetric selection in the esterification reactions
of chiral acid salts and achiral alkyl halides in the presence of

these polymers, and vice versa. On the other hand, tertiary amines are also known to act as phase-transfer catalysts[13].

The object of this work is to combine the two ideas of activation of the reaction and asymmetric selection in the presence of optically active polymers. For this purpose, optically active polyepichlorohydrin with optically active amine side chain, and optically active linear polyalkylenimines were chosen, and the activity was studied.

EXPERIMENTAL

Optically Active Epihalohydrins

Optically active epihalohydrins were synthesized starting from (S)-glycerol acetonide which was obtained from D-mannitol.

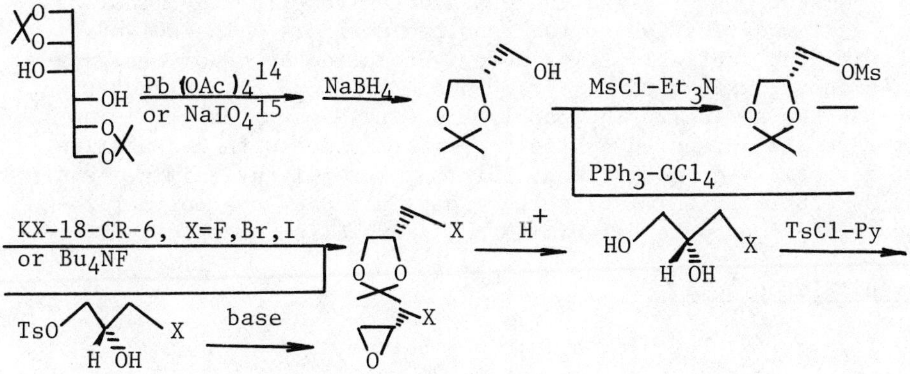

Scheme I. Routes to (R)-Epihalohydrins.

The synthetic scheme of (S)- and (R)-epichlorohydrins from (S)-glycerol acetonide via tosylation and chlorination was reported by McClure.[16] We explored the general synthetic scheme to (R)-epihalohydrins.

Optically Active Diamines

Three kinds of optically active diamines with differing degrees of substitution were synthesized starting from L-valine. The synthetic scheme is shown in Scheme II. By selective methylation at different stages of the reaction, the three diamines could be synthesized without any problem.

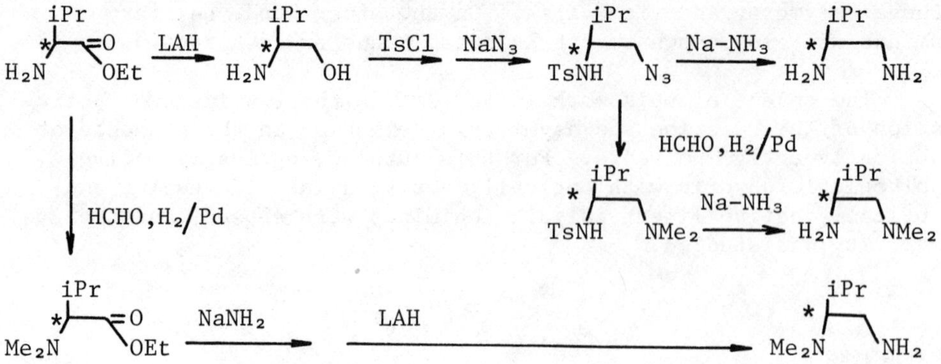

$$
\underset{H_2N}{\overset{iPr}{\overset{|}{*{\bigwedge}}}}=O \over OEt \xrightarrow{LAH} \underset{H_2N}{\overset{iPr}{*{\bigwedge}}}OH \xrightarrow{TsCl} \xrightarrow{NaN_3} \underset{TsNH}{\overset{iPr}{*{\bigwedge}}}N_3 \xrightarrow{Na-NH_3} \underset{H_2N}{\overset{iPr}{*{\bigwedge}}}NH_2
$$

HCHO, H$_2$/Pd

$$
\underset{TsNH}{\overset{iPr}{*{\bigwedge}}}NMe_2 \xrightarrow{Na-NH_3} \underset{H_2N}{\overset{iPr}{*{\bigwedge}}}NMe_2
$$

HCHO, H$_2$/Pd

$$
\underset{Me_2N}{\overset{iPr}{\overset{|}{*{\bigwedge}}}}=O \over OEt \xrightarrow{NaNH_2} \xrightarrow{LAH} \underset{Me_2N}{\overset{iPr}{*{\bigwedge}}}NH_2
$$

Scheme II. Synthesis of Optically Active Diamines.

Poly(N-methyl-L-isopropylethylenimine)
_____

     L-4-Isopropyl-2-oxazoline was synthesized from L-valinol by
the slight modification of the reported method.[17] The polymers
with different optical purity were synthesized by copolymerizing
the L-monomer with the racemic monomer, followed by the reduction
with LAH(PDL series). The copolymers with N-methylethylenimine
unit were also prepared by the copolymerization with 2-oxazoline
(PUL series). For comparison, poly(L-isopropylethylenimine) and
poly(N-methylethylenimine) were also prepared by the polymerizations of L-isopropylethylenimine and 2-oxazoline, respectively.

Polyepichlorohydrin
_____

     (S)- or (R)-Epichlorohydrin was polymerized by a Et$_2$AlCl-
Tetraphenylporphin system[18].

Introduction of Diamine on Polyepichlorohydrin
_____

     The chlorine atoms of the polyepichlorohydrin were replaced
by primary amino groups of the diamine similarly to the reported
method[19]. The full characterization of these reactions is now in
progress.

RESULTS AND DISCUSSION

     The synthesis of optically active epichlorohydrins from D-
mannitol was reported by McClure. The stereochemistry of each
transformation is well defined, however, the chemical yield is
usually poor depending on the reaction conditions. We tried to
improve the yield by inverting the order of tosylation and chlorination, and at the same time searched for a general method to

synthesize the optically active epihalohydrins using potassium halide-18-crown-6 systems, which we found effective in halogenation reactions, as shown in Scheme I. For fluorination, bromination, and iodination, the combination of potassium halide-18-crown-6 systems developed by us were effective. Furthermore, tetrabutylammonium fluoride was found very effective in fluorination. The stereochemistry of each transformation was studied by the help of selectively deuterated analogues. The stereochemistry of fluorination, bromination, and iodination was well defined, however, that of chlorination was very much dependent on the reaction conditions. The selectivity of the chlorination of glycerol 1,2-acetonide-3,3-d$_2$ is shown in Table 1.

Table 1. Stereochemistry of Chlorination of Glycerol
          1,2-Acetonide-3,3-d$_2$

No.	reagent	mol%*	yield(%)	A(%)	B(%)
1	PCl$_3$,Py	–	trace	–	–
2	POCl$_3$,Py	–	trace	–	–
3	POCl$_3$,DMF	–	63	47	53
4	Ph$_3$P,CCl$_4$	–	59	49	51
5	Ph$_3$P,CCl$_4$,DMF	–	66	56	44
6	Ph$_3$P,CCl$_4$,Py	1	24	80	20
7	Ph$_3$P,CCl$_4$,Py	5	37	87	13
8	Ph$_3$P,CCl$_4$,Et$_3$N	1	53	85	15
9	Ph$_3$P,CCl$_4$,Et$_3$N	5	61	87	18
10	Ph$_3$P,CCl$_4$,iPr$_2$EtN	1	53	86	14
11	Ph$_3$P,CCl$_4$,iPr$_2$EtN	5	67	88	12
12	Ph$_3$P,CCl$_4$	–	46	90	10
13	Ph$_3$P,CCl$_4$,Et$_3$N	100	27	92	8
14	Ph$_3$P,CCl$_4$,DBU	3	39	93	7
15	Ph$_3$P,CCl$_4$,DBU	30	49	93	7
16	Ph$_3$P,CCl$_4$,DBU	100	36	97	3
17	Ph$_3$P,CCl$_4$,DBP	3	29	86	14
18	Ph$_3$P,CCl$_4$,DBP	15	55	89	11
19	Ph$_3$P,CCl$_4$,DBP	100	55	93	7

DBU = 1,8-Diazabicyclo[5.4.0]-7-undecene
DBP = 2,6-Di-t-butylpyridine

drying agent of CCl$_4$: 4-11 : P$_2$O$_5$, 12-19 : CaH$_2$
* mol% of base to glycerol 1,2-acetonide-3,3-d$_2$

A: (structure with CD$_2$Cl)   B: (structure with D$_2$ and CH$_2$Cl)

CaH$_2$ dried CCl$_4$ and addition of non-nucleophilic tertiary amine
are quite effective in getting high selectivity. CaH$_2$ dried CCl$_4$
and equimolar amount of DBU gave almost pure compound without
racemization.

The optically active epihalohydrins prepared by our methods
are tabulated in Table 2.

Table 2.   Optically Active 3-Halogeno-1,2-propanediol Acetonides
           and (R)-Epihalohydrins

X	agent	yield (%)	$[\alpha]_D$	yield (%)	$[\alpha]_D$	e.e. (%)
		3-Halogeno-1, 2-propanediol acetonide				
F	MsCl-Bu$_4$NF	40	+12.7	12	- 6.1	97
Cl	Ph$_3$P,CCl$_4$	53	+35.2	26	-32.8	--
Br	MsCl-KBr,18-CR-6	46	+35.8	7	-14.6	--
I	MsCl-KI,18-CR-6	43	+34.5	9	- 8.7	--

The optical purity of the products is considered high. The chiral
purity of (R)-epifluorohydrin determined by ^{1}H-NMR in the presence
of 2 mol % of Tris[3-(2,2,2-trifluoro-1-hydroxyethylidene)-d-
camphorato]europium was 97$\pm$3 %. The optically active epifluoro-
hydrin is a new compound as far as we know.

Unsymmetrically substituted diamines were synthesized according to Scheme II without any difficulty.

The efficiency of various tertiary amines in the esterification
reaction (1) is shown in Fig. 2. Polytertiary amine showed almost
the same activity as 18-crown-6. The stereochemistry of the esterification reaction was studied by the use of optically active
sec-butyl bromide, and was proved to be basically inversion(80%)
at the carbon linked to bromine. These activity and selectivity
in the reaction may open the possibility of asymmetric selection
by optically active amines from chiral acid salts and achiral
bromides, and vice versa.

The following three reactions were carried out in order to
test the stereoregulation by optically active polyamine.

$$r\text{-}R^1CO_2K + a\text{-}R^2Br \xrightarrow{\text{optically active polyamine}} R^1CO_2R^2 + KBr$$

(a-)   (r-) $\cdots$(2)

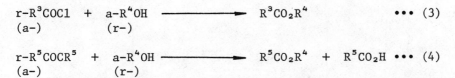

$$r-R^3COCl + a-R^4OH \longrightarrow R^3CO_2R^4 \qquad \bullet\bullet\bullet \ (3)$$
$$(a-) \qquad\quad (r-)$$

$$r-R^5COCR^5 + a-R^4OH \longrightarrow R^5CO_2R^4 + R^5CO_2H \ \bullet\bullet\bullet \ (4)$$
$$(a-) \qquad\qquad (r-)$$

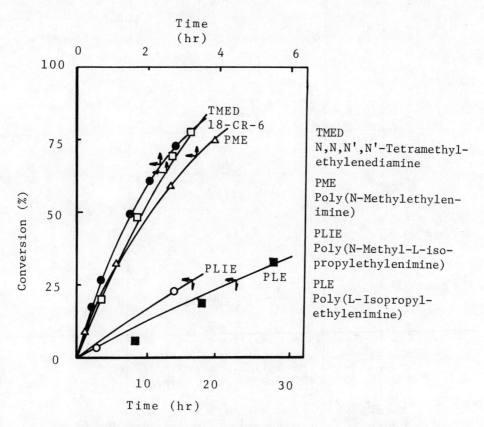

TMED
N,N,N',N'-Tetramethyl-
ethylenediamine

PME
Poly(N-Methylethylen-
imine)

PLIE
Poly(N-Methyl-L-iso-
propylethylenimine)

PLE
Poly(L-Isopropyl-
ethylenimine)

Fig.2. Activation of the Esterification
(1) by Amines.
$[CH_3CO_2K]=[BuBr]= 4$ m mol, $[Cat]=$
1.4 m mol in 5.6 ml Benzene at
80°C.

The results of the enantiomer selection in reaction (2) in the
presence of PLIE and PLE are shown in Table 3. Potassium acetate
does not react with the bromides in the absence of PLIE or PLE.
This is also true in the cases of potassium propionate or iso-
butyrate. 1.5-4.2% enantiomer selection was observed. The products
from control reactions showed no optical activity. Thus the optical
activity in the ester can be attributed to the presence of
optically active polyamines. The polymers can be repeatedly used.

Table 3. Enantiomer-Selective Ester Synthesis from Acid Salts and Alkyl Bromides in the Presence of Optically Active Polymers at 80°C in Benzene

No.	Salt (R^1, mole/l)		Bromide (R^2, mole/l)		PLIE or PLE (unit mole/l)		Time (hr)	Conversion (% on r-component)	Optical Purity (%)
1	CH_3	0.7	$CH_3CH_2CHCH_3$	0.7	PLIE	0.30	108	38	2.4
2	CH_3	1.0	α-Phenethyl	1.0	PLIE	0.30	11	45	1.5
3	CH_3	1.0	α-Phenethyl	1.0	PLIE[a]	0.22	19.5	51	1.6
4	CH_3CH_2	1.0	$CH_3CH_2CHCH_3$	1.0	PLIE	0.50	95	44	4.2
5	$(CH_3)_2CH$	1.0	$CH_3CH_2CHCH_3$	1.0	PLIE	0.50	108	44	(+0.16)[b]
6	$(CH_3)_2CH$	1.0	$CH_3CH_2CHCH_3$	1.0	PLE	0.50	166	42	(+0.78)[b]
7	$(CH_3)_2CH$	1.0	α-Phenethyl	1.0	PLIE	0.52	19	44	(−1.89)[b]
8	$(CH_3)_2CH$	1.0	α-Phenethyl	1.0	PLIE[a]	0.25	20	51	(−1.92)[b]
9	$(CH_3)_2CH$	1.0	α-Phenetyl	1.0	PLE	0.50	23	52	(+0.78)[b]
10	$CH_3CH_2CHCH_3$	1.0	$CH_3CH_2CH_2CH_2$	0.5	PLIE	0.25	110	41	2.5
11	$CH_3CH_2CHCH_3$	1.0	$CH_3CH_2CH_2CH_2$	0.5	PLE	0.52	68	36	2.5
12	CH_3	1.0	$CH_3CH_2CHCH_3$	1.0	PME	0.30	5	49	0.0
13	CH_3	1.0	$CH_3CH_2CHCH_3$	1.0	TMED	0.30	3	42	0.0

a Recovered polymers were used.

b $[\alpha]_D$ value of the product.

The optically active secondary polyamine(PLE) seems to have a weaker selecting ability compared with linear tertiary polyamine PLIE. An optical purity of 2.5% was observed in racemic salt and achiral bromide. Thus optical activity was observed in both the product esters from the reaction of racemic salt and achiral bromide and vice versa. This fact seems to suggest that the enantio-selection has occurred by the activation of carboxylate in the reaction of racemic salt and achiral bromide by PLIE or PLE, and by quaternization of alkyl bromide in the reaction of achiral salt and racemic bromide.

It was considered that the optically active polyamines interact more strongly with the acid chloride or anhydride than acid salt. Accordingly higher optical yield was anticipated in the products. However, only slight enantiomer selection was observed in reactions (3) or (4). The diastereomeric energy difference is apparently too small to differenciate the enantio-meric acid chloride or anhydride.

The effect of optical purity of PLIE(PDL series) and sequence length of the L-unit(PUL series) is shown in Fig.3. High selectiv-ity is achieved only by the copolymers of high L-unit content. This fact suggests that rather long N-methyl-L-isopropylethylen-imine unit is necessary for the effective enantiomer selection and that secondary structure of the polymer might be playing important role in the stereoselection.

Although as discussed above, some information was obtained in the assymetric selection by polyamines, the optical yield was not high enough to be used as a synthetic reaction. In order to improve the stereoselection, stereoregular polymer which is capable of complexing with cation, such as isotactic poly-epichlorohydrin was chosen as the polymer back-bone, and optically active tertiary amine as the catalytic site on the side chain. This type of the polymer may act as an effective complexing agent with cation by the polymer back-bone, and may exert a strong stereo-regulation by the isotactic structure of the polymer and by the asymmetric environment produced by chiral amine functions. Preliminary results of the enantiomer selection in the reaction of potassium acetate and α-phenethyl bromide by the optically active isotactic polyepichlorohydrin functionalized by L-2-dimethylamino-3-methylbutylamine gave 5-7% of the stereoselection. Further efforts are needed to achieve high enantioselection.

CONCLUDING REMARKS

The study of phase-transfer catalysts has been rapidly ex-panding. Most interests have been given to the activation of reactions, but there have been only few works done in the area of asymmetric synthesis utilizing these systems.

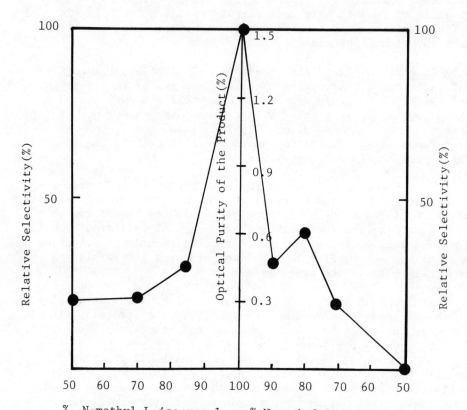

Fig. 3. Effect of N-methyl-L-isopropylethylenimine Unit(%)
on the Enantiomer Selection in the Reaction

$$CH_3CO_2K \quad + \quad PhCHBr \quad \xrightarrow{\quad\quad\quad} \quad CH_3CO_2CHPh \quad +KBr$$
$$\qquad\qquad\qquad\quad \underset{CH_3}{|} \qquad\qquad\qquad\qquad \underset{CH_3}{|}$$

Tsuboyama used optically active polyalkylenimine as catalysts for asymmetric addition of hydrogen cyanide to benzaldehyde,[20] however the optical yield was not so high. Inoue found cyclic dipeptide containing (S)-hystidine unit is a very effective catalyst achieving as high as 90% e.e. selectivity of the product in the same reaction.[21] This fact clearly indicates the importance of setting-up the rigid asymmetric environment during the reaction takes places (necessity of rigid transition state). Another important factor is the reaction temperature. The phase-transfer reactions are usually carried out at room temperature or higher temperature which makes it difficult to differentiate the enantiomers having only slight difference in the activation energy.

In order to overcome these disadvantages of phase-transfer systems in asymmetric selection or asymmetric synthesis, the catalytic function must be put in an organized phase. The use of block or graft polymers with micro phase separation will be one solution, but in this case care must be given to maintain the surface area of the catalysts. Recent progress of the chemistry in organized phase is expected to contribute very much in this field in near future.

REFERENCES

1. Y. Imai, J. Macromol Sci.-Chem., A15, 833 (1981).

2. J. M. J. Frechet, J. Macromol. Sci.-Chem., A15, 877 (1981).

3. T. D. N'Guyen, A. Deffieux, S. Boileau, Polymer, 19, 423 (1978).

4. H. Molinari, F. Montanali, S. Quici, and P. Tundo, J. Am. Chem. Soc., 101, 3920 (1979).

5. S. L. Regen, Angew. Chem. Int. Ed., 18, 421 (1979).

6. T. Balakrishnan, and W. T. Ford, Tetrahedron Lett., 22, 4377 (1981).

7. Y. Kawakami, T. Sugiura, and Y. Yamashita, Bull. Chem. Soc. Jpn., 51, 3053 (1978).

8. Y. Kawakami and Y. Yamashita, J. Org. Chem., 45, 3930 (1980).

9. Y. Kawakami, T. Asai, K. Umeyama, and Y. Yamashita, J. Org. Chem., in press.

10. H. Lehmkull, F. Rabet, and K. Hauschild, Synthesis, 184 (1977).

11. E. Santaniello, A. Manzocchi, and P. Sozzani, Tetrahedron
     Lett., 4581 (1979).

12. J. M. Parker, P. V. Wright, and C. C. Lee, Polymer, 22, 1305
     (1981).

13. H. Normant, T. Cuvigny, and P. Savignac, Synthesis, 805
     (1975).

14. J. J. Baldwin, A. W. Raab, K. Mensler, B. H. Arison, and
     D. E. McClure, J. Org. Chem., 53, 4876 (1978).

15. H. Eibl, Chem. and Phys. Lipids, 28, 1 (1981).

16. D. E. McClure, E. L. Engelhardt, K. Mensler, S. King,
     W. S. Saari, J. R. Huff, and J. J. Baldwin, J. Org. Chem.,
     44, 1826 (1979).
     also ref. 14.

17. Y. Ito, I. Ito, I. Hirao, and T. Saegusa, Synth. Comm., 4,
     97 (1974).

18. T. Aida, and S. Inoue, Macromolecules, 14, 1162, 1166 (1981).

19. E. Schacht, D. Bailey, O. Vogel, J. Polym. Sci.-Polym. Chem.
     Ed., 16, 2343 (1978).

20. S. Tsuboyama, Bull. Chem. Soc. Jpn., 35, 1004 (1962).

21. J. Oku and S. Inoue, J. Chem. Soc.-Chem. Comm., 229 (1981).

     J. Oku, N. Ito, and S. Inoue, Makromol. Chem., 183, 579
     (1982).

SYNTHETIC REACTIONS BY GAS-LIQUID PHASE-TRANSFER CATALYSIS

Pietro Tundo and Paolo Venturello

Istituto di Chimica Organica dell'Università

Torino, Italy

INTRODUCTION

Gas-liquid phase-transfer catalysis (GL-PTC)[1] is a new
synthetic organic method that has similarities both with phase-
transfer catalysis (PTC)[2] and with gas-liquid chromatography (GLC)
in that anion transfer processes and partition equilibria between
gaseous and liquid phases both take place and affect the synthesis.
Using GL-PTC, nucleophilic substitution reactions have been so far
carried out under operative conditions and with synthetic results,
making this method different from the well known liquid-liquid
(LL-) and solid-liquid (SL-) phase-transfer catalysis. As regards
these latter, phase-transfer catalysts (onium salts, crown ethers
and cryptands) transfer the reactive anion from an aqueous liquid
(LL-PTC) or a solid salt (SL-PTC) phase into the organic one in which
the substitution reaction occurs. In the case of GL-PTC, where no
solvent is used, the catalyst always acts as an anion transfer
(between solid and liquid phases) but, as it works in the molten
state it also constitutes the medium in which the reaction proceeds.

Experiments with molten salts were carried out by passing an
alkyl halide over a molten layer of an alkaline salt.[3] Notwith-
standing the presence of eutectic mixtures the temperature remained
fairly high and therefore numerous byproducts were obtained.
syntheses are carried out, but it is necessary to use phosphonium
salts as they can be kept at 150 °C for several days without

275

decomposition while ammonium salts decompose at this temperature.[4]

Owing to the experimental conditions of GL-PTC (high temperatures, lack of anions stabilizing interactions with protic solvents)
weak complexing agents as polyethylene glycol (Carbowax 6000) and
polyoxyethylene-lauryl ether (Brij 35) as well as classical PT
catalysts have proved effective catalysts providing a wider range
of catalysts that can be varied, affecting reaction results
differently.

HOW A GL-PTC REACTION IS CARRIED OUT

The reactions under GL-PTC conditions are carried out in a
thermostatized glass column containing a solid bed consisting of
a solid salt of the nucleophile or of a base able to generate it,
and by the catalyst adsorbed on the solid, that must have a large
surface area. The organic substrates are continuously introduced
into the column where temperature and pressure should maintain
both reagents and products in the gaseous state. The reaction
mixture is continuously collected by condensation at the column
outlet (Figure 1).

In order to prepare the catalytic bed, the catalyst is dissolved
in a low-boiling solvent (methylene chloride, methanol), the
nucleophile salt (alkaline halide or carboxylate) or the inorganic
base ($NaHCO_3$ or K_2CO_3) or a porous inorganic solid (alumina or silica
gel), according to the reaction that must be carried out, is added
to the solution, the solvent is removed and the resulting solid
dried in a stove. Subsequently the catalyst is finely dispersed on
the surface of the solid support and the mass is introduced into
the column. The temperature of the column must keep the catalyst
molten, which is not a problem (unless thermofailing reagents are
used), because the high temperature increases the reaction rate and
generally no byproduct is detected as the apparatus is free from
air pollution.

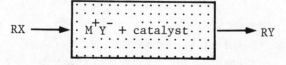

RX → M^+Y^- + catalyst → RY

Fig. 1.   GL-PTC scheme.

The bed, that remains solid in the column during the synthesis, always actively takes part in the reaction and is not a simple support for the catalyst; as regards this, two situations can be shown. In the first case the solid bed takes part in the reaction consuming itself as it proceeds according to equations 1a and 1b.

$$RX_{gas} + M^+Y^-_{solid} \xrightarrow{\text{cat.}} RY_{gas} + M^+X^-_{solid} \qquad (1a)$$

$$RX_{gas} + AH_{gas} + NaHCO_{3solid} \xrightarrow{\text{cat.}} RA_{gas} + CO_2 + H_2O + NaX \quad (1b)$$

In the second case the solid bed supplies the reactive sites in which the substitution reaction takes place and since it does not undergo irreversible chemical transformations, consequently it does not consume during the synthesis and the overall process becomes effectively catalytic: one can operate for a indefinite period, introducing fresh organic reagents without interruption, according to equation 2.

$$A + B \xrightarrow{\text{cat.}} products \qquad (2)$$

Below are reported some reactions carried out under GL-PTC conditions that indicate the possibility of synthetic applications, and show mechanistic aspects and differences with LL- and SL-PTC.

SYNTHESES WITH SOLID SALTS PLACED IN THE COLUMN

Alkyl Iodides[5]

We started the GL-PTC process synthesizing alkyl halides from gaseous alkyl chlorides or bromides and solid potassium iodide, as anion source according to equation 3.

$$RX_{gas} + KI_{solid} \xrightarrow{\text{cat.}} RI_{gas} + KX_{solid} \qquad (3)$$

In a first approach it was possible, using as entering group a 'soft' anion like I$^-$, to specify the type of catalyst necessary for the gas-phase reaction to proceed and moreover to investigate analogies between the reactions carried out on the solid bed and the well known substitution reactions[6] conducted with the salts immobilized on a solid support.

Table I.   Conversion of 1-bromobutane into 1-iodobutane.[a]

Solid bed	Catalyst[b]	Conversion %
KI (10.3 meq with respect to 1-bromobutane)	None	~ 2
	$Bu_4P^+I^-$	96
KI (1.5 meq with respect to 1-bromobutane)	Silica gel	50
	Silica gel + NaLS[c]	46
	Silica gel + $Bu_4P^+I^-$	93

[a] 1-Bromobutane, 0.21 moles; flow rate (liquid), 40 ml/h; T = 160 °C.
[b] 0.07 meq with respect to 1-bromobutane.
[c] Sodium laurylsulphate.

Results reported in Table I show that the presence of both a phosphonium salt and of a porous support like silica gel are important: when 1-bromobutane is allowed to flow in the absence of a catalyst through ground potassium iodide, no 1-iodobutane is pratically obtained; on the contrary, the conversion obtained in the presence of potassium iodide + silica gel clearly shows that the inorganic porous material plays the role of a 'solid solvent' and actives iodide anion by interactions which diminish the bond strength of the $K^+ I^-$ ion pair. Furthermore catalytic amounts of the phosphonium salt lead to higher conversion while the presence of the anionic surfactant sodium laurylsulphate (NaLS) on the surface of silica gel does not yield different results from those obtained with silica gel alone.

With a column 50 cm in length and 2.5 cm in diameter it is possible to obtain in about 35 min 0.2 moles of alkyl iodides (propyl, allyl, butyl, octyl and benzyl) with high conversions (74%, 78%, 93%, 82% and 81% respectively). In the case of benzyl iodide the reaction was carried out under vacuum and no bad effect due to the lachrymatory properties of both reagent and product were experienced.

Carboxylate Esters[7]

The synthesis of carboxylate esters can be carried out according to equation 4

$$\underset{gas}{RX} + \underset{solid}{R'COONa} \xrightarrow{\text{cat.}} \underset{gas}{R'COOR} + \underset{solid}{NaX} \qquad (4)$$

in which the alkaline salt of the acid is present in the column both
as reagent and catalyst-supporting solid, or alternatively according
to equation 5 in which the salt is generated in situ by reaction
of the acid with the base.

$$\underset{gas}{RX} + \underset{solid}{R'COOH} \xrightarrow[\text{cat.}]{K_2CO_3} \underset{gas}{R'COOR} \qquad (5)$$

When the solid reagent's melting point is lower than the
temperature at which the reaction is carried out it must
be supported on a solid material; this end can be achieved both
dispersing the low melting reagent on silica gel (in the case of
alkaline carboxylate) and on K_2CO_3 or K_2CO_3 + silica gel (in the
case of carboxylic acid).

In the Table II are reported some results obtained in the
carboxylate esters synthesis employing both procedures.

Table II.   Syntheses of Esters Under GL-PTC Conditions.[a]

Starting Compounds		Products	Conversion %
$n\text{-}C_3H_7CO_2Na$	$CH_3(CH_2)_7Cl$	$n\text{-}C_3H_7CO_2(CH_2)_7CH_3$	61
$2\text{-}ClC_6H_4CO_2Na$[b,c]	C_2H_5Br	$2\text{-}ClC_6H_4CO_2C_2H_5$	97
$4\text{-}ClC_6H_4CO_2H$[d]	C_2H_5Br	$4\text{-}ClC_6H_4CO_2C_2H_5$	100
$2,4,6\text{-}Me_3C_6H_2CO_2H$[e]	C_2H_5Br	$2,4,6\text{-}Me_3C_6H_2CO_2C_2H_5$	98

[a] Catalyst, $Bu_4P^+Br^-$, 0.05 molequiv. with respect to the carboxylate;
  alkyl halide, 0.67 molequiv. with respect to the carboxylate; flow
  rate (liquid), 40 ml/h; T = 150 °C.
[b] According to eq. 4.
[c] Bed mixed with silica gel, 1/1, w/w.
[d] Bed prepared with potassium carbonate, according to eq. 5; 0.4 Torr.
[e] Bed prepared with potassium carbonate and silica gel, according
  to eq. 5.

The GL-PTC synthesis of carboxylate esters is justified by some advantages: aprotic polar solvents in which the synthesis is conducted with classical nucleophilic substitution methods need not be used and moreover, in comparison with analogous reactions carried out under PTC conditions[8], the use of either a large carboxylate excess (LL-PTC) or of expensive catalysts like 18-crown-6 and cryptands[9] (SL-PTC) is avoided.

The reaction between sodium acetate and ethyl bromide (equation 6) carried out with different catalysts explains some peculiarities of GL-PTC; the results are reported in Table III.

$$CH_3CH_2Br + CH_3COONa \xrightarrow[150\ °C]{cat.} CH_3COOCH_2CH_3 + NaBr \qquad (6)$$

Table III.   Ethyl Acetate Synthesized from Sodium Acetate and Ethyl Bromide, as a Function of the Catalyst Used under GL-PTC Conditions.[a]

Catalyst[b]	m.p.(°C)	Conv. %	Catalyst[b]	m.p.(°C)	Conv. %
$Bu_4\overset{+}{P}\ \overset{-}{Br}$	87	96	18-crown-6	38	32
$C_{16}H_{33}\overset{+}{P}\ Bu_3\overset{-}{Br}$	54	86	Carbowax 6000[c]	55–62	4
$Bu_3\overset{+}{P}\ \overset{-}{MeI}$	135	59	Brji 35[c]	40	3
$C_{16}H_{33}\overset{+}{P}\ Et_3\overset{-}{Br}$	145	86	NaLS[c]	205	0(2)[d]
$C_{16}H_{33}\overset{+}{P}\ Me_3\overset{-}{Br}$	203	0	Silica gel[e]		20
$Et_3\overset{+}{P}\ \overset{-}{MeI}$	324	0			

[a] Sodium acetate, 0.62 mol.; ethyl bromide, 0.42 mol.; flow rate (liquid), 40 ml/h; T = 150 °C.
[b] 0.02 molequiv. with respect to the sodium acetate.
[c] 10 % by weight with respect to sodium acetate.
[d] T = 210 °C.
[e] Silica gel and sodium acetate, 1/1, w/w.

The acetate anion is certainly 'harder' than other carboxylate anions, and so its reactivity results are particularly remarkable. The reaction does not proceed at all in the absence of a catalyst

and silica gel alone, differently from alkyl iodides synthesis, is
not able to lead to relevant conversions. In this synthesis moreover,
18-crown-6 a cyclic polyether with strong complexing properties,
does not provide high conversion; still less so Carbowax 6000 and
Brij 35. Phosphonium salts, that clearly attack the crystalline
nucleophile more than a crown ether, prove highly active in this
reaction.

The presence of a liquid phase is needed for the reaction to
proceed. The function of this phase is twofold: firstly, it is the
liquid medium through which the alkyl halide diffuses and reacts
with the carboxylate anion, counterion of the phosphonium salt;
secondly, it attacks the crystal below and exchanges the halide
anion from alkyl halide with the reactive nucleophile.

Figure 2 shows that in GL-PTC onium salts bearing iodide as
counterion, normally a PTC poison, can also be used.

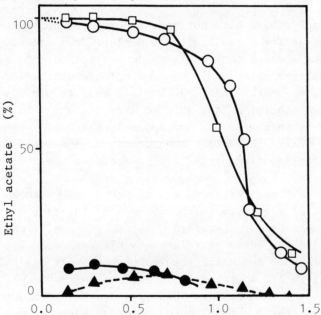

Molequiv. of EtX introduced (referred to AcONa)

Fig. 2. Plot of the conversion of EtX into EtOAc as a function of
the catalyst and of the alkylating agent; ($\square$ $Bu_4P^+Br^-$,
EtBr); ($\bullet$ $Bu_4P^+I^-$, EtI); ($\bigcirc$ $Bu_4P^+I^-$, EtBr). Under these
latter conditions EtI was also produced: its % amount is
shown as $\blacktriangle$. See Table III, footnote 'a' for reaction cond.

Under GL-PTC conditions, the iodide ion reacts with the alkyl halide
and is removed from the column as alkyl iodide (ethyl iodide in
Fig. 2). Moreover, Figure 2 shows that this is a real phase-transfer
process; in fact if the only function of the catalyst were to supply
a liquid phase to act as solvent for the nucleophile salt it would
not account for the fact that the activity of $Bu_4P^+Br^-$ + EtBr is
decidedly greater than that of $Bu_4P^+I^-$ + EtI.

Synthesis of Phenyl Ethers and Sulphides[10]

The synthesis of phenyl ethers and sulphides is carried out
according to GL-PTC, introducing the mixture of the phenol (or thiol)
and of the alkylating agent into the solid bed composed of K_2CO_3
(or $NaHCO_3$) supporting Carbowax 6000 as schematized in equation 7.

$$PhOH(R'SH) + RX \xrightarrow[K_2CO_3(NaHCO_3)]{cat.} PhOR(R'SR) \qquad (7)$$

Acyclic polyethers have not been reported to be phase-transfer
catalysts, except for a few limited examples. They are only shown
to be catalysts when present in large concentration and, since they[11]
are very soluble in water, only under SL-PTC conditions. Under
GL-PTC conditions on the contrary, they are able to mimic the functions of crown ethers: in the absence of water and with relatively
'soft' anions as phenoxide (or mercapdide), they can show
effective complexing properties toward cations and consequently to
increase the nucleophilic activity of reacting anions.

The basic strength of K_2CO_3 and of $NaHCO_3$ is enhanced by
several orders of magnitude in the absence of interactions with[12]
polar protic solvents; therefore under GL-PTC conditions they are
able to generate the anion of relatively high pK_a phenols, like
2-acetylphenol. Some synthetic results with phenols and thiols are
reported in Table IV. As reported in the case of carboxylate ester
synthesis, the reaction between phenols and alkyl halide (eq. 7)
also proceeds in the liquid film constituted by the molten catalyst:
the reaction products prove this statement, as only the O-alkylation
product is present when the Carbowax is used. On the contrary, when
no catalyst is present, in the presence of $NaHCO_3$ (68 % conv.) or
K_2CO_3 (3 % conv.) 13% or 12 % respectively of C-alkylation products
are obtained in the reaction of phenol with 1-bromobutane under
GL-PTC conditions. These data show that the reaction occurs in two

Table IV.   Synthesis of Phenyl Ethers and Thioethers, with K_2CO_3
or $NaHCO_3$ as a Base and Carbowax 6000 as Phase-Transfer
Catalyst.[a]

Starting Compounds			Products	Conversion %
$2\text{-}OHNp$[b]	EtBr	$NaHCO_3$	$2\text{-}NpOEt$	92
$3,5\text{-}Me_2C_6H_3OH$	BuBr	K_2CO_3	$3,5\text{-}Me_2C_6H_3OBu$	91
$2\text{-}AcC_6H_4OH$	BuBr	K_2CO_3	$2\text{-}AcC_6H_4OBu$	96
C_6H_5SH	$CH_3(CH_2)_7Cl$	K_2CO_3	$C_6H_5S(CH_2)_7CH_3$	92
$HSCH_2CH_2OH$	BuBr	K_2CO_3	$HOCH_2CH_2SBu$	85

[a] Catalyst, 5% by weight with respect to the base; flow rate (liquid),
40 ml/h; T = 150 °C.
[b] Np = naphtyl.

different environments; the first (the molten catalyst) aprotic
apolar, the second (the surface defects of the solid support)
highly polar.

Mechanistic Aspects

When the reaction proceeds stoichiometrically consuming the bed
placed in the column (eq. 1), the equations 8-11 illustrate its
mechanism.

$$RX_{gas} + Q^+Y^-_{liquid} \rightleftharpoons (RX + Q^+Y^-)_{liquid} \tag{8}$$

$$(RX + Q^+Y^-)_{liquid} \rightleftharpoons (RY + Q^+X^-)_{liquid} \tag{9}$$

$$(RY + Q^+X^-)_{liquid} \rightleftharpoons RY_{gas} + Q^+X^-_{liquid} \tag{10}$$

$$Q^+X^-_{liquid} + M^+Y^-_{solid} \rightleftharpoons Q^+Y^-_{liquid} + M^+X^-_{solid} \tag{11}$$

Equations 8-10 concern the nucleophilic substitution reaction that
takes place in the organic phase: the reaction rate is controlled
by the intrinsic reactivity of the electrophile and of the
nucleophile (eq. 9), by the diffusion rate and by the partition

equilibria of the reagents and products between gaseous and liquid phases (eq. 8 and 10), similarly to what occurs in GLC, but with the difference that in the former case there is a chemical reaction. The same consideration should be taken into account when the nucleophile is produced by reaction of an acidic compound (carboxylic acids, phenols, thiols, active methylene compounds) with the base (K_2CO_3, $NaHCO_3$) present in the bed; in this case the generation of the reactive anion may be promoted by the catalyst either at the solid-liquid interface (in comparison with what occurs in LL-PTC) or less probably in the organic phase through the HCO_3^- or CO_3^{--} anion-transfer from the solid phase to the liquid catalyst as counterion. Equation 11 represents the catalyst regeneration and is required in order that the reaction may occur. Strictly comparable with the analogous equation for SL-PTC, it represents the real function of the catalyst that is able to promote the exchange between its anion and the nucleophile salt.

The way and the rate of such an exchange in GL-PTC are under consideration;[13] in this connection the crystallographic and morphologic study on the salts (M^+X^- in eq. 1) produced in the syntheses is facilitated by the fact that the reactions are carried out without stirring and consequently crystalline habits are well defined. For example, when an ester synthesis is carried out according to equation 12, scanning electron micrographies showed small sodium chloride crystals of 1-10μ sizes, having a cubic habit in the case of R = $PhCH_2$ and an octahedral one in the case of R = Bu (Figures 3 and 4).

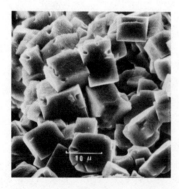   

Fig. 3.   Electron micrography of   Fig. 4.   Electron micrography of
           NaCl crystals produced   NaCl crystals produced
           in eq. 12 when R = $PhCH_2$.   in eq. 12 when R = Bu.

$$RCl_{gas} + AcONa_{solid} \xrightarrow[160\ °C]{Bu_4\overset{+}{P}\ \overset{-}{X}} ROAc_{gas} + NaCl_{solid} \quad (12)$$

Interestingly, the larger samples of these crystals present a round shape cavity on one face only of the cube or only on one corner of the octahedron. These phenomena are interpreted on the basis of the Schemes 2 and 3 that from another point of view explain the catalytic process.

The sodium chloride crystals are nucleated on the solid sodium acetate surface always according to the $\{100\}$ direction. Subsequently they grow as cubes or as octahedrons as a function of the ratio between the activities of the supersaturated and saturated NaCl solutions or/and of the medium in which the crystallization occurs. As the reaction proceeds, the solid bed is consumed and the crystals produced grow in the direction of the liquid phase but not towards the solid bed whose 'imprint' remains in the $\{100\}$ direction.

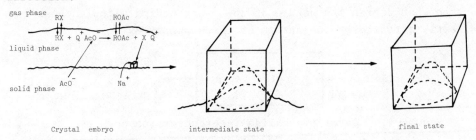

Scheme 2.   Nucleation and growth on the (100) face of the cubes of the alkaline halides produced during a GL-PTC reaction.

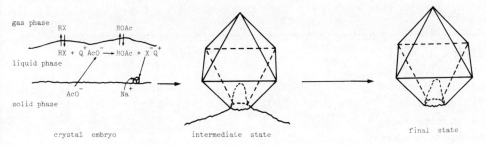

Scheme 3.   Nucleation and growth on the (100) face of the octahedrons of the alkaline halides produced during a GL-PTC reaction.

SOLID BED ACTING AS A CATALYST

Catalytic Interconversion of Alkyl Halides[14]

When a mixture of alkyl halides is allowed to pass through a solid bed containing a PT catlyst, rapid halide interconversions take place according to equation 13

$$RHal + R'Hal' \quad \xrightleftharpoons{cat.} \quad RHal' + R'Hal \quad (13)$$

The reaction between CH_2Cl_2 and C_2H_5Br (2/1, m/m) represented in Scheme 4 achieves thermodynamic equilibrium when the same statistic distribution of the halides is present in all the compounds. The equilibrium is achieved more or less rapidly, as a function of the catalyst and of the supporting solid. As shown in Table V, the nature of the supporting solid is a determining factor in promoting the reaction, even when a phosphonium salt is present: this points out to what degree the nature of the 'surrounding' liquid phase is directly affected by the nature of the support.

$$CH_2Cl_2 + EtBr \longrightarrow \boxed{\text{solid bed}} \longrightarrow \begin{cases} CH_2Br_2 + CH_2BrCl+ \\ \\ EtCl+CH_2Cl_2+EtBr \end{cases}$$

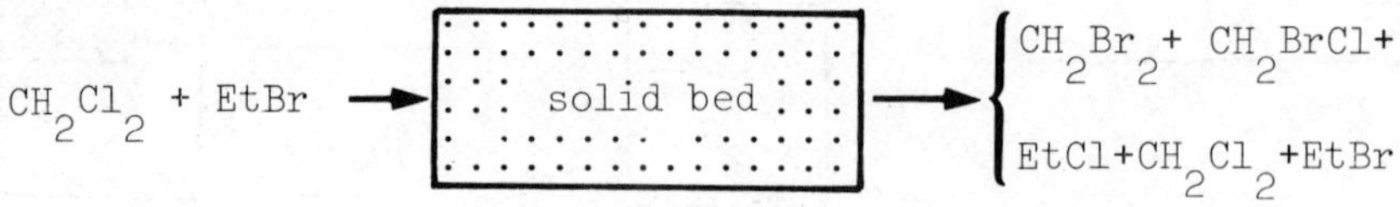

Scheme 4.   Catalytic halide interconversion between CH_2Cl_2 and C_2H_5Br under GL-PTC conditions.

Table V.   Catalytic Exchange between CH_2Cl_2 and C_2H_5Br (1/2, m/m).[a]

Solid bed			Product Yields %	
			as CH_2ClBr	as CH_2Br_2
Silica gel			1	0
Basic alumina			33	2
Acidic alumina			48	6
Silica gel	+ $Bu_4P^+Br^-$ (10%, w/w)[b]		25	1
Basic alumina	+ "	"[b]	50	24
Acidic alumina	+ "	"[b]	50	12

[a] Silica gel, 110 g; aluminas, 200 g; T = 170 °C; flow rate, 28 ml/h.
[b] Flow rate, 200 ml/h.

Figure 5 shows the conversions obtained as a funtion of the
temperature, when the organic reagents are introduced in a constant
manner into a bed constituted by basic alumina, in the presence
and in the absence of $Bu_4P^+Br^-$. In the presence of the catalyst
the reaction easily attains thermodynamic equilibrium and shows a
lower activation energy (the flex point occurs at a lower
temperature) than that observed in its absence. We are in the
presence of two different types of anionic activation; when a liquid
phase is adsorbed on the porous support, equations 14-19 count and
equation 20 represents the whole process.

$$RX_{gas} + Q^+Y^-_{liquid} \rightleftharpoons (RX + Q^+Y^-)_{liquid} \quad (14)$$

$$(RX + Q^+Y^-)_{liquid} \rightleftharpoons (RY + Q^+X^-)_{liquid} \quad (15)$$

$$(RY + Q^+X^-)_{liquid} \rightleftharpoons RY_{gas} + Q^+X^-_{liquid} \quad (16)$$

$$R'Y_{gas} + Q^+X^-_{liquid} \rightleftharpoons (R'Y + Q^+X^-)_{liquid} \quad (17)$$

$$(R'Y + Q^+X^-)_{liquid} \rightleftharpoons (R'X + Q^+Y^-)_{liquid} \quad (18)$$

$$(R'X + Q^+Y^-)_{liquid} \rightleftharpoons R'X_{gas} + Q^+Y^-_{liquid} \quad (19)$$

$$RX_{gas} + R'Y_{gas} \rightleftharpoons RY_{gas} + R'X_{gas} \quad (20)$$

In the absence of the liquid phase, the reaction proceeds via
adsorption of the alkyl halide on the support (similarly to GC)
and is promoted by sodium halides rather than by onium salts.
Both in the presence and in the absence of onium salt, sodium halides
were confirmed to be present on the support by titrating the solid
beds after the reactions.

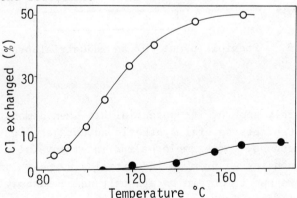

Figure 5.   % Cl exchanged in the reaction between CH_2Cl_2
and EtBr (scheme 4), as a function of the
temperature; in the presence (○) and in the
absence (●) of the phosphonium salt.

POTENTIAL INDUSTRIAL EXTENSION

The halide-exchange reaction can have industrial applications, since various and fine halides can be obtained, without using solvents and with high reaction rates. Particularly, starting from CH_2Cl_2 and CH_3I (1/1, m/m), CH_2ClI was obtained at thermodynamic equilibrium with a 34 % conversion; the collected reaction mixture was then fractionally distilled and CH_3Cl and CH_2ClI were separated. A subsequent reaction carried out on the remaining products (CH_2Cl_2, CH_2I_2 and CH_3I) enriched with fresh reagents, yielded the same conversion into CH_2ClI.

Consequently one reasonably expects that the plant scheme in Figure 6 may work. The coupling of a catalytic column and of a distillation column produces the required alkyl halide in a continuous manner.

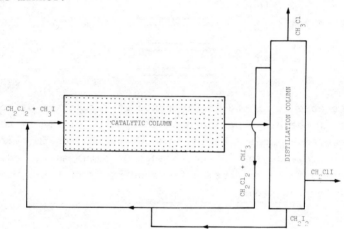

Figure 6.   Possible plant for an industrial process.

CONCLUSION

Although only some of its potential has been demonstrated to date, GL-PTC offers several synthetic advantages with respect to classical methods. For example halide interconversion can be interestingly considered from an industrial view-point and ester synthesis allows one to obtain esters only using carboxylic acids, alkyl halides, K_2CO_3 and the catalyst, without having to produce the acyl chloride employing thionyl chloride.

Furthermore some general advantages of GL-PTC are: (a) no

solvent is used; (b) large quantities of products can be obtained
with a small solid bed; (c) the reaction product can be collected
at the column outlet without extraction or distillation; (d) the
catalyst can be easily recovered; (e) the handling of the reagents,
products and solid bed is simple, and when dangerous compounds
are involved safety precautions can be easily provided.

REFERENCES

1. P.Tundo, J. Org. Chem., 44, 2048(1979).

2. (a) W. P. Weber and G. W. Gokel, "Phase Transfer Catalysis in
   Organic Synthesis", Springer-Verlag, West Berlin, 1977;
   (b) C. M. Starks and C. Liotta, "Phase Transfer Catalysis",
   Academic Press, New York, 1978; (c) E. V. Dehmlov and S. S.
   Dehmlov, "Phase Transfer Catalysis", Verlag Chemie, Weiheim, 1980;
   (d) F. Montanari, D. Landini and F. Rolla, PT Catalyzed Reactions
   in "Topics in Current Chemistry", Springer-Verlag, West Berlin,
   101, 147(1981).

3. D. I. Packham and F. A. Rackley, Chem. Ind.(London), 899(1966).

4. J. E. Gordon, J. Org. Chem., 30, 2760(1965).

5. P. Tundo and P. Venturello, Synthesis, 952(1979).

6. (a) S. L. Regen and C. Koteel, J. Am. Chem. Soc., 99, 3837(1977);
   (b) S. L. Regen, S. Quici and S. J. Liaw, J. Org. Chem., 44,
   2029(1979); (c) G. Bram and T. Fillebeen-Khan, J. Chem. Soc.,
   Chem. Commun., 522(1979).

7. E. Angeletti, P. Tundo and P. Venturello, J. Chem. Soc., Perkin I,
   993(1982).

8. Reference 1c, Chpt. 3, pp. 69-77.

9. A. Knöchel, J. Oehler and G. Rudolph, Tetrahedron Lett., 3167
   (1975).

10. E. Angeletti, P. Tundo and P. Venturello, J. Chem. Soc., Perkin I,
    1137(1982).

11. D. Balasubramanian, P. Sukumav and B. Chandani, Tetrahedron
    Lett., 3543(1979).

12. M. Fedorynsky, K. Wojciechowski, Z. Matacz and M. Makosza,
    J. Org. Chem., 43, 4682(1978).

13. E. Angeletti, M. Rubbo, P. Tundo and P. Venturello, submitted for publication.

14. E. Angeletti, P. Tundo and P. Venturello, J. Chem. Soc., Chem. Commun., 1127(1980).

15. P. Tundo, P. Venturello and E. Angeletti, submitted for publication.

ION-BINDING POLYESTERS AND POLYAMIDES CONTAINING THF RINGS

J. A. Moore* and E. M. Partain, III[†]

Department of Chemistry
Rensselaer Polytechnic Institute
Troy, New York 12181

INTRODUCTION

In 1955, Corbaz and coworkers[1] isolated an antibiotic from Streptomyces viridochromogenes and Streptomyces olivochromogenes which was named nonactin. The antibiotic was unusual because it was a neutral, optically inactive solid. In addition to nonactin, three homologs named monactin, dinactin, and triactin were subsequently isolated.[2] This class of antibiotics is called the actins (Fig. 1).

The structure of nonactin was elucidated by Gerlach and coworkers,[3] who found that nonactin was a macrocyclic ester consisting of four alternating units of levorotatory and dextrorotatory nonactic acid, a hydroxy acid containing a cis-tetrahydrofuran ring. In 1975, the structure of nonactin was confirmed by total syntheses by Gerlach[3] and by Schmidt.[4]

The ion-binding activity of nonactin was noted in its ability to stimulate the uptake of potassium ions into mitochondria.[5] Nonactin shows a high degree of cation specificity in regulating metabolic behavior. The translocation of sodium ion by nonactin is not as effective as that of potassium, and lithium is not translocated across mitochondrial membranes at all. The actins are very effective at enhancing the transport of cations across membranes, and are even able to mediate cation transport in carbon

---

*To whom correspondence should be addressed
†Present address: Union Carbide Corp., Bound Brook, N.J. 08805

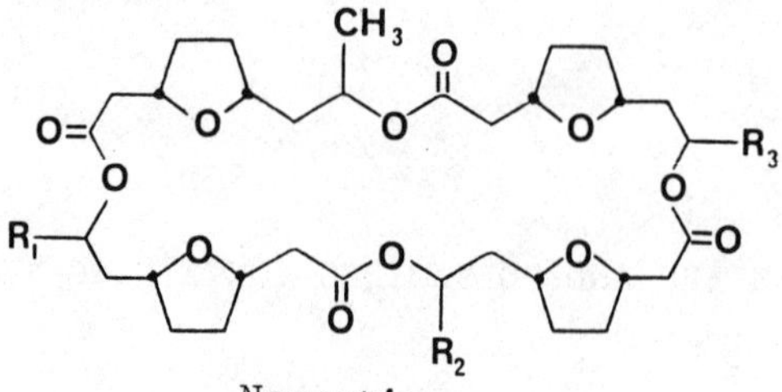

Nonactin

$R_1 = R_2 = R_3 = Me$

Fig. 1. The Actins

tetrachloride.[6] Later, through careful extraction studies with
alkali metal picrates, the salt-extraction equilibrium constants
were obtained for nonactin and its homologs.[7] These data clearly
demonstrated the selectivity of the ion-binding behavior of the
actins.

Naturally occurring ion-binding materials which are able to
translocate cations across a lipid barrier are referred to as
ionophores. For the purposes of this discussion, any compound
that binds cations will be considered an ionophore, and the terms
ion-binding material and ionophore will be considered synonomous.

The actins form lipid-soluble alkali ion complexes. Several
studies have been conducted to elucidate the mechanism of action
of these ionophores.[8,9] The complexed ion is bound to the ionophore by induced-dipole interactions with the oxygen atoms of carbonyl, ether, and/or hydroxyl groups. The ionophore encloses the
ion so that polar groups are directed toward the interior, while
the lipophilic hydrocarbon groups are exposed on the exterior.
This model of the ionophore/cation interaction is confirmed by
X-ray analyses of the potassium/nonactin complex. In nonactin,
the potassium ion is complexed through the four tetrahydrofuran
oxygen atoms and four carbonyl oxygen atoms to form a complex
folded to resemble the seam of a tennis ball.[10] No binding was
noted through the ester oxygen atoms.

In 1967, while these structural studies were underway,
Pedersen accidentally discovered a class of macrocyclic polyethers
which he called crown ethers[11,12] (Fig. 2). Before 1967, the literature on macrocyclic polyethers was scant, and their complexing
ability was not recognized. Pedersen understood the utility of

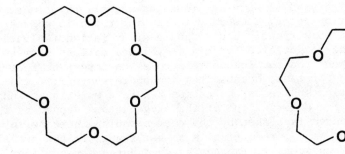

Fig. 2.   18-Crown-6 and 15-Crown-5

these compounds as models in membrane transport experiments. He
prepared a large number of these compounds, and proposed a mechanism of formation of salt-polyether complexes similar to the one
proposed for the macrotetrolides such as the actins.[11]

The relative stability of a cation-polyether complex depends
on four factors: (1) the basicity of the oxygen atoms of the polyether, (2) the charge of the cation, (3) the size of the cation,
and (4) the solvent in which the complex is prepared. In the presence of a cation, the electrons on the oxygen atoms in the polyether ring are polarized toward the cation forming coordinate
bonds. Polyether complexes become more stable as the constituent
oxygen atoms become more basic. This stability was demonstrated
early by Pedersen,[11] who noted that the less basic oxygen atoms
adjacent to an aromatic carbon atom were poorer cation ligands.
Studies of ion-binding to electron deficient oxygen atoms by
Izatt,[13] Cram[14] and Gray[15] confirmed the inability of weakly basic
oxygen atoms to function as efficient ligands. Later, when researchers began using heteroatoms other than oxygen, they found
that larger, more polarizable atoms such as sulfur were poorer
ligands for alkali and alkaline earth metal cations, but are effective ligands for small cations such as copper (II) and mercury
(II).

A more highly charged cation should polarize the oxygen atoms
more readily, making the electron pairs more available for coordinate bonding. Potassium and barium cations are approximately
the same size, and yet the barium complex of 18-crown-6 has a
stability constant greater than that for the potassium complex.
The higher stability constant is attributed to greater polarization of the oxygen atoms in the polyether by the bivalent barium
cation.[16] Thus, mono-, bi- and trivalent cations should have progressively greater ion-dipole interactions with the oxygen atoms,
resulting in increasing stability constants. However, increasing
the cationic charge results in a more tightly solvated cation in

solution. In highly polar solvents (such as water), this increase in cation solvation energy caused by increasing cationic charge leads to an even more tightly bound sphere of hydration (solvation). In this case, the competition for higher valence cations between the (polar) solvent and the ionophore becomes an important factor in the complexing process, because the ionophore must first displace the solvation sphere of the cation before complexing. Thus the increase in the polarization of the oxygen atoms caused by the higher cation charge is offset by the corresponding increase in cation solvation energy, resulting in no net change in the stability constant. Consequently, increasing the charge on a cation only results in higher stability constants in weakly polar or nonpolar solvents.[16,17]

The ability of a cation to polarize the polyether oxygen atoms is also a function of its size: larger cations have lower charge densities, and do not polarize atoms as well as small cations with high charge densities. Further, a stable complex is not formed if the ion is too large to fit in the "hole" of the polyether ring. The hole of the polyether ring is defined by the coplanar system of oxygen atoms in the ring. An oxygen atom is considered to be coplanar if it lies in the same plane as all the other oxygen atoms in the ring. X-ray studies of the potassium complex of dibenzo-30-crown-10 showed that strict coplanarity was not a prerequisite, and that large crown ethers wrap themselves around the cation in the same manner as nonactin.[18] However, in most cases, it was found that the oxygen atoms of cation-polyether complexes were, in fact, coplanar.

The structures of cyclic polyether-cation complexes postulated by Pedersen,[11] and later confirmed by X-ray studies[18,19] do not require removal of the entire solvation shell, because coplanar complexing of the cation with a polyether still allows solvent contacts perpendicular to the plane of the ring.[20] Perpendicular solvent contact was noted experimentally when X-ray studies of the sodium dibenzo-18-crown-6 complex revealed two molecules of water complexed above and below the plane of the polyether ring.[18] These results were noteworthy, because the complex had been prepared in methanol, and no water had been deliberately added to the crystallizing solvent, indicating the preference of the cation to bind to water rather than to methanol.

In contrast to these polyethers, macrocyclic antibiotics such as the actins complex cations by much larger rings which wrap themselves completely around the cation. The wrapping of the ionophore requries complete removal of the solvent shell of the cation, and this solvent stripping was demonstrated experimentally by Prestegard and Chan[21] in NMR studies of nonactin. Because polyether complexation requires at least partial displacement of the solvent shell surrounding the cation, the complex formation is minimized if the ion is strongly bound to the solvent.[11] Therefore, larger cations have lower stability constants because they are unable to fit with the cavity of the cyclic polyether, and smaller cations also have lower stability constants because they are too strongly solvated for the cyclic polyether to compete successfully for them.[20]

Since Pedersen first reported the synthesis and cation-binding behavior of cyclic polyethers, literally hundreds of additional crown compounds have been reported. Several reviews have appeared.[17]

A new class of multidentate compounds, the polyether-esters, has recently been reported.[22] By condensing pentaethylene glycol with malonyl chloride, Bradshaw and coworkers were able to prepare compound 1. Compound 2 was prepared in an analogous manner (Fig. 3). Space-filling models of 1 and 2 indicated that both carbonyl groups cannot be directed into the cavity of the macrocycle. Binding of magnesium to the macrocycle was observed.[23] Polyether-ester 3 was prepared as a simple model of actin antibiotics, but no ion-binding properties were reported[24] (Fig. 3).

Heterocyclic rings such as pyridine have been incorporated into the polyether ring as shown in compound 4.[25,26] Reinhoudt and Gray[27] incorporated furan rings into cyclic polyethers of type 5 (Fig. 4). Complexing of the macrocycle with Zeise's salt (potassium trichloro(ethylene)platinate (II), $K[PtCl_3(C_2H_4)]$) was noted, as indicated by downfield shifts in the NMR signals of the furan and furyl protons. Other researchers[15,28] prepared structurally similar crown ethers substituted with furan, thiophene, and various aromatic systems.

Tetrahydrofuran should be an excellent ionophore component because of its balance between hydrophilic and lipophilic character. Cram and coworkers[14] prepared a series of tetrahydrofuran containing crown ethers (6), and found that these substituted cyclic polyethers were equally as effective in cation binding as the corresponding monocyclic crown ethers (Fig. 5).

Kobuke and coworkers[29] prepared a series of macrocycles consisting entirely of tetrahydrofuran rings (7, 8). These materials did bind alkali metal cations, but the larger macrocycles were not as effective as their corresponding crown ethers. The smaller macrocycles were as effective in their cation-binding capacity as crown ethers of the same size (Fig. 5).

Bradshaw and coworkers[13] prepared macrocycles that combined the ester and tetrahydrofuran elements into crown ethers (9, 10, 11; Fig. 5). These materials also bound cations, but were not as effective or as selective as the corresponding crown ethers. Spectroscopic studies of the carbonyl group in both the infrared and

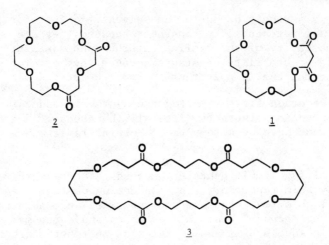

Fig. 3.   Polyether-ester Macrocycles

ultraviolet regions indicated that it was not involved in complexation.

     The discovery of the cation-binding properties of crown ethers
renewed interest in studying the ion-binding behavior of linear polyethers and their homologs.   Linear oligoethylene glycol dimethyl
ethers (glymes) bind cations more weakly by two or three orders of

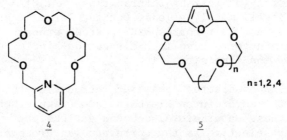

Fig. 4.   Macrocycles 4 and 5

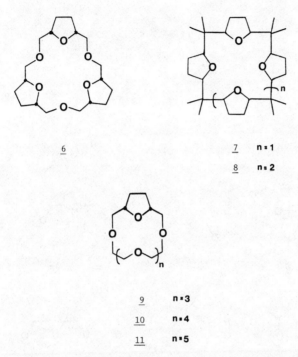

6

7   n = 1

8   n = 2

9   n = 3

10   n = 4

11   n = 5

Fig. 5.   Synthetic Tetrahydrofuran-Containing Macrocycles

magnitude compared to crown ethers.[30,31]   Japanese workers[32] studied
the cation complexation of glymes by NMR spectroscopy. Downfield
shifts for the methylene protons were noted when a complex was formed. However, if aromatic groups were incorporated at the ends of
the polyether chains, the cation-binding ability was dramatically
improved.[31,33]   Vogtle and coworkers prepared several polyethers
of different chain lengths with terminal quinoline groups.[31,34,35]
They found that compounds 12 and 13 were very flexible as shown by
the lack of specificity between sodium and potassium. In contrast,
the introduction of a bridging pyridine group renders the ligand

more rigid, and causes preferential binding of sodium to compound
14^{34} (Fig. 6). Ligands containing five heteroatoms or less form
circular complexes, while larger arrays of ligands (15, Fig. 6)
lead to helical structures, as determined by X-ray crystallography.[33]

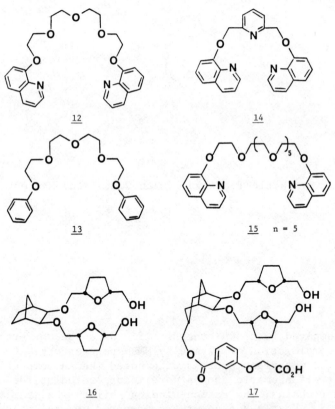

Fig. 6. Synthetic Acyclic Ionophores

Wierenga and coworkers[36,37] prepared acyclic ionophores with specificity for calcium. These ionophores consisted of furan or tetrahydrofuran ligands pendent from bicyclo[2.2.1]heptane (16) or bicyclo[3.2.1]octane (17) systems (Fig. 6).

Finally, Kirsch and Simon[38] prepared two different, acyclic ionophores (18, 19; Figure 7), studied the ion-selectivity, and determined the stability constants. They found high selectivity for alkaline-earth metals, with respect to alkali metal cations.

As soon as the novel ion-binding character of crown ethers was discovered, researchers sought ways to incorporate these character-istics into polymeric structures.[39] An ion-binding site could be incorporated into the polymer backbone, or the site could be pendent from the polymer backbone. The ion-binding site could be either a macro-cyclic or acyclic multidentate ligand. A review of crown ethers, including polymer-bound macrocyclic polyethers, has recently appeared.[40]

The first reported incorporation of a crown ether into a polymer backbone was the preparation of structure 22 by condensing isophthal-oyl chloride with 4,4'diaminodibenzo-18-crown-6.[41] Later workers[42] were able to prepare higher molecular weight polymer 22, and found that they could cast strong, highly permeable membranes. The ion-selectivity of the polymeric crown ether was retained. Blasius and coworkers[43] prepared a number of polymers with either cryptates (24) or crown ethers (23) as integral parts of the polymer backbone. Some of these materials are listed in Fig. 9. Blasius found that

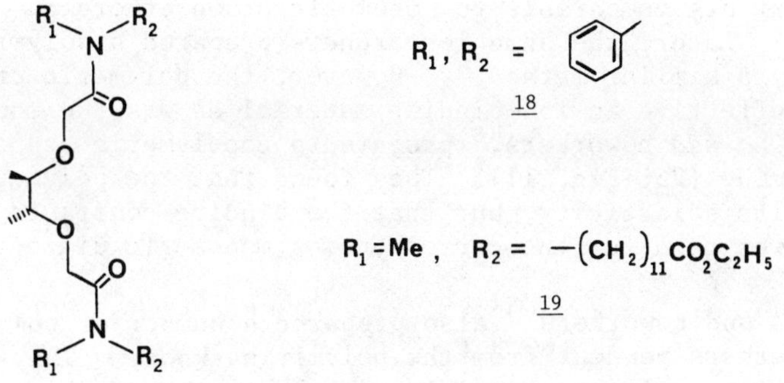

$$R_1, R_2 = \text{(phenyl)}$$

18

$$R_1 = Me, \quad R_2 = -(CH_2)_{11}CO_2C_2H_5$$

19

Fig. 7.   Ionophores 18 and 19

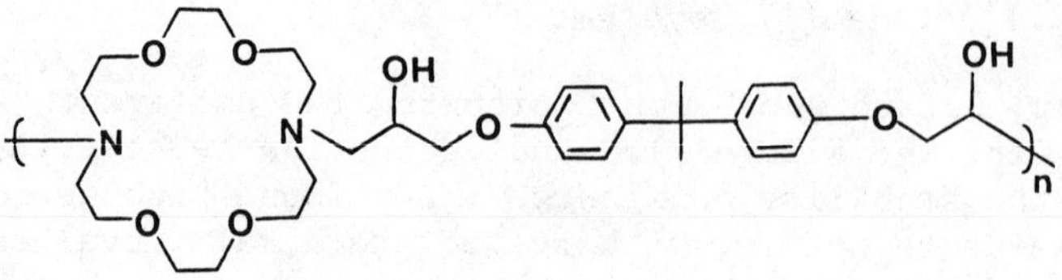

20

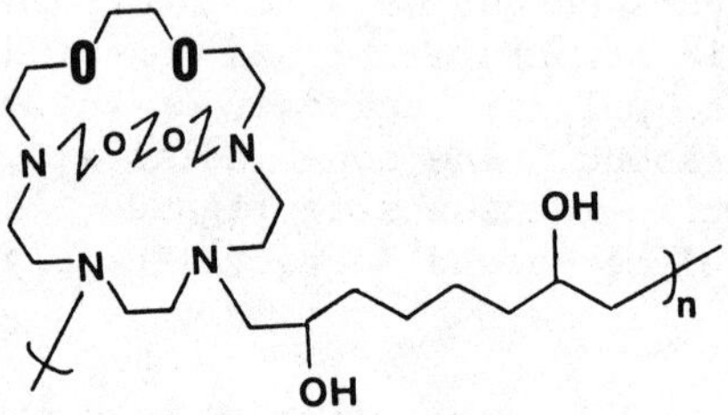

21

Fig. 8.   Ion-Binding Polymers 20 and 21

these polymers were useful in ion-exchange chromatography. A
cryptate is a bicyclic species with two terminal nitrogen atoms
bridged by three polyether chains (see Fig. 10). Gramain and
Frere[44] also prepared an ion-binding polymer containing diaza-crown
ether moieties as integral portions of the polymer backbond (20;
Fig. 8). The polymer was an effective ion-binding material, with
binding constants comparable to monomeric crown ethers on a per
crown basis. Later, the same researchers prepared a polymric cryptate (30) by a similar method.[45] However, the polymeric cryptate
was not as effective an ion-binding material as was the monomeric
cryptate. Cho and coworkers[46] prepared a copolymeric aza crown
ether/polyimide (25; Fig. 11). They found that the polymer retained most of its selectivity, but that the binding constants were
somewhat lower than for the corresponding monomeric diazacrown ether.

Blasius and coworkers[43] also prepared a number of polymers
with crown ethers pendent from the polymer backbone. They noted
the effectiveness of these materials in ion-exchange chromatography,
but did not conduct specific ion-binding measurements.

The most successful effort to date to incorporate pendent
cyclic polyethers onto a polymer backbone was accomplished by Smid

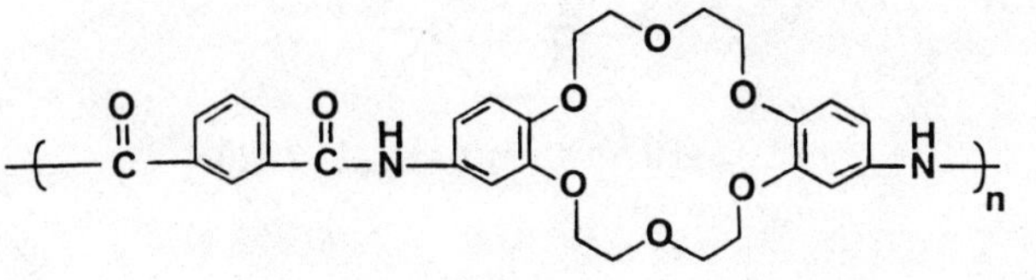

Fig. 9. Polymers Containing Macrocyclic Ionophores

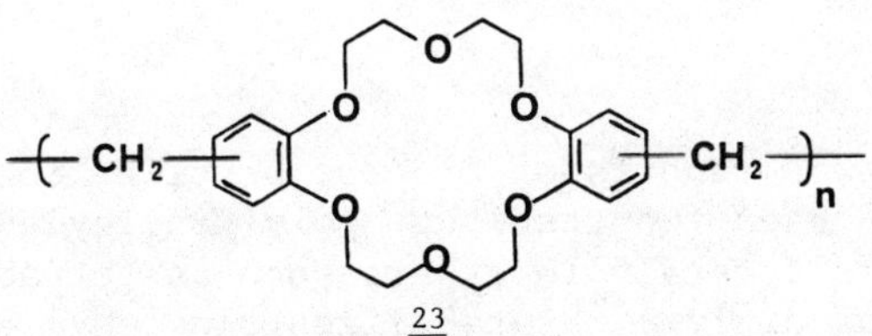

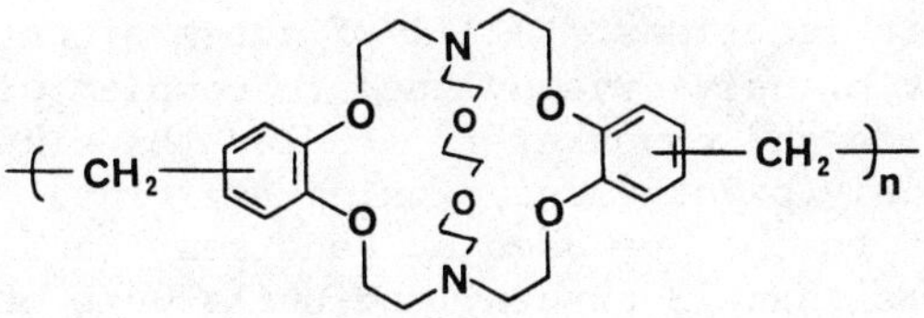

and coworkers.[47,48] They prepared a polystyrene backbone with pendent crown ethers (26) by polymerizing 4'-vinylbenzocrown ethers (Fig. 12). The polymers obtained were found to be as effective or more effective in binding cations than the corresponding monomeric macrocyclic polyethers. Later, Smid and Sinta[69] reported the preparation of poly(vinylbenzoglymes) (27, 28): polystyrene derivatives with one or two $O(CH_2CH_2O)_n CH_3$ (n=2, 3, or 7) substituents. These materials bound cations weakly compared to the corresponding polymeric crown ethers. Yagi and coworkers[49] prepared a series of polymethacrylamides with pendent crown ethers (29) similar to those prepared by Smid, with similar ion-binding properties (Fig. 12).

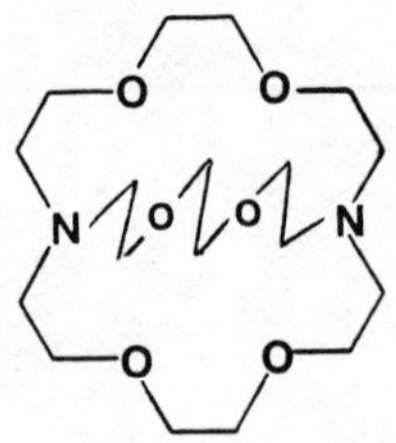

Fig. 10.   Cryptate [2.2.2]

There are fewer reports of linear, acyclic, ion-binding
polymers. It has been reported that poly(oxyethylene) improves the
solubility of alkali metals in ethers such as tetrahydrofuran, di-
methoxyethane, and diglyme,[50] stabilizes fluorenyl alkali metal compounds,[51] accelerates Williamson reactions[52] and accelerates several
other nucleophilic reactions.[52]   All of these effects were attributed to the ability of poly(oxyethylene) to complex with cations in
solution.   Yanagida and coworkers[53] studied the alkali metal cation
complexation of poly(oxyethylene), using a picrate salt extraction
technique similar to the one used by Pedersen[11] and Frensdorff.[20]
Polymers with more than 23 oxyethylene units were effective ionophores for potassium, with degrees of extraction (percent extracted)
comparable to crown ethers.   The extractability per oxyethylene unit
was nearly constant, and the complex stability increased linearly
with increasing numbers of repeating oxyethylene units.   Seven
oxyethylenes were the minimum number of repeat units necessary to
bind potassium ion effectively in the aqueous phase.   The less efficient extraction of short-chain poly(oxyethylene) is apparently
caused by its hydrophilic character.

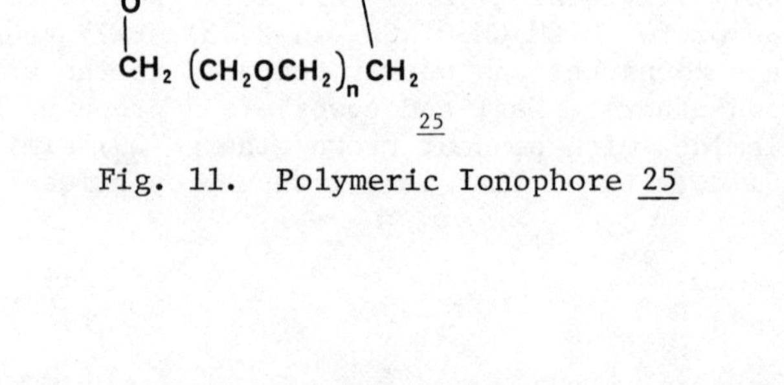

25

Fig. 11.   Polymeric Ionophore 25

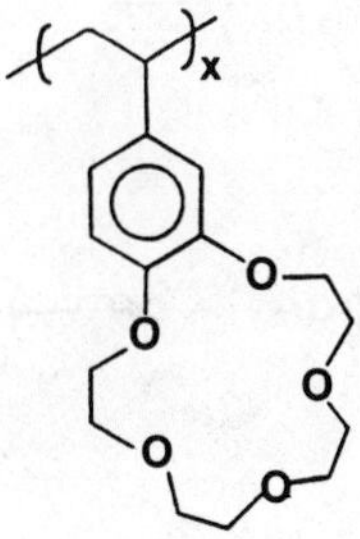

Fig. 12.   Polymers Containing Pendent Ionophores

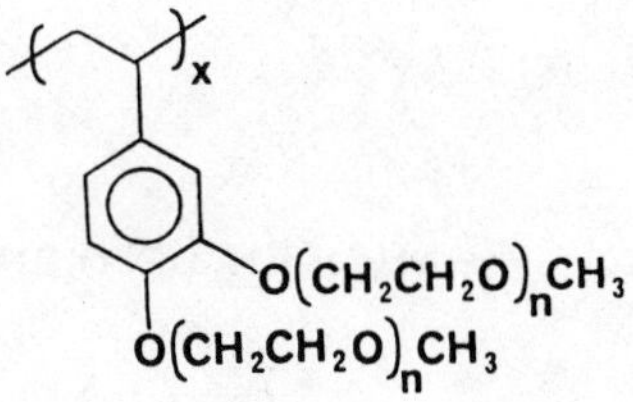

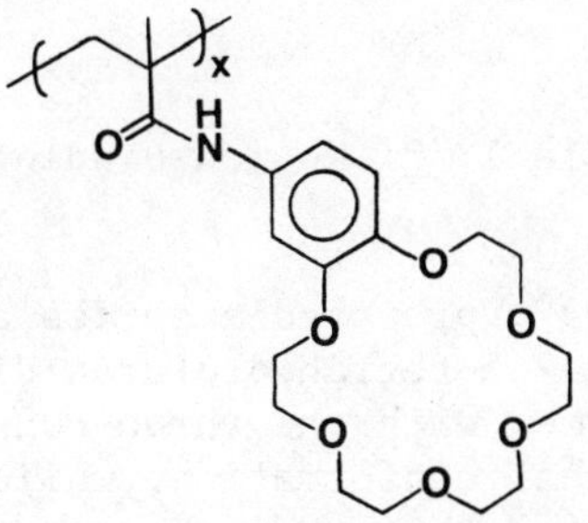

Tadokoro[54] established the structure of poly(oxyethylene) in the crystalline state as a distorted helical structure containing seven oxyethylene units in two turns of the helix. Molecular models of the polymers indicated that with a slight amount of bond rotation, a cavity similar in size to 18-crown-6 is formed. Yanagida[53] postulated that poly(oxyethylene) may exist in solution in a helical conformation, or that the addition of cationic species would induce a helical structure.

Böhmer recently prepared a series of polyamides containing specific numbers of oxyethylene units (30). By reducing the polyamide with lithium aluminum hydride, the corresponding polyamine was prepared (31; Fig. 13). Preliminary reports on the polyamide indicate that it does bind cations,[55] and preliminary ion-binding

$$\left(\!-NHCH_2CH_2\left(OCH_2CH_2\right)_x NHCO\left(CH_2OCH_2\right)_2 -\overset{\overset{\displaystyle O}{\displaystyle \|}}{C}\!\right)_{\!n}$$

<u>30</u>

$$\left(\!-NHCH_2CH_2\left(OCH_2CH_2\right)_x NHCH_2\left(CH_2OCH_2\right)-CH_2\!-\!\right)_{\!n}$$

<u>31</u>   x = 3
<u>32</u>   x = 4
<u>33</u>   x = 5

Fig. 13.   Polymers <u>30</u> and <u>31</u>

data is included in Table 1.[56] No ion-binding data is available
for the polyamines.

Smith and coworkers[57] prepared a series of ion-binding polymers
consisting of recurring 2,5-tetrahydrofuran diyl units. They found
that models of the polymer with the threo-configuration (<u>34</u>) were
able to exist in a helical conformation, while the erythro-configuration (<u>35</u>) experiences enough crowding among the methine hydrogen
atoms that the erythro-configuration favors an extended chain conformation (Fig. 14). One would expect the threo-polymer (<u>34</u>) to
have a helical cavity with the oxygen atoms on the inside of the
coil. A variable-sized cavity with several coordinating oxygen
atoms is available for binding. No conformation of the erythro-
polymer (<u>35</u>) presents any reasonable geometry for multidentate co-
ordination of the oxygen atoms with any cation. Consistent with
these models, the polymer with the threo-configuration is an effective binder of ions, while the erythro polymer does not exhibit
any ion-binding character.

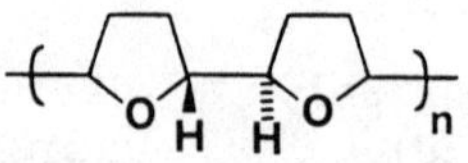

<u>34</u> 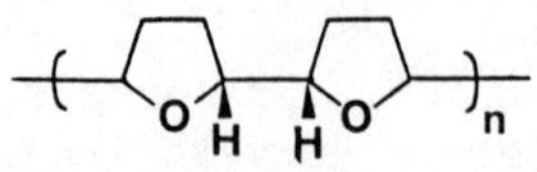   <u>35</u>

Fig. 14.   Polymers <u>34</u> and <u>35</u>

Sumitomo and coworkers prepared 6,8-dioxabicyclo[3.2.1]octan-7-one (37) and 6-aza-8-oxabicyclo[3.2.1]octan-7-one (36) from 3,4-dihydro-2H-pyran-2-carbaldehyde (acrolein dimer), and polymerized these monomers by ring-opening methods[58] (Fig. 15). The lactam (36) was polymerized to a high molecular weight polyamide (39). Films of the polyamide showed good permeation characteristics. The diffussion coefficient of potassium ion in films of this material was greater than that of sodium ion, and both coefficients were higher in polymer 39 than in cellulosic membranes. No specific ion-binding studies were conducted. Later, optically active 6-aza-8-oxabicyclo[3.2.1]octan-7-one (36) was resolved using dehydroabietyl-amine, and polymerized to yield optically active polymer.[59] Diffusion coefficients for the optically active polymer were similar to those of the racemic polymer. No ion-binding data were reported. Cationic polymerization of lactone (37) gave linear polyester or cyclic oligomers with degrees of polymerization of two, four, or six, depending on the reaction conditions. Complexation of the cyclic hexamer and metal thiocyanates was noted. In the infrared spectrum, the carbonyl stretching vibration was shifted to lower frequency, indicating metal bonding to the nonbonding electrons on the carbonyl oxygen atoms. Methine protons adjacent to the carbonyl groups showed pronounced downfield shifts in the NMR spectrum, and ^{13}C NMR spectroscopy showed large shifts in the carbonyl carbon atom signals. Sumitomo concluded that the cation interaction is primarily with the carbonyl oxygen atom.[59] Sumitomo recently reported the preparation of an optically active cyclic tetramer of 37, and reported that an acetonitrile molecule can be held in the macrocyclic cavity.[60] No further ion-binding data was reported.

In earlier work, Moore and Kelly[61] reported the preparation of polyesters based on 2,5-disubstituted furans in various states of reduction. As an approximation to the structure of nonactin, lactone 45 was prepared and polymerized with tetra-t-butoxytitanate.[62] In preliminary measurements, significant ion-binding ability was not noted.

The object of this research is to incorporate the structural elements of the macrotetrolide antibiotic nonactin into a linear polymer. These structural elements consist of a cis-tetrahydrofuran ring, an ester linkage and a variable number of methylene carbon atoms between the ester and the tetrahydrofuran ring. The ion-binding behavior of the polymer is being studied to try to determine which structural and stereochemical features are of importance in modifying binding capacity. In addition to ultimately preparing a series of polyesters, it was also decided to study the corresponding polyamides. Models of the polyamides and polyesters containing cis-tetrahydrofuran units are able to form helical structures similar to those postulated by Smith[57] for polymer 34, and Yanagida[53] for poly-(oxyethylene). If such helical structures do form, the oxygen atoms should be on the inside of the coil, thus creating an ion-binding

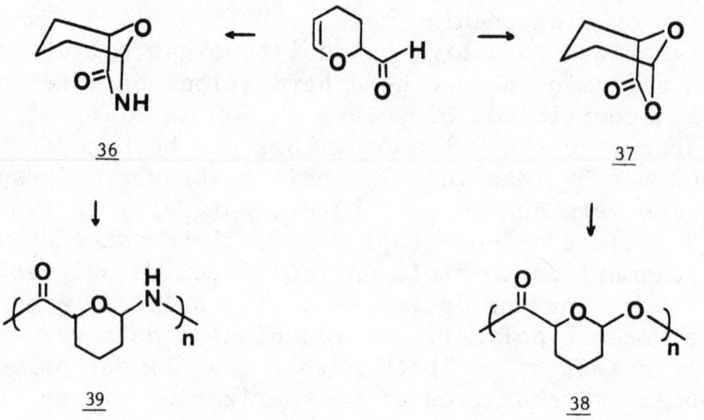

Fig. 15.   Preparation of Polymers 38 and 39

cavity similar to 18–crown–6.   If a helical structure forms and preferentially binds only specific–sized ions, then it would be reasonable to assume that the cavity formed by the polymer is not variable in its diameter, and is fairly rigid in solution.   If, on the other hand, the binding of ions shows no size preference, then this fact would imply either a helical structure with a cavity of variable diameter, or a non–helical, flexible polymer which is able to bind different size ions in an unspecified, multidentate manner.

Ion–binding polymers might have several potential uses, including ion–selective electrodes and reverse–osmosis or dialysis membranes.   In addition to cation selectivity, good film–forming and water permeation characteristics would be necessary.

As initial synthetic targets, polyester 40 and polyamide 41 (Fig. 16) were chosen, as representatives of a series of tetrahydro-furan–containing polymers, because of their structural simplicity

Fig. 16.   Polymers Prepared in This Work

and similarity to nonactin. The desired polymers were prepared by
ring-opening polymerization.

RESULTS AND DISCUSSION

A. Preparation of Monomers

      2-Oxo-3,8-dioxabicyclo[3.2.1]octane (45) was prepared by the
method of Moore and Kelly[62] using the process outlined in Fig. 17.
3-Aza-8-oxabicyclo[3.2.1]octan-2-one (49), the lactam analog of (45)
was prepared[63] by the procedure outlined in Fig. 18.

B. Polymerization of Monomers

      3,8-Dioxabicyclo[3.2.1]octane-2-one (45)

      After purification by molecular distillation, the monomer was
polymerized in bulk, at slightly elevated temperatures. Typically,
250 to 300 mg of monomer were placed in a dry test tube under nitrogen, sealed with a serum cap, and the initiator was added as a solution in diethyl ether. The polymerization was conducted under a
positive pressure of dry, deoxygenated nitrogen, and excess diethyl
ether was entrained. After the specified period of time had elapsed, the reaction mixture was dissolved in chloroform and the polymer
was precipitated from petroleum ether F as a sticky, resinous solid.
Viscosity measurements were made on chloroform solutions at 25°.
The apparent number- and weight-average molecular weights of polyester 40, with an intrinsic viscosity of 0.082, were found to be
11,000 and 45,000 by gel-permeation chromatography, respectively.

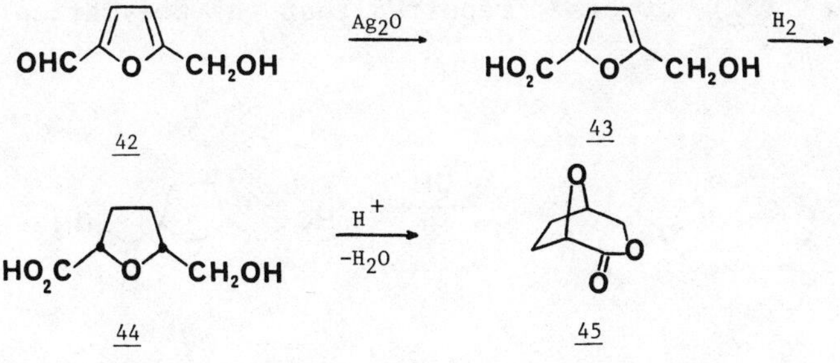

Fig. 17. Synthesis of 3,8-Dioxabicyclo[3.2.1]octan-2-one (45)

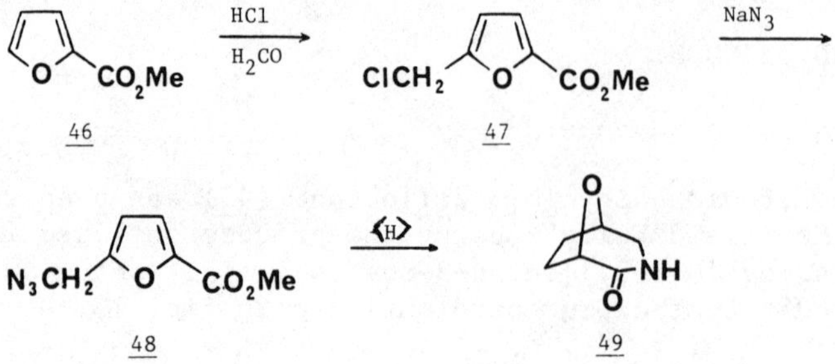

Fig. 18.   Preparation of 3-Aza-8-oxabicyclo[3.2.1]-
octan-2-one (49)

To determine the structure of the polyesters, they were
hydrolyzed by refluxing in dilute, aqueous sodium hydroxide solution (Fig. 19).   The mixture was acidified with Amberlyst-15 ion-
exchange resin, and the water was removed in vacuo.   For the polymer
prepared from tetraisopropyl titanate (Tyzor TPT), the hydrolysis
product was identified as hydroxy acid 44. This result confirmed the
work of Moore and Kelly.[62]   However, for the polymer prepared with
triisobutyl aluminum as catalyst, hydrolysis gave a white solid.
One possible explanation would be that the aluminum alkyl initiator
polymerized lactone 45 through the oxygen bridgehead atom to give a
polyether rather than a polyester (Fig. 20).   Another possible explanation is the epimerization of the lactone during the polymerization, giving a trans hydroxy acid repeat unit in the polymer.

The ring-opening polymerizations of oxetanes (1,3-epoxides)
with Lewis acids such as boron trifluoride etherate and aluminum
alkyls are well documented.[64]   Hall and coworkers[65] reported the
ring-opening polymerization of 7-oxabicyclo[2.2.1]heptane (50) with
Lewis acids such as boron trifluoride etherate to give poly(cyclo-
hexene oxide) (51).   Ogata[66] reported that the polymerization of

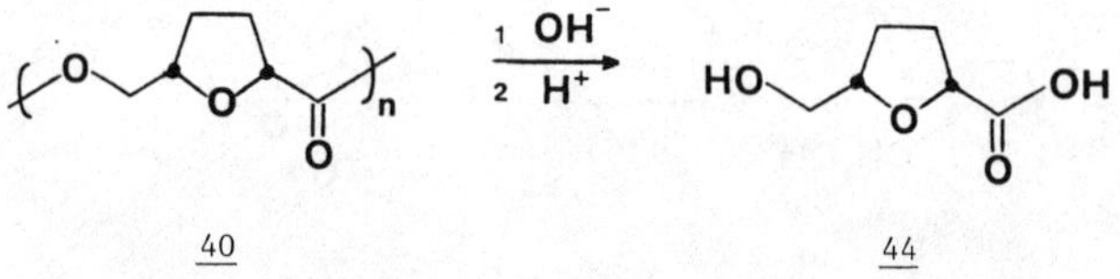

Fig. 19.   Hydrolysis of Polyester 40

Fig. 20.   Possible Route of Polyether Formation

3-aza-10-oxabicyclo[4.3.1]decan-4-one (52) proceeds through opening
of the ether linkage (Fig. 21).

However, spectral data do not support the formation of polyether
units. The NMR and IR spectra of the polyester prepared with triisobutyl aluminum were identical to those of the polyester prepared
with Tyzor TPT or TBT. The NMR spectrum of the hydrolysis product
was identical to that of cis-hydroxy acid 44, except that the proton
absorption for the hydroxyl methylene group was observed at 3.82 ppm,
rather than at 3.93 ppm. It is likely that the solid isolated is
trans-5-hydroxymethyl-2-tetrahydrofuroic acid.

3-Aza-8-oxabicyclo[3.2.1]octan-2-one (49)

After purification by recrystallization from toluene, monomer
49 was polymerized hydrolytically.[79] Polymerization of lactam 49,
either cationically or anionically, might result in epimerization
of the cis-tetrahydrofuran ring junction which would result in a
polymer containing a mixture of cis and trans tetrahydrofuran rings
(Fig. 22). The importance of tetrahydrofuran ring geometry has

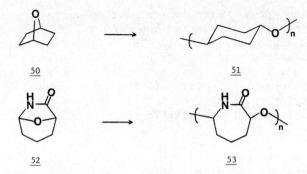

50   51

52   53

Fig. 21.   Polyethers Prepared by Ring-Opening

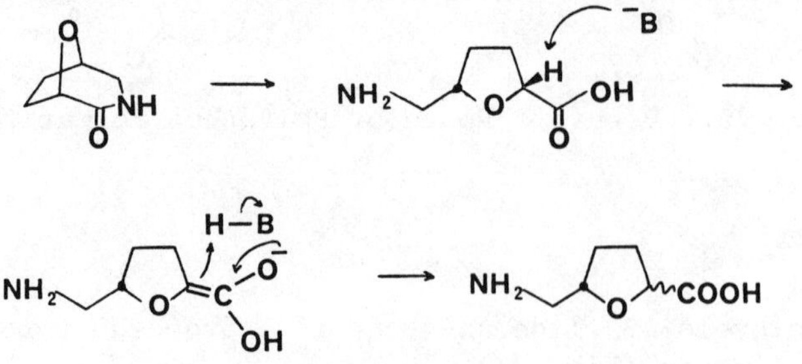

Fig. 22.   Anionic Epimerization of Lactam (49)

already been demonstrated by Smith[57] with polymers 34 and 35. Because we desired only the cis tetrahydrofuran polyamide, the polyamide used in the ion-binding studies was prepared by hydrolytic polymerization. Water is a weak acid and base, and it is unlikely that epimerization would occur under hydrolytic polymerization conditions. After a specified period of time had elapsed, the reaction mixture was dissolved in chloroform, and the polymer was precipitated from diethyl ether as a white, somewhat fibrous solid. Viscosity measurements were made on solutions in either chloroform or hexafluoroisopropanol at 25°. Films of the polymer cast from chloroform were optically clear, but very brittle.

The proton nuclear magnetic resonance spectrum of polyamide 41 was not very helpful in structural determination because of excessive peak broadening. The infrared spectrum of polyamide 41 showed two carbonyl stretching frequencies at 1655 cm^{-1} and 1532 cm^{-1}. The absorption at 1532 cm^{-1} was absent in monomer[49] and this band is referred to as the amide II band.[58] It is characteristic of a trans amide group, which indicates that polymerization occurred through opening of the lactam bond and not through opening of the ether bridgehead oxygen bonds. The infrared data were interpreted as indications that the structure of polymer 41 is a polyamide and not a polyether. Similar results were obtained by Sumitomo[58] with 6-aza-8-oxabicyclo[3.2.1]octan-7-one (36), who noted a carbonyl stretching frequency in the monomer at 1721 cm^{-1}, and two carbonyl absorptions at 1678 cm^{-1} and 1528 cm^{-1} in the polymer (39).

A sample of the polyamide with [η] = 0.3, had a glass transition temperature (T$_g$) by differential scanning calorimetry (DSC) of 118°. Other structurally similar polyamides such as Nylon 4

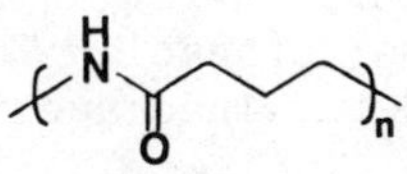

Fig. 23.   Structurally Similar Polyamides

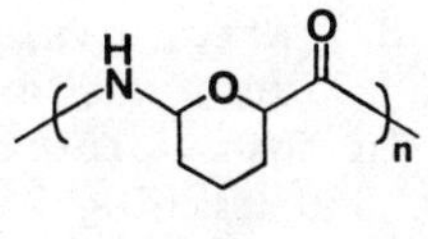

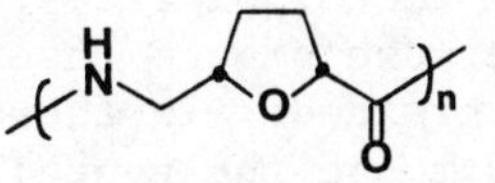

(54) [poly(imino-1-oxotrimethylene)] and Sumitomo's polyamide 39 have values of T_g of 111°[67] and 130°[58], respectively (Fig. 23). No melting transition was observed in the DSC trace below 350°; either the polymer is completely amorphous, or the melting point is higher than 350°. Examination of the polymer samples after DSC showed that they were partially charred, indicating that the decomposition temperature was close to 350°, and a higher melting point probably did not exist.

The number-average molecular weight ($\overline{M}_n$) was determined by vapor-pressure osmometry (VPO) in 1,2-dichloroethane solution, and was found to be 4000; $\overline{M}_n$ (GPC) = 12,400 ± 200; $\overline{M}_w$ (GPC) = 48,000 ± 200.

C.   Ion-Binding Measurements

The combination of small molecules or ions with macromolecules is governed by the usual laws of equilibrium. Strictly speaking, the equilibrium expressions are based on activities and not on concentrations. However, the correction from concentration to activity is rarely justified when dealing with polymer systems because of the relatively high uncertainty in the determination of exact binding constants.[68]

The general ion-binding technique   used in this work was similar to that employed by Smid,[47,48,69] Frensdorff,[20] Pedersen[11] and Frere.[44]   In a typical ion-binding experiment, equal volumes of a polymer solution in chloroform and an alkali metal picrate solution in water were shaken at regular intervals (approximately 5 minutes) in a thermostated water bath for one hour. The concentration of the polymer solution was usually 2 x 10^{-2} M (as a concentration of monomer repeat units), and the picrate solution varied between

10^{-2} and 10^{-5} M.   After mixing for one hour, the organic layer was removed and an aliquot was diluted with 95% ethanol. The picrate concentration in this solution was then determined spectrophotometrically, at 360 nm (log ε = 4.15).

Originally, the alkali metal picrates were employed in ion-binding experiments because of the ease with which the cation concentration could be determined. In these cases, the picrate anion is simply considered to be the gegenion of the alkali metal cation, and as a spectrophotometric label by which the cation concentration could be determined. Smid[69] later found that some polymeric systems with pendent ionophores not only bind cations in the usual way, but that these systems also bind the picrate anion directly. In this instance we are dealing with alkali metal cations binding to the ionophore, using the picrate anion only as a spectrophotometrically detectable gegenion.

Most other researchers[11,20,44,47,69] who used alkali metal picrates to study ion-binding prepared the alkali metal picrates in situ by titrating aqueous picric acid with the desired alkali metal hydroxide. In this work, the alkali metal picrate salts were prepared in crystalline form by reacting picric acid with alkali metal hydroxides or carbonates in water, followed by three recrystallizations from water.[70] The picrate salts were then dried in vacuo, and aqueous solutions were prepared from the crystalline salts. This method was judged superior because no competing ionic species such as unreacted picric acid or alkali metal hydroxide would be present.

In extraction experiments with chloroform and aqueous alkali metal picrate solutions, no picrate anion was detected in the organic phase. Within the limits of detection of the spectrophotometer (4 x 10^{-7} M), the picrate anion was insoluble in chloroform. All of the picrate solutions obeyed Beer's law over the absorbance range (0.15 to 0.65) used in the spectral measurements.

One hour was allowed for the extraction mixture to reach equilibrium. Pedersen,[11] Frensdorff,[20] Frere[44] and Smid[47,69], using similar extraction methods, found that equilibrium was rapidly achieved, usually after a few minutes. With the polymers prepared in this work, prolonged extraction experiments failed to increase the concentration of picrate in the organic phase.

In the absence of polymer, none of the picrates are extracted into the chloroform phase. In the presence of either polyester 40 or polyamide 41, a portion of the alkali metal picrate is extracted into the organic phase. It is reasonable to assume that the alkali metal picrate that is extracted into the organic phase in the presence of polymer is being bound to the polymer in some way. It is assumed that only polymer-bound alkali metal picrate will exist in

the organic phase. Consequently, the concentration of picrate in
the organic phase is equal to the concentration of bound solute.
Conversely, because all of the alkali metal picrate in the organic
phase is bound solute, the alkali metal picrate remaining in the
aqueous phase is unbound, or free solute.

Log-log plots of bound solute versus free solute for polyester
40 are nearly linear. At high free and bound solute concentrations,
some deviation from linearity is noted (see Fig. 27, 28). Similar
log-log plots for polyamide 41 are virtually flat; changing the
free solute concentration by two or three orders of magnitude fails
to change the bound solute concentration by more than a factor of
two (Fig. 29). The result is apparent solute saturation of the
polyamide. A possible cause of this behavior may be the mode of
binding of cations to polyamides, and the resulting generation of
positively charged nitrogen atoms.

It has been found that binding of alkali metal cations,
particularly lithium cations, to poly(vinylpyrrolidone),[70] poly-
(hydroxyethyl aspartamide),[70] and poly(hydroxyalkyl glutamines)[71]
resulted in expansion of the neutral polymers in a pseudopolyelec-
trolyte fashion. The spectroscopic evidence suggests that the binding of metal cations to synthetic poly(α-amino acids) such as poly-
(1-lysine),[72] poly(1-serine),[73] and poly(1-proline)[74] occurs through
the peptide linkage.[75] The metal cation is bound through the carbonyl oxygen atom, resulting in a positive charge on the nitrogen
atom as illustrated in Fig. 24. The generation of cationic nitrogen atoms on the polymer backbone as a result of metal cation binding could cause uncoiling of the polymer through electrostatic repulsion among the cationic centers on the polymer. The generation
of such cationic centers in a polymer would repel additional metal
cations, and would supress further ion-binding. Consequently, increasing the number of bound metal cations would increase the cationic charge on the nitrogen atoms of the amide groups of the polymer. The increasing charge on the polymer decreases the affinity
of the remaining amide groups for ions. This mode of binding is
manifested in anti-cooperative behavior and apparent solute saturation, as observed with polyamide 80.

Fig. 24. Mode of Binding of Metal Cations to Amides

Another factor contributing to the difference in ion-binding between polyamide 41 and polyester 40 is the inherent rigidity of polyamides. Flexibility in polymer chains arises from rotation around single bonds in the polymer backbone. Less flexible polymers have higher melting temperatures (T_m). The T_m of poly(caprolactam) is 223-228°,[76] while the T_m of poly-(caprolactone) is 55-60°.[77] The difference in T_m is attributed to the higher rigidity of the polyamide. This reduced flexibility arises from the partial double bond character of the carbon-nitrogen bond in the polyamide. Delocalization of the non-bonding pairs of electrons on the nitrogen atom (Fig. 25) through the carbonyl group imparts some sp^2 character to the carbon-nitrogen bond. The result is a partial carbon-nitrogen double bond, which restricts free rotation about the polymer chain, and makes it more rigid. If similar effects are at work in polyamide 41, then the polymer may be too rigid to bend around a cation to bind it in a multidentate fashion. The lower molecular weight polyamide (based on viscosity measurements) seems to be a slightly better ionophore. This effect may be colligative, i.e., caused by the larger number of polymer chains in solution with equal weights of samples of different average sizes. If the anti-cooperative behavior postulated is occurring, then lower molecular weight material might be a better ionophore, because the electrostatic repulsion caused by bound cations would be distributed over a larger number of independent polymer chains, resulting in a smaller incidence of intrapolymeric repulsion.

The polymers most similar to polyester 40 and polyamide 41 are the poly(tetrahydrofuran diyls) (34 and 35) prepared by Smith and coworkers,[57] and the polyamides 30, 32, 33, 56, 57 and 58 prepared by Böhmer and coworkers[55,56] (see Table 1 and Fig. 13 and 26). Using a picrate extraction method both research groups reported similar ion-binding behavior. At roughly the same polymer and picrate concentrations, both polyester 40 and polyamide 41 are similar in their extraction capacity to Böhmer's polyamides and to Smith's poly(tetrahydrofuran diyls). Plots of percent solute extracted versus the logarithm of the total solute concentration for lithium, potassium, and cesium picrate and polyamide 41 show that the percentage varies widely over solute cencentration (Fig. 30-32). A similar curve for polyester 40 and potassium picrate is included in Fig. 31 for comparison. Its curvature is not nearly as sharp

Fig. 25.   Delocalization of Amide Electron Pair

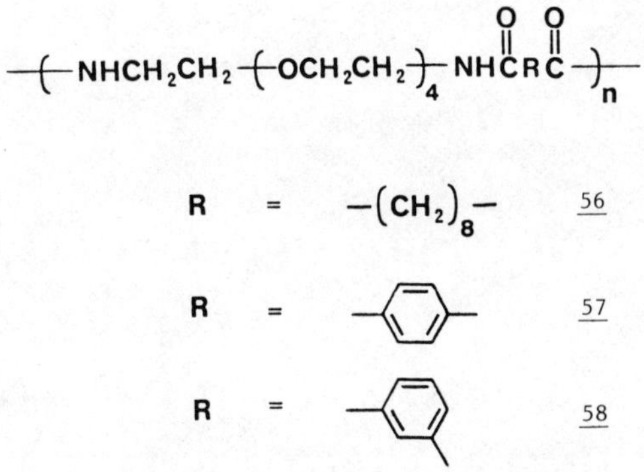

$$-\left(-NHCH_2CH_2-\left(OCH_2CH_2\right)_4-NH\overset{O}{\overset{\|}{C}}R\overset{O}{\overset{\|}{C}}-\right)_n$$

R = $-(CH_2)_8-$ <u>56</u>

R = —⟨ ⟩—   <u>57</u>

R = —⟨ ⟩   <u>58</u>

Fig. 26. Polyamides <u>56</u>, <u>57</u> and <u>58</u>

as that for polyamide <u>41</u>. As have many other researchers, Böhmer and Smith reported ion-binding data at only one solute concentration, and have not yet reported the intrinsic binding constants, making it difficult to draw comparisons to other polymer systems. It is still clear however that these polymers, with structures most similar to

Table 1. Ion-Binding Properties of Selected Ionophores

Ionophore	Percent Solute Extracted			Ref.
	Li	K	Cs	
18-crown-6	63	74	--	57
Polymer <u>34</u>	37	53	--	57
Polymer <u>35</u>	--	0	--	57
Polymer <u>30</u>	3	2	--	56
Polymer <u>36</u>	5	14	--	56
Polymer <u>37</u>	10	21	--	56
Polymer <u>56</u>	15	31	28	55
Polymer <u>57</u>	14	27	26	55
Polymer <u>58</u>	10	32	29	55
Polyamide <u>41</u>	30	32	34	a
Polyester <u>40</u>	33	33	--	a,b,c

The ionophore concentration used was 10^{-2}M. The solute concentration used was 10^{-4}M, except where noted.
    a. These values were determined in this work.
    b. Polyester <u>40</u> was prepared with i-Bu$_3$Al.
    c. The solute concentration used was 10^{-3}M.

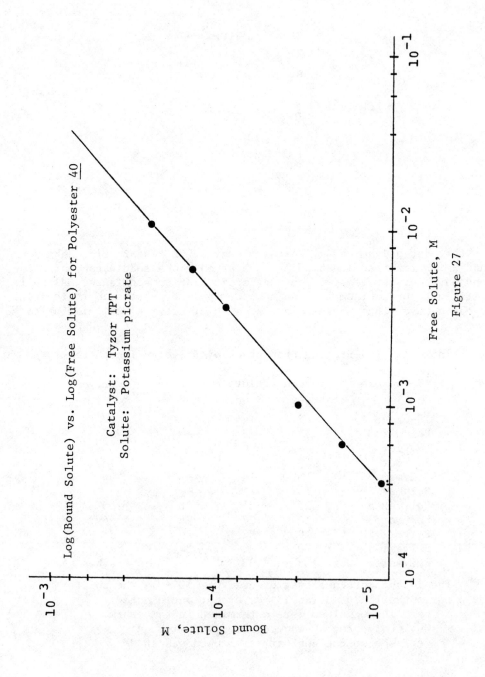

Figure 27

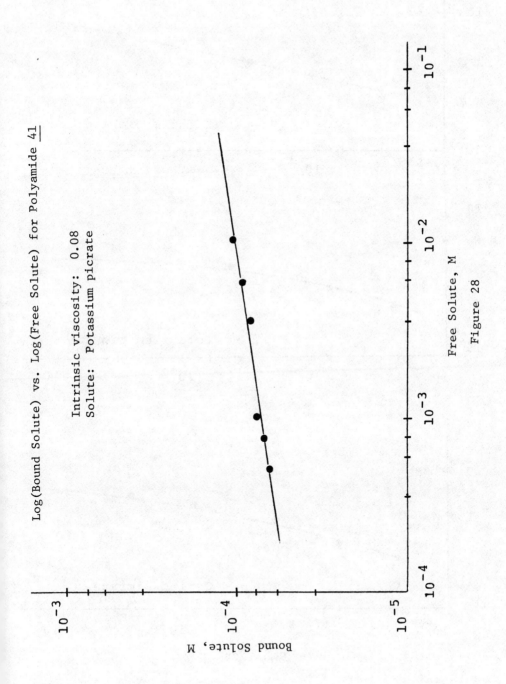

Log(Bound Solute) vs. Log(Free Solute) for Polyamide 41

Intrinsic viscosity: 0.08
Solute: Potassium picrate

Figure 28

Log(Bound Solute) vs. Log(Free Solute) for Polyamide 41

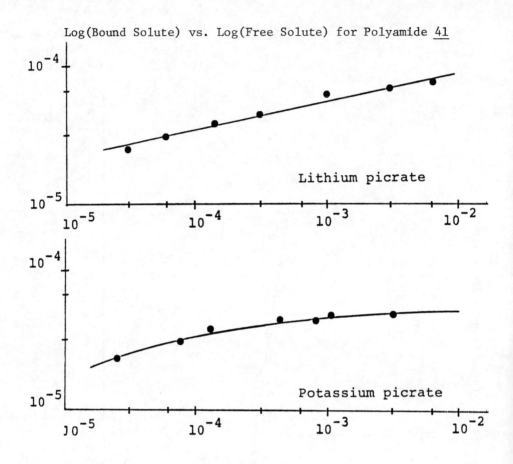

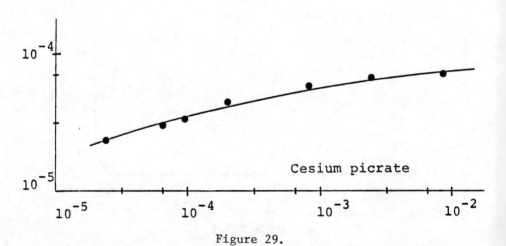

Figure 29.

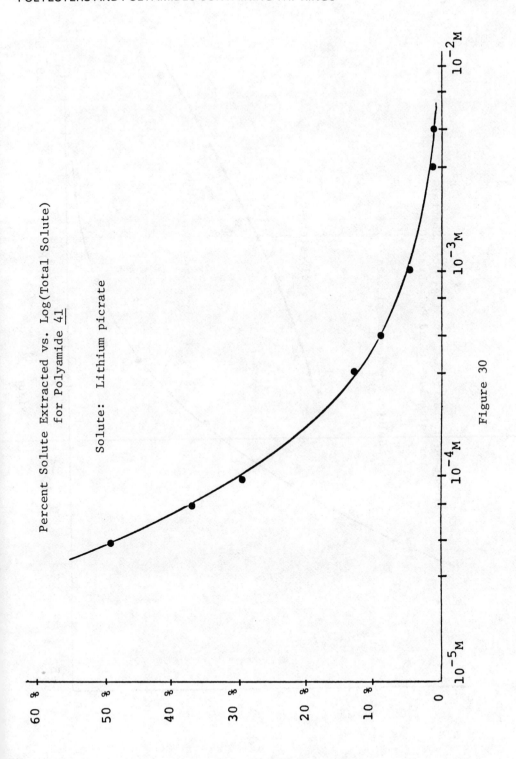

Percent Solute Extracted vs. Log(Total Solute) for Polyamide 41

Solute: Lithium picrate

Figure 30

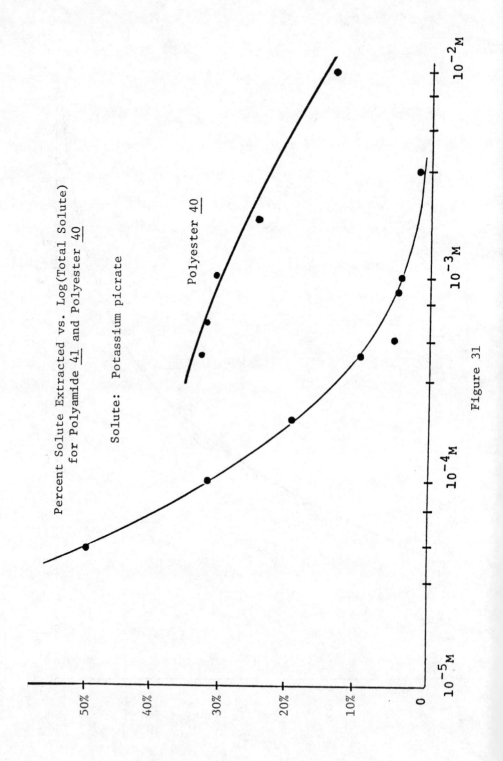

Percent Solute Extracted vs. Log(Total Solute)
for Polyamide 41 and Polyester 40

Solute: Potassium picrate

Polyester 40

Figure 31

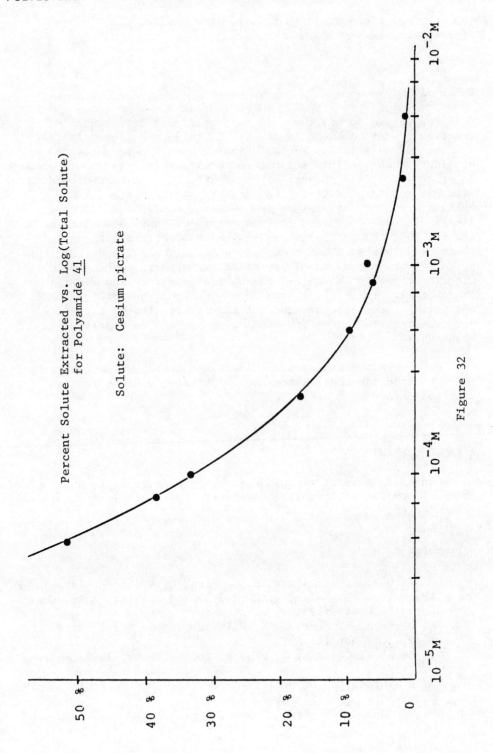

Percent Solute Extracted vs. Log(Total Solute)
for Polyamide 41

Solute:　Cesium picrate

Figure 32

the polymers prepared in this work, have similar ion-binding characteristics.

CONCLUSIONS

For polyamide 41, there is no significant difference in the fraction of cation bound in any of the three alkali metals tested. The lack of selectivity seems to indicate that some mechanism of binding without a fixed polymer conformation in solution is at work. In other words, the polymer is apparently free to wrap around any size cation, to bind it in the required multidentate fashion. If a helical structure exists, it is of variable size and diameter. Similar results were obtained for polyester 40. Note also that no selectivity between potassium and cesium was observed with any of Böhmer's polymers. This observation seems to indicate that those polymers were not particularly selective either, perhaps also indicating either a variable-sized helical cavity, or a flexible random coil. It is clear from this work, and the work of Böhmer[55,56] and Smith,[57] that the presence of macrocyclic ion-binding moieties is not necessary to confer ion-binding properties on a polymer in solution.

We are continuing to probe these and other systems in an effort to understand and control the mode of binding of ions to these flexible polymers.

ACKNOWLEDGEMENT

Financial support of this work by a Goodyear Fellowship and an Andrew P. Dunlop Fellowship (Quaker Oats Foundation), is gratefully acknowledged.

REFERENCES

1.   R. Corbaz, L. Ettlinger, E. Gaumann, W. Keller-Schierlein, F. Krandolfer, L. Neipp, V. Prelog, and H. Zahner, Helv. Chim. Acta, 38, 1445 (1955).

2a. H. Gerlach, K. Oertle, A. Thalmann, and S. Servi, Helv. Chim. Acta, 58, 2036 (1975).

 b. J. Beck, H. Gerlach, V. Prelog, and W. Voser, Ibid., 45, 620 (1962).

3a. J. Dominguez, J. D. Dunitz, H. Gerlach, and V. Prelog, Helv. Chim. Acta, 45, 129 (1962).

 b. H. Gerlach and V. Prelog, Ann., 669, 121 (1963).

4a. J. Gombas, E. Haslinger, H. Azk, and U. Schmidt, Tetrahedron
     Letters, 3391 (1975).
4b. J. Gombas, E. Haslinger, A. Nikiforov, H. Zak, and U. Schmidt,
     Monatsh. Chem., 106, 1043 (1975).
 c. U. Schmidt, J. Gombas, E. Haslinger, and H. Zak, Chem. Ber.,
     109, 2628 (1976).
5a. B. C. Pressman, E. J. Harris, W. S. Jagger, and J. H. Johnson,
     Proc. Natl. Acad. Sci., 58, 1949 (1967).
 b. B. C. Pressman, ibid., 53, 1076 (1965).
6a. S. N. Graven, H. A. Lardy, and S. Estrada-O, Biochemistry, 6
     365 (1967).
 b. S. N. Graven, H. A. Lardy, D. Johnson, and A. Rutter, ibid.,
     5, 1729 (1966).
7.   G. Szabo, G. Eisenman, and S. Ciani, J. Membrane Biol., 3,
     346 (1969).
8.   B. C. Pressman, Fed. Proced., 27, 1283 (1968).
9a. A. Agtarap, J. W. Chamberlin, M. Pinkerton, and I. Steinrauf,
     J. Am. Chem. Soc., 89, 5737 (1967).
 b. L. A. R. Pioda, H. A. Wachter, R. E. Dohner, and W. Simon,
     Helv. Chim. Acta, 50, 1373 (1967).
10. B. T. Kilbourn, J. D. Dunitz, L. A. R. Pioda, and W. Simon,
     J. Mol. Biol., 30, 559 (1967).
11a. C. J. Pedersen, J. Am. Chem. Soc., 89, 7017 (1967).
 b. idem., ibid., 92, 391 (1970).
12a. idem., Fed. Proced., 27, 1305 (1968).
 b. C. J. Pedersen and H. K. Frensdorff, Angew. Chem., Int. Ed.,
     11, 16 (1972).
13. J. S. Bradshaw, S. L. Baxter, J. D. Lamb, R. M. Izatt, and
     J. J. Christensen, J. Am. Chem. Soc., 103, 1821 (1981).
14. J. M. Timko, S. S. Moore, D. M. Walba, P. C. Hiberty, and
     D. J. Cram, ibid., 99, 4207 (1977).
15. R. T. Gray, D. N. Reinhoudt, C. J. Smit, I. Veenstra, Rec.
     trav. Chim. Pays-Bas, 95, 258 (1976).
16. J. D. Lamb, R. M. Izatt, J. J. Christensen, and D. J. Eatough,
     in "Coordination Chemistry of Macrocyclic Compounds", G. A.
     Melson, ed., Chap. 3, Plenum Press, New York, 1979.
17a. J. J. Christensen, D. J. Eatough, and R. M. Izatt, Chem. Rev.,
     74, 351 (1974).
 b. J. J. Christensen, J. O. Hill, and R. M. Izatt, Science, 174,
     459 (1971).
 c. R. M. Izatt, D. J. Eatough, and J. J. Christensen, Structure
     and Bonding, 16, 161 (1973).
 d. P. N. Kapoor and R. C. Mehrotra, Coord. Chem. Rev., 14, 1
     (1974).
 e. J.-M. Lehn, J. Simon, and J. Wagner, Angew. Chem., Int. Ed.,
     12, 578 (1973).
 f. D. J. Cram and J. M. Cram, Science, 183, 803 (1974).
 g. J. S. Bradshaw, J. Y. Hui, B. I. Haymore, J. J. Christensen,
     and R. M. Izatt, J. Heterocycl. Chem., 10, 1 (1973).

h.   G. R. Newkome, J. D. Sauer, J. M. Roper, and D. C. Hager, Chem. Rev., 77, 513 (1977).

17i. R. M. Izatt and J. J. Christensen, "Synthetic Multidentate Macrocycle Compounds", Academic Press, New York, 1978.

j.   R. M. Izatt and J. J. Christensen, "Progress in Macrocyclic Chemistry", Volumes 1 and 2, J. Wiley, New York, 1980 and 1981.

k.   D. A. Laidler and J. F. Stoddart (Chapter 1), and F. Vogtle and E. Weber (Chapter 2), in "The Chemistry of Ethers, Crown Ethers, Hydroxyl Groups, and their Sulfur Analogs", Supplement E, Part 1, S. Patai, ed., J. Wiley, New York, 1980.

18a. D. Bright and M. R. Truter, J. Chem. Soc., B, 1544 (1970).

b.   M. A. Bush and M. R. Truter, ibid., 1440 (1971).

19.   M. R. Truter, Struct. and Bonding, 16, 71 (1973).

20a. H. K. Frensdorff, J. Am. Chem. Soc., 93, 600 (1971).

b.   idem., ibid., 93, 4684 (1971).

21a. J. H. Prestegard and S. I. Chan, Biochemistry, 8, 3921 (1969).

b.   idem., J. Am. Chem. Soc., 92, 4440 (1970).

22a. J. S. Bradshaw, L. D. Hansen, S. F. Nielsen, M. D. Thompson, R. A. Reeder, R. M. Izatt, and J. J. Christensen, J. Chem. Soc., Chem. Comm., 874 (1975).

b.   J. S. Bradshaw, G. E. Maas, R. M. Izatt, and J. J. Christensen, Chem. Rev., 79, 37 (1979).

23a. J. S. Bradshaw and M. D. Thompson, J. Org. Chem., 43, 2456 (1978).

b.   R. M. Izatt, J. D. Lamb, G. E. Maas, R. E. Asay, J. S. Bradshaw, and J. J. Christensen, J. Am. Chem. Soc., 99, 2365 (1977).

24.   A. Samat, J. Elgnero, and J. Metzger, J. Chem. Soc., Chem. Comm., 1182 (1979).

25a. G. R. Newkome, G. L. McClure, J. B. Broussard, and F. Danesh-Khoshboo, ibid., 97, 3232 (1975).

b.   M. Newcomb, G. W. Gokel, and D. J. Cram, ibid., 96, 6810 (1974).

c.   G. W. Gokel, J. M. Timko, and D. J. Cram, J. Chem. Soc., Chem. Comm., 444 (1975).

26a. J. S. Bradshaw, G. E. Maas, J. D. Lamb, R. M. Izatt, and J. J. Christensen, ibid., 102, 467 (1980).

b.   R. M. Izatt, J. D. Lamb, R. E. Asay, G. E. Maas, J. S. Bradshaw, and J. J. Christensen, ibid., 99, 6134 (1977).

27.   D. N. Reinhoudt and R. T. Gray, Tetrahedron Letters, 2105, 2109 (1975).

28.   G. R. Newkome, J. D. Sauer, J. M. Roper, and D. C. Hager, Chem. Rev., 77, 513 (1977).

29.   Y. Kobuke, K. Hanji, Kl Horiguchi, M. Asada, Y. Nakayama, and J. Furukawa, J. Am. Chem. Soc., 98, 7414 (1976).

30a. L. L. Chan and J. Smid, J. Am. Chem. Soc., 89, 4547 (1967).

b.   L. L. Chan, K. H. Wong, and J. Smid, ibid., 92, 1955 (1970).

31.   B. Tummler, G. Maass, F. Vogtle, H. Sieger, U. Heimann, and E. Weber, J. Am. Chem. Soc., 101, 2588 (1979).

32. S. Yanagida, K. Takahashi, and M. Okahara, Bull. Chem. Soc. Jap., 51, 3111 (1978).

33. W. Saenger, I. H. Suh, and G. Weber, Israel J. Chem., 18, 253 (1979).

34a. E. Weber and F. Vogtle, Tetrahedron Letters, 2415 (1975).

 b. B. Tummler, G. Maass, E. Weber, W. Wehner, and F. Vogtle, J. Am. Chem. Soc., 99, 4683 (1977).

35a. F. Vogtle and H. Sieger, Angew. Chem., Int. Ed., 16, 396 (1977).

 b. idem., ibid., 17, 198 (1978).

 c. W. Rasshofer, G. Oepen, and F. Vogtle, Chem. Ber., 111, 419 (1978).

36. W. Wiereng, B. R. Evans, and J. A. Woltersom, J. Am. Chem. Soc., 101, 1334 (1979).

37. see also J. O. Gardner and C. C. Beard, J. Med. Chem., 21 357 (1978).

38. N. N. L. Kirsch and W. Simon, Helv. Chim. Acta, 59, 356 (1976).

39a. E. Blasius and P. G. Maurer, Makromol. Chem., 178, 649 (1977).

 b. G. Dotsevi, Y. Sogah, and D. J. Cram, J. Am. Chem. Soc., 98, 3038 (1976).

 c. W. M. Feigenbaum and R. H. Michel, J. Polym. Sci., A-1, 9, 817 (1971).

 d. E. Shchori and J. Jagur-Grodzinski, J. Appl. Polym. Sci., 20, 773 (1976).

40. J. S. Bradshaw and P. E. Stott, Tetrahedron, 36, 461 (1980).

41. W. M. Feigenbaum and R. H. Michel, J. Polym. Sci., A-1, 9, 817 (1971).

42. E. Shchori and J. Jagur-Grodzinski, J. Appl. Polym. Sci., 20, 773, 1665 (1976).

43a. E. Blasius, K. P. Jangen, W. Adrian, G. Klantke, R. Lorscheider, P. G. Maurer, V. G. Nguyen, T. Nguyen Tien, G. Scholten, and J. Stockerer, Z. Anal. Chem., 284, 337 (1977).

 b. E. Blasius, K. P. Jangen, and W. Neumann, Mikrochim. Acta, 2, 279 (1977).

 c. E. Blasius and P. G. Maurer, Makromol. Chem., 178, 649 (1977).

 d. E. Blasius, W. Adrian, K. P. Jangen, and G. Klautke.

44a. P. Gramain and Y. Frere, Mactomolecules, 12, 1038 (1979).

 b. S. Bormann, J. Brossas, E. Franta, P. Gramain, M. Kirsch, and J. M. Lehn, Tetrahedron, 31, 2791 (1975).

45. P. Gramain and Y. Frere, Makromol. Chem., Rapid Comm., 2, 161 (1981).

46. I. Cho and S. K. Chang, ibid., 2, 155 (1981).

47a. K. H. Wong, G. Konizer, and J. Smid, J. Am. Chem. Soc., 92, 666 (1970).

 b. S. Kopolow, T. E. Hogen-Esch, and J. Smid, Macromolecules, 6, 133 (1973).

 c. S. Kopolow, Z. Machacek, U. Takaki, and J. Smid, J. Macromol. Sci., Chem., A7, 1015 (1973).

 d. S. C. Shah, S. Kopolow, and J. Smid, J. Polym. Sci., Polym. Chem., 14, 2023 (1976).

48.   K. H. Wong, K. Yagi, and J. Smid, J. Membrane Biol., 18, 379
      (1974).
49a. K. Yagi, J. A. Ruiz, and M. C. Sanchez, Makromol. Chem., Rapid
      Comm., 1, 263 (1980).
  b. K. Yagi and M. C. Sanchez, ibid., 2, 311 (1981).
50.   I. M. Panayotov, C. B. Tsvetanov, I. V. Berlinova, and R. S.
      Velichkova, Makromol. Chem., 134, 313 (1970).
51.   I. M. Panayotov, C. B. Tsvetanov, and D. K. Dimov, ibid., 177,
      279 (1976).
52.   A. Hirao, S. Nakahama, M. Takahashi, and N. Yamazaki, ibid.,
      179, 1735 (1978), and 179, 915 (1978).
53.   S. Yanagida, K. Takahashi, and M. Okahara, Bull. Chem. Soc.
      Jap., 50, 1386 (1977).
54a. Y. Takahashi and E. Tadokoro, Macromolecules, 6, 672 (1973).
  b. H. Tadokoro, Y. Chatani, T. Yoshihara, S. Tahara, and S. Mura-
      hasi, Makromol. Chem., 73, 109 (1964).
55.   S. Iwabuchi, T. Nakahira, A. Tsuchiya, K. Kojima, and V.
      Böhmer, Makromol. Chem., in press.
56a. W. Kern, S. Iwabuchi, H. Sato, and V. Bohmer, Makromol. Chem.,
      180, 2539 (1979).
  b. H. Sato, S. Iwabuchi, V. Bohmer, and W. Kern, ibid., 182, 755
      (1981).
  c. W. Kern, M. M. Bhagwat, and V. Bohmer, Makromol. Chem., Rapid
      Comm., 2, 557 (1981).
57.   W. J. Schultz, M. C. Etter, A. V. Pocius, and S. Smith, J. Am.
      Chem. Soc., 102, 7981 (1980).
58a. H. Sumitomo and K. Hasimoto, Macromolecules, 10, 1327 (1977).
  b. H. Sumitomo, K. Hasimoto, and T. Ghyamo, Polym. Bull., 1, 133
      (1978).
  c. H. Sumitomo, K. Hasimoto, and M. Ando, J. Polym. Sci., Polym.
      Lett., 11, 635 (1973).
59.   K. Hashimoto and H. Sumitomo, Macromolecules, 13, 786 (1980).
60a. I. Tajima, M. Okada, and H. Sumitomo, Makromol. Chem., Rapid
      Comm., 1, 197 (1980).
  b. I. Tanaka, I. Tajima, Y. Hayakawa, M. Okada, M. Bitoh, T.
      Ashida, and H. Sumitomo, J. Am. Chem. Soc., 102, 7873 (1980).
61.   J. E. Kelly, Ph.D. Dissertation, Rensselaer Polytechnic In-
      stitute, 1975.
62.   J. A. Moore and J. E. Kelly, J. Polym. Sci., Polym. Lett. Ed.,
      13, 333 (1975).
63.   J. A. Moore and E. M. Partain, III, ibid., submitted for pub-
      lication.
64a. T. Saegusa, H. Imai, and J. Furukawa, Makromol. Chem., 65, 60
      (1963).
  b. K. Weissermel and E. Nolken, ibid., 68, 140 (1963).
  c. I. Penczek, and S. Penczek, ibid., 94, 228 (1966).
  c. V. A. Kropachev, L. V. Alferova, and B. A. Dolgoplosk, Polym.
      Sci., USSR, 5, 46 (1964).
65.   E. L. Wittbecker, H. K. Hall, and T. W. Campbell, J. Am. Chem.
      Soc., 82, 1218 (1960).

66a.   N. Ogata, T. Asahara, and S. Tohyama, J. Polym. Sci. A-1, $\underline{4}$, 1359 (1966).

  b.   N. Ogata and S. Tohyama, Bull. Chem. Soc. Jpn., $\underline{39}$, 1556 (1966).

67.   H. Wexler, Makromol. Chem., $\underline{115}$, 262 (1968).

68.   C. Tanford, "Physical Chemistry of Macromolecules", Chap. 8, J. Wiley, New York, 1961.

69.   R. Sinta and J. Smid, Macromolecules, $\underline{13}$, 339 (1980).

70.   O. Silberrad and H. A. Phillips, J. Chem. Soc., $\underline{93}$, 474 (1908).

71.   D. Balasubramanian and B. C. Misra, "Metal-Ligand Interactions in Organic Chemistry and Biochemistry, Part 1", B. Pullman and N. Goldblum, eds., D. Reidel Comp., New York, 1977, pg. 159.

72.   N. Lotan, J. Phys. Chem., $\underline{77}$, 242 (1973).

73.   M. L. Tiffany and S. Krimm, Biopolymers, $\underline{8}$, 347 (1969).

74.   F. Quadrifoglio and D. W. Urry, J. Am. Chem. Soc., $\underline{90}$, 2760 (1968).

75.   W. L. Mattice and L. Mandelkern, Biochemistry, $\underline{8}$, 1049 (1969).

76a.   D. Balasubramanian, A. Goel, and C. N. R. Rao, Chem. Phys. Lett., $\underline{17}$, 482 (1972).

  b.   M. E. Noelken and S. N. Timasheff, J. Biol. Chem., $\underline{242}$, 5080 (1969).

  c.   B. Rode and R. Fussenegger, J. Chem. Soc., Faraday Trans. II, $\underline{71}$, 1958 (1975).

77a.   K. Dachs and E. Schwartz, Angew. Chem., Int. Ed., $\underline{1}$, 430 (1962).

  b.   G. B. Gechele and L. Crescentini, J. Appl. Polym. Sci., $\underline{7}$, 1349 (1963).

78a.   F. J. Van Natta, J. W. Hill, and W. H. Carrothers, J. Am. Chem. Soc., $\underline{56}$, 455 (1934).

  b.   V. Crescenzi, G. Manzini, G. Calzolari, and C. Borri, Eur. Polym. J., $\underline{8}$, 449 (1972).

79.   J. A. Moore and E. M. Partain, III, Macromolecules, submitted for publication.

80.   It should be noted that repulsion cannot be invoked in the instance of polyaminde $\underline{41}$ at the concentration we have investigated so far. At a repeat unit concentration of 10^{-4}M we have, on the average, only one ion pair per hundred repeat units, an ion concentration which is probably not sufficient to cause repulsion effects.

ION-BINDING PROPERTIES OF NETWORK POLYMERS

WITH PENDANT CROWN ETHER LIGANDS

Johannes Smid and Roger Sinta

Polymer Research Institute
Chemistry Department
College of Environmental Science and Forestry
State University of New York
Syracuse, NY 13210

INTRODUCTION

The macrocyclic crown ethers and cryptands and the acyclic podands have been incorporated into polymeric structures either as part of the backbone or anchored as pendant ligands.[1] Considerable effort has been focussed on attaching these cation-chelating agents to insoluble supports in the form of polymeric networks, gels or glass beads[2-8]. This immobilization facilitates of course the recovery of these materials and prevents contamination of reaction products. The polymeric ligands have been tested extensively as chromatographic stationary phases for the separation of ionic and neutral solutes and as heterogeneous anion-activating catalysts. In all such applications the cation interaction with the polymers depends not only on the ligand structure and type of cation but is also sensitive to the nature of the counterion, solvent and temperature. Other important factors affecting the binding are the spacing between the ligands along the polymer chain, the length and structure of the spacer connecting the ligand with the polymer backbone, the crosslinking density and the presence of comonomer substituents which either may hinder or cooperate in the binding process.

Most studies with the immobilized crown ethers and other ligands have been concerned with determining their effectiveness in separating ionic or neutral solutes and with their role as phase transfer catalysts in modifying reaction rates, yields or distribution of products. However, quantitative information on the actual binding of ionic solutes as a function of the many

variables listed above is scarce. We have recently embarked on a
more detailed study of the binding of salts to network polymers
containing crown ether and oligo-oxyethylene (glyme) ligands as
pendant groups. Since most anion-activating reactions involving
crown ether phase transfer catalysts have been carried out in low
polarity media, we have carried out the investigations in solvents
such as dioxane, tetrahydrofuran and toluene. The work discussed
here deals with the binding of alkali picrates to network polymers
containing benzo-15-crown-5 and benzo-18-crown-6 ligands as pendant
moieties. The picrate anion serves as a sensitive spectrophotometric
probe, and permits the use of salt concentrations as low as 10^{-5}
to 10^{-6} M. This minimizes the formation of ionic aggregates higher
than ion pairs. Binding constants of alkali picrates to networks
of different crown content were measured in both dioxane and
tetrahydrofuran. The immobilized crown ethers were also used in
competition with soluble ligands to arrive at formation constants
of the latter with ionic solutes in apolar media.

BINDING OF PICRATE SALTS TO CROWN ETHER NETWORKS

The general structure of the crown ether network polymers is
depicted below. The polymers were synthesized from 2% chloromethylated polystyrene and the 4'-hydroxymethyl derivatives of
benzo-15-crown-5 and benzo-18-crown-6. The low capacity network
polymers R15C5 and R18C6 were derived from a polystyrene resin

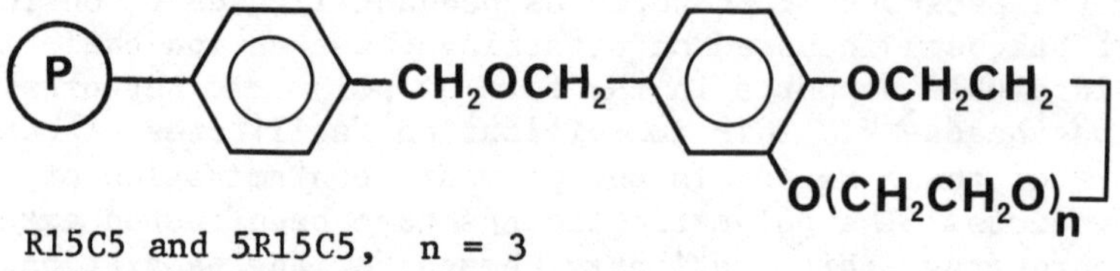

$$\text{P} - \bigcirc - CH_2OCH_2 - \bigcirc - OCH_2CH_2 - \left[O(CH_2CH_2O) \right]_n$$

R15C5 and 5R15C5,   n = 3
R18C6 and 5R18C6,   n = 4

containing 0.9 meq Cl per gram of polymer. They possess on the
average one crown ligand per eleven monomer units. The high crown
content polymers 5R15C5 and 5R18C6 contain on the average close
to one crown ether per two monomer units and were synthesized
from a chloromethylated polystyrene of 5 meq Cl per gram of polymer.

The cation binding to the crown ether resins was measured
spectrophotometrically by monitoring the disappearance of the
alkali picrates from solution. Besides their high molar absorp-
tivities in a convenient spectral range (340-380 nm), the picrate
absorption maxima are also sensitive to the interionic ion pair
distance. This makes it often possible to draw important conclusions regarding the structure of the alkali picrate ion pairs or
their ligand complexes in solution. In both THF and dioxane the
picrate salts are externally solvated tight ion pairs Pi^-, M^+, S_n,

where S denotes a solvent molecule. The dissociation constants in THF at 25ºC are less than 10^{-7} M. Hence, above 10^{-5} M salt concentrations the fraction of free ions remains small. The presence of a significant fraction of free cations complicates the experiments since ligand binding constants to ion pairs and free ions can be vastly different[9]. Dissociation constants of complexes between crown ethers and alkali picrates are usually substantially higher[9]. For this reason, competition experiments with soluble crown ethers (see below) were carried out in dioxane or toluene.

Picrate binding to a pre-swollen network in THF, dioxane or toluene reaches equilibrium within half an hour. The heterogeneous reaction can be written as

$$Cr* + Pi^- \, M^+ \; \underset{\longleftarrow}{\overset{K_N}{\longrightarrow}} \; Pi^- \, M^+ \, Cr* \tag{1}$$

where Cr* denote the uncomplexed crown ligands in the network, $Pi^- \, M^+ \, Cr*$ is the network-bound picrate and $Pi^- \, M^+$ the free picrate concentration in solution. The latter is measured after each successive addition of the salt after equilibrium is reached. Equation (1) assumes a 1:1 crown-ion pair complex, but complexes may also contain two crown ligands. The intrinsic binding constant, K_N, can be evaluated from the rearranged form of the Langmuir adsorption isotherm

$$1/R = 1/n + 1/nK_N A \tag{2}$$

In this expression, $1/R$ equals $Cr_0*/Pi^- \, M^+ \, Cr*$, A is the free picrate concentration and $1/n$ denotes the number of crown ligands in the $Pi^- \, M^+ \, Cr*$ complex. Some examples of $1/R$ versus $1/A$ plots are shown in Figures 1 and 2. They graphically depict the binding of picrate salts to the networks R18C6 in THF and 5R15C5 in dioxane.

Inspection of Figures 1 and 2, and those not shown here, lead to the conclusion that linear plots with intercepts $1/n$ equal to unity are obtained for those crown-cation combinations which are known to give stable 1:1 complexes only. This behavior is found for $Pi^- \, Na^+$ with all four networks, for $Pi^- \, K^+$ with R18C6 and 5R18C6, and for Cs^+ with R15C5 in THF. Linearity implies that the ion pair binding to these networks is solely governed by statistical factors. Hence, there is no evidence for electrostatic attraction or repulsion between bound ion pairs even under conditions where nearly all crown units contain a cation, that is, at low $1/A$ values. This is not only the case for the low crown content networks, but also for those where every other monomer unit contains a ligand (see Fig. 2).

Deviations from linearity frequently occur in systems where

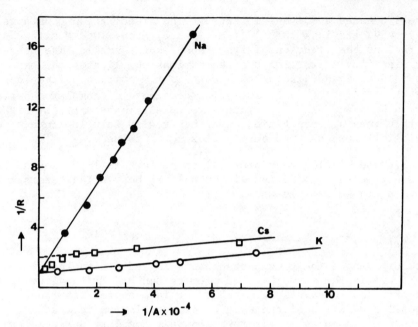

Fig. 1. Plots of 1/R versus 1/A for the binding of alkali picrates
to R18C6 in THF at 25°C.   (●) Na⁺ ; (○) K⁺; (□) Cs⁺.

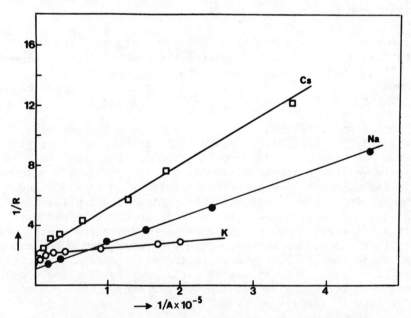

Fig. 2. Plots of 1/R versus 1/A for the binding of alkali picrates
to 5R15C5 in dioxane at 25°C.   (●) Na⁺; (○) K⁺; (□) Cs⁺.

the cation is known to form stable 1:1 and 2:1 crown-cation-com-
plexes. Examples pertinent to our work are $Pi^- K^+$ with R15C5 and
5R15C5, and $Pi^- Cs^+$ with all four networks. In most of these
systems the linear portion of the curve, at low cation loading
(high $Cr_0^*/Pi^- M^+ Cr^*$ values) can be extrapolated to yield a 1/n
value equal or close to two (see plots for $Pi^- Cs^+$/R18C6 in Fig. 1
and $Pi^- K^+$ for 5R15C5 in Fig. 2). This clearly suggests the
formation of stable 2:1 crown-complexes in these networks. Most
of these lines eventually curve downward at high picrate concentration and yield an intercept 1/n equal or close to one. Apparently
as more and more ion pairs are forced into these networks, the
2:1 complexes change into 1:1 crown-cation complexes (equation 3).
This conclusion was previously arrived at from viscosity

$$Pi^- Cr^* M^+ Cr^* + Pi^- M^+ \rightleftarrows 2 Pi^- M^+ Cr^* \qquad (3)$$

measurements of salt-containing solutions of poly(crown ether)s[10].

Cation Selectivities and Effect of Solvent

Intrinsic binding constants calculated from 1/R versus 1/A
plots are collected in Table 1. Starred values were derived from

Table 1. Intrinsic Binding Constants, K_N, of Alkali Picrates
to Crown Ether Networks in Tetrahydrofuran and
Dioxane at 25°C.

	$K_N \times 10^{-3} M^{-1}$					
	Sodium		Potassium		Cesium	
Network	THF	DIOX	THF	DIOX	THF	DIOX
R15C5	1.66	15.5	20.8	220*	1.3	(2.0)
5R15C5	2.2	58	79*	350*	(55*)	66*
R18C6	3.41	25.8	54	172	125*	(150)
5R18C6	6.8	97	100	258	(96*)	410*

linear plots with an intercept 1/n = 2.0, or from curved plots
containing a distinctly linear portion at high 1/A values and
which extrapolates to 1/n = 2 (e.g., R18C6/$Pi^- Cs^+$/THF and
5R15C5/$Pi^- K^+$/dioxane, see Figs. 1 and 2). It is assumed that in
the linear regions most complexes have a 2:1 stoichiometry and

K_N was calculated by taking $1/n = 2$. In some of the cesium picrate systems (e.g., 5R15C5/Pi⁻ Cs⁺/THF) the entire plot is slightly curved, or the intercept is between 1 and 2. The K_N values derived for those systems have been bracketed in Table I. They were calculated by using either $1/n = 1$ or $1/n = 2$, depending on whether the best straight line through the points intercepts close to 1 or close to 2. They are average values and the true K_N depends somewhat on the fraction of crown units complexed to ion pairs. These networks most likely contain variable mixtures of 1:1 and 2:1 complexes depending on the salt content.

Inspection of Table I reveals higher K_N values for all networks when THF is replaced by dioxane, the effect being largest for sodium picrate (as much as a factor 25) and smallest for the cesium salt (about a factor 2). Complexation of crown to an externally-complexed tight ion pair is accompanied by release of most or all of the bound solvent molecules (equation 4). The cation-solvent interaction is considerably stronger with THF than with dioxane,

$$Pi^- M^+ S_n \; + \; Cr* \; \rightleftharpoons \; Pi^- M^+ Cr* \; + \; nS \qquad (4)$$

and decreases rapidly in the order $Na^+ > K^+ > Cs^+$. Hence, the observed changes in K_N with solvent are reasonable. Also, the release of solvent molecules is suggested by the positive entropy change of about 8 eu found from temperature dependent studies of the equilibrium represented by equation 4.

The selectivity pattern for the four crown ether networks in terms of their K_N values parallel that found for the corresponding soluble crown ether ligands. The selectivity for the R15C5 networks is $K^+ \gg Na^+ \geqslant Cs^+$ and for the two R18C6 networks $Cs^+ \geqslant K^+ \gg Na^+$. Although the orders are not reversed, the actual selectivity does depend significantly on solvent and on the crown content.

Effect of Crown Content

K_N values for the 5R15C5 and 5R18C6 networks exceed those of the corresponding resins of low crown content by a factor of about two to four (See Table I). The increase is more than a factor thirty for cesium picrate when 5R15C5 is used instead of R15C5. The enhanced binding is also found in systems where only 1:1 complexes are formed. Assuming a swelling factor of ten, the local crown concentration in the low capacity network compares with a crown concentration of roughly 0.04 M. This increases to about 0.2 M in the high crown content polymers. This change may exert a marked influence on the swelling properties and the nature of the microenvironment within the solvent channels of the network, factors that could influence the binding constants of the ion pairs. Also, the chains in a solvent-swollen 2% crosslinked network have considerable mobility, and the high crown content

networks may be considered to have a microenvironment consisting of
a high concentration of potential oxygen binding sites. Even in
a 1:1 complex (Pi^-Na^+ with R15C5 and R18C6, or Pi^- K^+ with R18C6),
additional oxygen atoms from neighboring crown ligands may be
utilized to form a more effective solvation shell for the cation.
This type of effect was also found for 1:1 complexes with linear
poly(crown ether)s, where homopolymers of vinylbenzocrown ethers
are often more effective salt binders than the corresponding
styrene copolymers[11].

The effective binding exhibited by the 5R resins for salts
which form stable 2:1 crown cation complexes can be expected on
the basis of a closer spacing between crown ligands. It is well
recognized that bis-crown ethers of benzo-15-crown-5 (that is,
two crown units connected at their 4' position by a chain) are
effective complexing agents for potassium picrate[9]. They form
crown-separated ion pair complexes in THF, and the binding constant
depends on the length and structure of the chain connecting the
crown units. The same occurs for cesium picrate with bis(crown
ethers) containing the benzo-18-crown-6 moiety. Such 2:1 complexes
are commonly found for Pi^- K^+ with linear homopolymers and styrene
copolymers of vinylbenzo-15-crown-5 and for Pi^- Cs^+ with those of
vinylbenzo-18-crown-6[10,12]. A high crown content increases the
probability of formation of intramolecular 2:1 crown-cation
complexes. The effect appears to be especially large for Pi^- Cs^+
with R15C5 and 5R15C5, probably because benzo-15-crown-5 itself
is a poor complexing agent for Cs^+, while two crown units can
form a more complete solvation shell. The Cs^+ cation can ac-
commodate at least twelve oxygen atoms as suggested by the high
efficiency of the 5R18C6 networks. Intermolecular 2:1 complexes
between a cation and two crown units coming from different chains
also increases at higher crown content. This should lead to a
higher crosslinking density, similar to what has been reported
for linear poly(crown ether)s[10].

Increasing the picrate concentration appears to convert part
or all of the 2:1 complexes into 1:1 complexes, except for 5R18C6/
Pi^- Cs^+/dioxane and possibly for 5R15C5/Pi^- Cs^+/dioxane. It is
likely that this reaction converts crown-complexed loose ion pairs
into crown-complexed tight ion pairs (see equation 4). Since in
many anion-activated reactions the tight ion pairs are considerably
less reactive than the loose ion pairs, conditions leading to high
ion pair loading may well result in a less effective use of the
crown catalyst.

STUDIES OF SALT BINDING TO SOLUBLE LIGANDS IN APOLAR MEDIA USING
CROWN ETHER NETWORK POLYMERS

Insoluble polymeric ligands such as the crown ether networks

described in the previous section can be used to great advantage
in the study of ion or ion pair interactions with soluble ion-
binding ligands in solvents of low polarity[8]. In the method that
we applied, a soluble ligand, L, and an immobilized crown ligand,
Cr*, are competing for the picrate salt in a solvent like dioxane
according to equation (5)

$$Pi^- \ M^+ \ Cr* \ + \ L \ \overset{K}{\rightleftarrows} \ Pi^- \ M^+ \ L \ + \ Cr* \tag{5}$$

In a typical experiment, a picrate salt was first solubilized
in dioxane ($\sim 10^{-4}$ M). Sufficient crown-network polymer was then
added to bind at least 95% of the salt. Picrate salt was then
released from the network by adding increasing amounts of the
ligand. Each time the soluble complex $Pi^- \ M^+$ L was measured
spectrophotometrically after equilibrium was reached. If f_1 is
the fraction of ligand bound to the picrate salt ($f_1 = Pi^- \ M^+ \ L/L_o$,
assuming a 1:1 complex), and f_2 is the fraction of network-crown
bound to the salt ($f_2 = Pi^- M^+ \ Cr*/Cr_o*$), then the following relationship can be derived

$$[1/f_2 - 1] \ = \ K \ [1/f_1 - 1] \tag{6}$$

Since the total concentration of picrate salt (P_o), ligand (L_o)
and network-crown units (Cr_o*, in equivalents) are known, the
parameters f_1 and f_2 can be obtained spectrophotometrically by
determining $Pi^- \ M^+ L$, since $Pi^- \ M^+ \ Cr_o* = Pi^- \ M^+_o - Pi^- \ M^+ \ L$
(both concentrations must be converted into equivalents). A plot
of $(1/f_2 - 1)$ versus $(1/f_1 - 1)$ should yield a straight line with
slope K. The equilibrium constant, K, is a measure of the binding
strength of a ligand, L, relative to that of the network-bound
crown ether.

Equilibrium (5) can be analyzed in terms of reactions (1) and
(7), where $K = K_L/K_N$. Hence, if the intrinsic binding constant, K_N,

$$Pi^- \ M^+ \ + \ L \ \rightleftarrows \ Pi^- \ M^+ \ L \tag{7}$$

of the salt to the crown ether network is known (see Table 1), a
measurement of K will yield the formation constant, K_L, of the
complex $Pi^- \ M^+$ L in the particular solvent used.

Two examples of plots of equation (6) are shown in Figs. 3
and 4. Fig. 3 refers to the release of potassium picrate from an
R18C6 network in dioxane using the ligands 18-crown-6 (18C6), di-
cyclohexyl-18-crown-6 (DCH 18C6), 4'-methylbenzo-18-crown-6
(MB18C6) and 4,4'-dimethyldibenzo-18-crown-6 (DMB18C6). Fig. 4
refers to the same systems but in toluene as solvent. Since alkali
picrates do not dissolve in toluene, it is necessary to first
solubilize the required amount of salt with a small amount of the

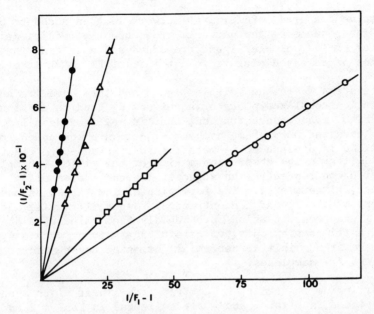

Fig. 3.   Release of R18C6-bound potassium picrate in dioxane on
addition of 18C6 (●), DCH18C6 (△), MB18C6 (□), and
DMB18C6 (○).

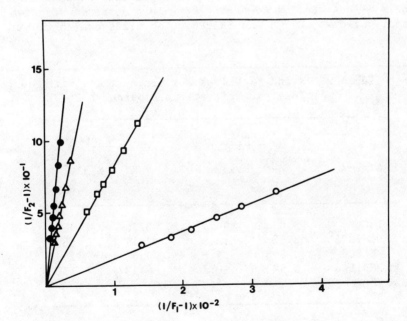

Fig. 4.   Release of R18C6-bound potassium picrate in toluene on
addition of 18C6 (●), DCH18C6 (△), MB18C6 (□) and
DMB18C6 (○).

ligand to be measured. Sufficient network is then added to bind
most of the picrate to the network. Subsequently, additional
quantities of the soluble ligand are added and each time the concentration of released picrate in the form of $Pi^- M^+ L$ is measured.

In systems where the soluble ligand can form both 1:1 and 2:1
ligand-cation complexes, plots of equation (6) tend to curve upward at high ligand concentration, that is, at increased $1/f_1$ values.
This was observed when 4'-methylbenzo-15-crown-5 was added to
$Pi^- K^+$/R18C6 in dioxane or 4'-methylbenzo-18-crown-6 to $Pi^- Cs^+$/
R18C6. Shifts in the absorption maxima of the $Pi^- M^+ L$ complexes
at higher crown concentration suggests the conversion of 1:1 complexes into 2:1 complexes. The formation of the latter depends on
the square of the ligand concentration. Hence, $1/f_2$ will increase
at high ligand concentration more rapidly than calculated on the
basis of a 1:1 complex only. A careful analysis of such a system
should make it possible to determine formation constants for both
the 1:1 and 2:1 complexes.

Equilibrium constants, K, derived from linear plots such as
those depicted in Figs. 3 and 4 are collected in Tables 2 and 3.

Listed are also the complex formation constants, K_L, calculated from the relationship $K_L = KK_N$ by using the appropriate K_N
values given in Table 1. The reliability of this competition
method is illustrated by comparing K_L values obtained with two
different networks, R18C6 and 5R18C6. Table 2 shows that while

Table 2.   K and K_L Values for Crown Ethers with
           Potassium Picrate in Dioxane at 25° C

| | R18C6[a] | | 5R18C6[b] | |
Ligand	K	$K_L \times 10^{-4} M^{-1}$	K	$K_L \times 10^{-4} M^{-1}$
18C6	5.62	96.7	3.76	96.7
DCH18C6[c]	3.13	53.8	2.19	56.5
MB18C6	0.908	15.6	0.60	15.5
DMB18C6	0.595	10.3	0.437	11.3

[a] 0.70 meq crown/g; $K_N = 1.72 \times 10^5$ M^{-1}
[b] 1.48 meq crown/g; $K_N = 2.58 \times 10^5$ M^{-1}
[c] Mixture of isomers of dicyclohexyl-18-crown-6

Table 3.   K and K_L Values for Crown Ethers with
Sodium Picrate in Dioxane and Toluene
at 25°C.   Network R18C6[a]

| | Dioxane | | Toluene |
Ligand	K	$K_L \times 10^{-4} M^{-1}$	K
18C6	1.34	3.46	1.26
DCH18C6	1.00	2.58	1.24
MB18C6	0.773	2.00	0.67
DMB18C6	0.191	0.472	0.124
15C5	2.80	7.22	---
MB15C5	0.739	1.91	---

[a]0.70 meq crown/g; $K_N = 2.58 \times 10^4 M^{-1}$ for
sodium picrate

the K_N and K values for these two networks differ substantially,
their product, $K_L = KK_N$ for a particular ligand is the same within
experimental error. The same observation was made when K_L values
were determined for several ligands using both R18C6 and 5R15C5.

Complex formation constants with potassium picrate decrease
in the order 18C6 > DCH18C6 > MB18C6 > DMB18C6. This parallels
the substituent effect on the basicity of the oxygen atoms nearest
the substituent. A similar trend is found with sodium picrate
although K_L values are lower by about a factor ten. The order in
decreasing K values is not altered when toluene is used as solvent
instead of dioxane. As long as the same network is involved, K is
proportional to K_L. Hence, a scale of equilibrium constants K for a
series of ligands with an ionic solute in toluene should in most
instances reflect the trend in the affinity of these ligands for
that solute in toluene.

In dioxane, the K value for the system $Pi^- K^+$/MB18C6/R18C6 is
0.908, not far from unity. R18C6 contains a 4'-alkyl-substituted
benzocrown ether, and should be comparable in its cation-binding
properties to 4'-methylbenzo-18-crown-6. Hence, the conclusion
may be drawn that in dioxane the cation affinity of a free crown
ether is not much different from the same crown anchored to a
network as long as the crown moieties are not too closely spaced.
With 5R18C6 the K for MB18C6 is 0.6, reflecting the enhanced
binding affinity of a network with higher crown content. In

toluene, the K for $Pi^- K^+$/MB18C6/R18C6 is only 0.60 (Table 2), and the value is 0.67 for the same system but with the sodium salt. The microenvironment for the ionic solute in the network is not the same as that in solution, and it is not expected that a bound ligand, even when isolated from other bound ligands, will behave exactly the same as in solution. The differences, however, appear not to be great, as long as the cation only forms 1:1 complexes with both free and bound ligand.

For sodium picrate (Table 3) the 15C5 ligand is more effictive than 18C6. The reverse is found for the free sodium ion in methanol[13]. MB18C6 and MB15C5 in dioxane are nearly identical in their complexing ability of $Pi^- Na^+$. It may be worth pointing out that equal affinity of crown ethers for an ionic solute does not necessarily mean that their catalytic effect will be the same. One factor determining the catalytic behavior of a crown ether is its ability to activate the anion in terms of increasing the interionic ion pair distance. This increase may be more pronounced in a complex of $Pi^- Na^+$ with MB18C6 than that with MB15C5 in spite of their identical complexing power.

The K values in toluene yield the same decreasing sequence for the four crown ligands (Table 2) as in dioxane, although the ratio K (dioxane)/K (toluene) can be quite different for each ligand. In nearly all systems, the equilibrium constants are lower in toluene than in dioxane, sometimes by as much as a factor three. This could be due to the much higher oxygen content inside the network, providing additional binding sites or a more polar environment for the ionic solute.

Effect of Counteranion

Formation constants of ion pair-ligand complexes will depend not only on the specific interaction between ion and ligand but also on the nature of the counterion. Other salts that can be conveniently studied are mono- or dinitrophenolates, dyes like methyl orange, or compounds with a distinct chromophore, for example pyrene carboxylates. For a salt such as sodium tetraphenylboron the K_N is more difficult to determine since it has a very high binding constant to crown ethers and its absorption maximum at 270 nm has a relatively low molar absorptivity. For solutes with less favorable spectra, a competitive binding method may be employed to arrive at their network binding constants, using a salt with a favorable spectrum and for which K_N is known, e.g., picrates. Assume a mixture of $Pi^- Na^+$ (P) and Na BPh4 (B) in dioxane in the presence of a network Cr*:

$$Cr^* + P \underset{}{\overset{K_{N,1}}{\rightleftarrows}} Cr^*, P \qquad\qquad (8)$$

$$Cr* \; + \; B \; \underset{\xrightarrow{\hspace{1cm}}}{\overset{K_{N,2}}{\rightleftarrows}} \; Cr* \, B \qquad\qquad (9)$$

Assuming that the fraction of free ions is neglibibly small, it can be shown that the ratio of the two binding constants $K_{N,1}$ and $K_{N,2}$ is given by

$$\frac{K_{N,2}}{K_{N,1}} = \frac{(B_o - B) P}{(Cr*, P) B} \qquad\qquad (10)$$

Since $Cr* = Cr_o* - (Cr*, B + Cr*, P)$ (only 1:1 complexes are assumed to form), the following relationship can be derived

$$\frac{B_o}{K_1 P_f (1/R - 1) - 1} = \frac{1}{K_{N,2}} + \frac{P_b}{K_{N,1} P_f} \qquad\qquad (11)$$

where $R = Cr_o*/Cr*$, P (subscripts zero refer to initial concentrations) and P_b and P_f denote the bound and free picrate concentrations, respectively. The latter is measured spectrophotometrically at different initial concentrations B_o of NaBPh$_4$. A plot of the left hand side of equation (11) versus P_b/P_f should yield a straight line with $K_{N,1}$ and $K_{N,2}$ being the reciprocals of the slope and intercept, respectively. Such a plot is shown in Fig. 5 for the system sodium picrate/sodium tetraphenylboron/ 5R18C6 in THF. The slope yields a value $K_{N,1} = 6.5 \times 10^3 \, M^{-1}$, close to the K_N value of $6.8 \times 10^3 \, M^{-1}$ obtained directly for Pi$^-$ Na$^+$ binding to 5R18C6 in THF (Table 1). Unfortunately, the intercept cannot be determined with any accuracy, but $K_{N,2}$ for NaBPh$_4$ appears to be in the order of $10^5 \, M^{-1}$, much larger than for sodium picrate. A large interionic ion pair distance as in NaBPh$_4$ will increase the electric field strength around the cation and enhance the ligand binding constant to that salt. More accurate intercepts can be achieved by choosing the proper combinations of salt and network so that $K_{N,1}$ and $K_{N,2}$ differ by a smaller factor.

The use of immobilized ligands such as crown ether network polymers to study the interaction of soluble ligands with ionic solutes constitutes a versatile method to arrive at complex formations constants of ion- and ion-pair ligand complexes or for obtaining a scale of ligand affinities for ionic solutes. The crown networks are less satisfactory for weaker cation-binding ligands such as glymes or sulfoxides. A high concentration of ligands must be added to release picrate from the crown networks, and this can change the bulk properties of media such as dioxane or toluene. In such cases it is more advantageous to use network polymers to which weaker ion-binding ligands are anchored. We

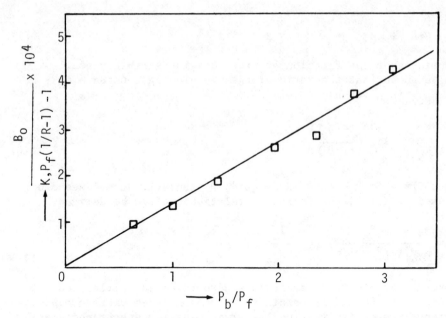

Fig. 5.   Competitive binding plot for sodium picrate and sodium
           tetraphenylboron to 5R18C6 in THF.

have recently synthesized a series of oligo-oxyethylene (glyme)
containing networks which are very effective in the study of interactions between picrate salts (including the lithium salt) with
ligands such as polyamines, sulfoxides and glymes in hydrocarbon
solvents.

Acknowledgement   We gratefully acknowledge the financial support
of this research by the National Science Foundation (Grant Che
7905890).

References

1.   J. Smid, "Progress in Macrocyclic Chemistry" (R.M. Izatt and
     J.J. Christensen, Eds.), Wiley-Interscience, New York, 1982,
     Chpt. 3.
2.   E. Blasius, K.P. Janzen, M. Keller, H. Lander, T. Nguyen-Tien
     and G. Scholten, Talanta, 27, 107 (1980).
3.   S.L. Regen, Angew. Chem. Intern. Edit. Engl., 18, 421 (1979).
4.   G. Manecke and A. Kramer, Makromol. Chem., 182, 3017 (1981).
5.   F. Montanari and P. Tundo, J. Org. Chem., 46, 2125 (1981).
6.   P. Kutchukov, A. Ricard and C. Quivoron, Eur. Polym. J., 16,
     753 (1980).
7.   A. Akelah and D.C. Sherrington, Chem. Rev., 81, 557 (1981).
8.   R. Sinta and J. Smid, J. Am. Chem. Soc., 103, 6962 (1981).

9. M. Bourgoin, K.H. Wong, J.Y. Hui and J. Smid, J. Am. Chem. Soc., 97, 3462 (1975).
10. S.C. Shah, S.L. Kopolow and J. Smid, Polymer, 21, 188 (1980).
11. K.H. Wong, K. Yagi and J. Smid, J. Membr. Biol., 18, 379 (1974).
12. S. Kopolow, Z. Machacek, U. Takaki and J. Smid, J. Macromol. Sci.-Chem., A7, 1015 (1973).
13. J.D. Lamb, R.M. Izatt, C.S. Swain and J.J. Christensen, J. Am. Chem. Soc., 102, 475 (1980).

SELECTIVE CATION BINDING BY POLYMERS WITH PENDANT CROWN ETHERS

Koji Yagi and Maria Cristina Sanchez

Departamento de Química, ICET, Universidad Autónoma de Guadalajara
Lomas del Valle, Guadalajara, Jalisco, México

INTRODUCTION

Polymers with pendant crown ethers show different selectivity from their monomers in complex formation with metal cations due to the ease of a 1:2 complex formation. They can be prepared either by the polymerization of the crown vinyl monomers or by the polymer reaction. The former method is preferable for studies on basic properties since it gives a regular polymer with pendant crown ether, however, the latter one is advantageous from an economic standpoint.

The first vinyl monomers containing crown ethers were the styrene derivatives prepared by Smid et al.[1,2] More recently, derivatives of acrylate[3,4] and acrylamide[5,6] have been prepared. These monomers can be polymerized by radical initiators and certain cation binding properties of these polymers have been reported.

With the object to obtain an effective and selective extracting material and also to investigate the dependence of the selectivity in the cation binding on the copolymer composition, we have prepared the methacrylamide derivatives of 15-crown-5 (1) and 18-crown-6 (2) and their homopolymers and copolymers with methyl methacrylate (MMA) with various crown monomer compositions. Cation binding properties of these polymers were evaluated by the extraction of alkali metal picrates into methylene chloride.

We also tried an immobilization of crown ether on the surface of a porous support since it is preferable from an economic standpoint as well as the ease of the control of the physical properties of the beads. In the actual application of the crown polymers for the separation of some salts by the column method, they have to be processed

345

as granules of a suitable size. The technique of suspension polymerization is hard to apply in this case, since the monomers are
very costly. Crown ethers have been introduced by the polymer
reaction to the crosslinked polystyrene backbone, and immobilized
on the surface of beads of silica gel.[9-11] Hiraoka et al. coated
inorganic supports with a kind of epoxy resin using diaminodibenzo-
18-crown-6 and bisepoxide.[12] We have developed a new method to
immobilize crown ethers using poly(glycidyl methacrylate) (3) and
aminobenzocrown ethers (4).

EXPERIMENTAL

Materials

The monomers (1, 2) were prepared as reported previously.[5,7]
Homo- and copolymerizations were carried out using AIBN as an
initiator. The copolymer compositions were determined by nitrogen
analysis. The results are shown in Tables 1 and 2.

Immobilization of 15-crown-5 on the surface of Chromosorb GNAW
(60-80 mesh), Gasukuro Kogyo Co. Ltd, was carried out by submerging
the Chromosorb (25g) in 50 ml of a THF solution of aminobenzo-15-
crown-5 (4a, 1.25g) and 3 (1.20g). The mixture was kept overnight
and then the solvent was evaporated at room temperature. The
residue was finally vacuum dried and heated at 150 °C for 2 hrs
under nitrogen. The resulting material was then washed in a Soxlet
extractor first with THF and then with methanol for 24 hrs each.
The product thus purified was vacuum dried (RC-12). The product
from 4b (RC-22) was prepared in a similar manner.

n=1 ; 1
n=2 ; 2

n=1 ; 4a
n=2 ; 4b

Evaluation of Cation Binding Properties

Extractions were carried out using 10 ml of aqueous picrate solution (7×10^{-5}M) with metal hydroxide (0.01 M) and the same volume
of crown solution in methylene chloride (approximately 3.5×10^{-4}M).
They were shaken well in a separating funnel and then separated.
Sometimes foam formation was observed at the interface. This
almost disappeared on standing about half an hour. The organic
phase was filtered. Picrate concentrations in the aqueous (357 nm,

ε=1.5 x 10^4 1/mol.cm) and the organic (378 nm, ε=1.8 x 10^4 1/mol.cm) phases were determined by spectrophotometry.

RESULTS AND DISCUSSIONS

Selective Extraction of Alkali Metal Picrates

The monomer reactivity ratios were calculated on the basis of the data shown in Tables 1 and 2 and were obtained as follows, M_1 being MMA; r_1=1.15, r_2=0.80 for 1 ; and r_1=1.0, r_2=1.0 for 2. In the latter case, they turned out to be unity by chance. These data show that the distribution of each monomer in the copolymer is almost random.

The percentage of the picrates extracted in methylene chloride are shown in Figs. 1 and 2. The polymers turned out to extract much better than the corresponding monomers. Moreover, the homopolymer of 1 (CP10) showed an especially high selectivity toward K and Rb, and that of 2 (CP20) toward Cs, as compared to the monomers. These are due to the ease of sandwich type 1:2 complex formation between 15-crown-5 and K or Rb, and between 18-crown-6 and Cs. In other words, it may be said to be an "entropy effect" or "neighboring group effect".

The data are qualitatively parallel with those reported in the case of polystyrene derivatives, however, they differ in that the methacrylamide derivatives extract much more effectively than the polystyrene derivatives. One of the reasons for the high extractability of the methacrylamide derivatives is ascribed to their higher partition coefficients in aqueous phase.[13] The extractions by the derivatives of polyacrylate[3,4] and polyacrylamide[5,6] showed the same tendencies, although the extractions were carried out in a chloroform-water system.

The data for Li seem to be higher than expected. However, the complex formation between Li and amide group, if anything, does not seem to affect the extraction since poly(methacrylanilide) did not show any significant extraction.

To shed more light on the cation binding properties of the polymeric crown ethers, the extractions were carried out with various copolymers with different crown monomer compositions and the results are shown in Figs. 1 and 2. In the case of the copolymers of the monomer 1, the percentage of the extraction decreased gradually on increasing the MMA fraction in the copolymer. This is especially noteworthy in the cases of K and Rb which form a strong 1:2 complex with 15-crown-5. It should be noted that the selectivity pattern finally approached that of the monomer.

Table 1. Copolymerization of 1 with MMA.

Run	Feed				Conditions		Results		
1m	1(mg)	MMA(mg)	AIBN(mg)	Solv.[a](ml)	Temp.(°C)	Time(hr)	Yield(%)	N(%)	[MMA]/[1][c]
10	500	0	5	A(1)	70	13	32[d]	3.91	0
11	277	78	6.9	A(2.5)	60	24	79	2.86	1.4
12	176	100	5.5	A(2)	60	24	48	2.39	2.4
13	249	245	6.9	A(1.3)	60	20	59	1.85	4.6
14	100	403	6.0	A(0.5)+B(1)	60	18	62	0.71	16
15	51	446	5.3	A(0.5)+B(1)	60	18	71	0.59	36

a A:DMF; B:THF. b After twice reprecipitation. c Molar ratio of each unit in the copolymer.
d [η]= 0.10 in DMF at 30°C.

Table 2. Copolymerization of 2 with MMA.

Run	Feed				Conditions		Results		
2m'	2(mg)	MMA(mg)	AIBN(mg)	Solv.[a](ml)	Temp.(°C)	Time(hr)	Yield(%)[d]	N(%)[b]	[MMA]/[2][c]
20	204	0	5	A(1)+B(0.5)	70	22	51[d]	3.52	0
21	170	43	2.1	A(0.5)	60	8	61	2.87	1.0
22	166	84	2.5	A(0.5)	60	12	44	2.36	2.1
23	105	107	3.1	A(0.2)+B(1)	60	12	53	1.71	4.3
24	100	400	5.1	B(1)	60	18	65	0.54	16
25	50	450	5.0	B(1)	60	18	71	0.38	33

a–c, The same as in Table 1. d [η]= 0.08 in DMF at 30°C.

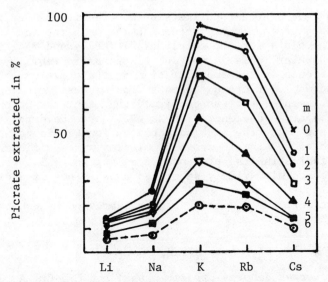

Fig. 1. Percent of picrates extracted in organic phase with polymeric 15-crown-5 (CP1m). See table 1 for numbering. [Cr]o/[A]o=
5.0, 4.3, 4.6, 4.6, 4.5, 4.8, 4.7 for n=0 to 6 in this order,
respectively. Dotted line: monomer 1.

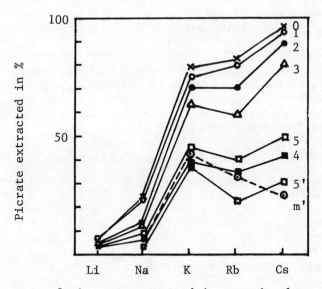

Fig. 2. Percent of picrates extracted in organic phase with polymeric 18-crown-6 (CP2m'). See table 2 for numbering. [Cr]o/[A]o=
4.9, 5.0, 5.2, 4.9, 3.8, 5.5 for m'=0 to 5 in this order, respectively, and 3.8 for 5'. Dotted line: monomer 2.

In the case of the copolymers of 2, the high selectivity for Cs also decreased gradually on increasing the MMA fraction in the co-polymer. The plots for the copolymer 24 (CP24) and 25 (CP25) are reversed. However, this is not due to the extractability but is due to the difference in the concentration of the crown ether. In the case of CP24, a lower crown concentration was used because we had insufficient sample. It was confirmed that the extraction by CP25 with the same crown concentration as CP24 gave the data lower than CP24. In this case, the high selectivity for Cs did not disappear completely although the 18-crown-6 moieties were sufficiently separated, indicating the presence of a long range 1:2 complex, i.e. a complex formed with widely separated crown ether ligands.

The extraction equilibrium can be expressed by equation 1.

$$M^+ \;+\; Cr^* \;+\; A^- \underset{\longleftarrow}{\overset{Ke}{\longrightarrow}} M^+Cr,A^-* \qquad (1)$$

where the asterisk denotes organic phase, and M^+, Cr, A^- denote metal cation, crown ether, picrate anion, respectively. The extrac-tion equilibrium constant can be calculated from equation 2.

$$Ke = \frac{\alpha}{a(1 - \alpha)([Cr]o/[A]o - m\alpha)[A]o} \qquad (2)$$

where α, a, [Cr]o, [A]o, denote the fraction of the picrate extracted in the organic phase, cation activity, initial crown and picrate con-centration, respectively, m being 1 for a 1:1 complex and 2 for a 1:2 complex.

Crown complexed picrate seems to be partially dissociated in this solvent, so the values of Ke obtained are apparent ones, however, the dissociation behavior may not be the same as in the case of the polystyrene derivatives since picrate anion may form a hydrogen bond with the hydrogen of the amide group.

Table 3. Values of Ke of the polymeric 15-crown-5 (x $10^4 l^2/m^2$).[a]

Crown ether	Li	Na	K	Rb	Cs
CP10	5.0	11	1010	490	43
CP11	5.6	8.7	550	150	22
CP12	5.3	6.7	220	120	22
CP13	5.5	9.0	150	81	16
CP14	5.5	8.0	64	31	11
CP15	4.1	6.6	25	15	5.7

[a] m=1 for Li and Na, and m=2 for K, Rb and Cs

Table 4. Values of Ke of the polymeric 18-crown-6 $(\times 10^4 1^2/m^2)^a$

Crown ether	Li	Na	K	Rb	Cs
CP20	2.0	11	160	190	1800
CP21	1.7	5.7	120	150	790
CP22	1.1	4.4	87	75	390
CP23	1.6	5.0	68	54	210
CP24	1.9	3.4	30	25	39
CP25	1.3	2.8	26	21	35

^{a}m=2 for Cs and m=1 for the other metals.

In Tables 3 and 4 are shown the values of Ke thus calculated. As expected, the decrease is noteworthy for the system where it forms a 1:2 complex, for example, polymeric 15-crown-5 and K or Rb, and polymeric 18-crown-6 and Cs. Slight decreases were also observed for the system where it forms a 1:1 complex probably due to the decrease in the partition coefficient (Pe).

The extraction equilibrium consists of the following three equilibria.

$$M^+ + Cr \underset{\overline{}}{\overset{Ks}{\rightleftharpoons}} MCr^+ \qquad (3)$$

$$Cr^* \underset{\overline{}}{\overset{Pe}{\rightleftharpoons}} Cr \qquad (4)$$

$$MCr^+ + A^- \underset{\overline{}}{\overset{Pc}{\rightleftharpoons}} MCr^+,A^- {}^* \qquad (5)$$

Ke is equal to the product of the three equilibrium constants, i.e. Ke=KsPePc. Crown monomers should be more hydrophilic than MMA and it is reasonable that the values of Pe and Pc depend on the co-polymer composition. Their dependence should be in the opposite direction and the dependence of Pe seems to contribute more to the overall extraction than that of Pc in this case judging from the results.

Accordingly, as a parameter to express the ease of a 1:2 complex formation, the relative selectivities normalized for Na, Ke(x)/Ke(Na), were calculated and are plotted against the molar ratio of MMA units to crown monomer units in the copolymer(Figs. 3 and 4).

In the case of the copolymers of 1, K and Rb gave rapidly and continuously decreasing curves while Cs gave a moderately decreasing one. These are reasonable since K and Rb form a strong 1:2 complex while Cs forms a weak one, if it forms one at all. Li gave almost

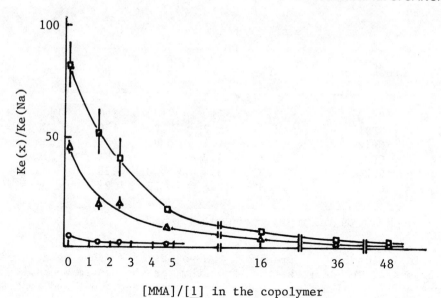

Fig. 3.   Dependence of the relative selectivities of the polymeric
15-crown-5 on the copolymer composition.   (o):Cs; (Δ):Rb; (□):K.
For some points, the extractions were repeated and the ranges are
shown by the perpendicular lines.

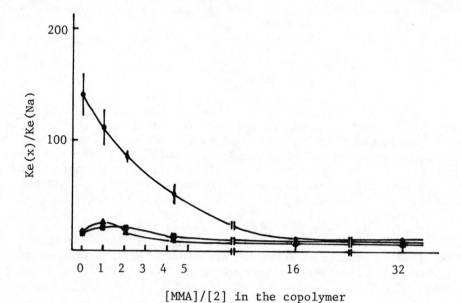

Fig. 4.   Dependence of the relative selectivities of the polymeric
18-crown-6 on the copolymer composition.   (●):Cs; (▲):Rb; (■):K.
Perpendicular lines show the ranges of the data.

constant values, which is also reasonable since both Li and Na form
1:1 complexes with 15-crown-5.

In the case of the copolymers of 2, Cs, which forms a strong
1:2 complex with 18-crown-6, shows a rapidly and continuously decreasing curve, whereas K and Rb, which form a 1:1 complex, show
almost constant values.

In both cases, the continuously decreasing curves indicate that
a 1:2 complex is most easily formed between two adjacent crown
ligands of the continuous crown monomer sequence and that it becomes
more difficult the more MMA units there are incorporated in between
two crown ligands. The decrease in Ke is due to the decrease in the
partition coefficient for the system where it forms a 1:1 complex,
whereas the marked decrease in Ke for the system where it forms a
1:2 complex is mainly due to the decrease in the binding constant.

To ascertain the results, we have tried the extraction at lower
crown and picrate concentrations and have obtained the same tendencies. The details will be published eleswhere.

Smid et al showed that two benzo-15-crown-5 ligands connected
by a $-CO_2(CH_2)_nO_2C-$ chain passed through a maximum complex formation
constant with K in THF for n=5 (i.e, 9 chain atoms in between the
ligands).[14] They also reported that in the case of the copolymers
of styrene derivative of 15-crown-5 and styrene, crown ligands
separated from another crown ligand by less than 7 to 9 chain carbon
atoms will preferencially form a cooperative 1:2 intramolecular
K/crown complex.[15] Biscrown ether of the glutaranilide type,
$(-NHCO(CH_2)_3CONH-)$, which can be considered as the most simple model
compound of polyacrylamide derivative[6] also showed a high 1:2 intramolecular complex formation constant.

It should be noted that in the poly(1) or poly(2), the adjacent
benzocrown ligands are separated from each other by seven atoms. Of
course, the number of atoms separating the two ligands is not the
only factor, the flexibility of the side chain as well as the main
chain will play an important roll in the proper alignment of the two
crown ligands for an effective 1:2 complex formation. In addition,
a comparison between the two copolymers may be difficult since they
differ in the side chain structure as well as the comonomer.
However, in the light of these previous reports, the results obtained here seem reasonable.

Immbilization of Crown Ethers on the Surface of Chromosorb and their Application.

From an economic standpoint, it is important to immobilize crown
ethers on the surface of porous support to make full use of the
costly crown ether. The scheme of our method can be shown as follows.

$$
\begin{array}{c}
\text{CH}_3 \\
\text{-(-C-CH}_2\text{-)}_{\overline{n}} \\
\text{CO}_2\text{CH}_2\text{CH-CH}_2 \\
\diagdown\!\diagup \\
\text{O}
\end{array}
\quad + \quad \text{H}_2\text{N-Crown} \quad \xrightarrow[\text{2hr}]{150\,^\circ\text{C}} \quad
\begin{array}{c}
\text{CH}_3 \\
\text{-(-C-CH}_2\text{-)}_{\overline{n}} \\
\text{CO}_2\text{CH}_2\text{CH-CH}_2 \\
\text{OH NH-Crown}
\end{array}
$$

$$\underline{3} \qquad\qquad\qquad\qquad\qquad \underline{4} \qquad\qquad\qquad\qquad\qquad\qquad \underline{5}$$

$$
\underline{3} \ + \ \underline{5} \ \longrightarrow \
\begin{array}{c}
\text{CH}_3 \\
\text{-(-C-CH}_2\text{-)}_{\overline{n}} \\
\text{CO}_2\text{CH}_2\text{CH-CH}_2\text{-N-CH}_2\text{-CHCH}_2\text{O}_2\text{C} \\
\text{OH} \qquad\qquad\qquad \text{OH}
\end{array}
\quad
\begin{array}{c}
\text{Crown} \\
\text{|}
\end{array}
\quad
\begin{array}{c}
\text{CH}_3 \\
\text{-(-C-CH}_2\text{-)}_{\overline{n}}\text{,} \\
\end{array}
$$

$$\underline{6}$$

The thermal treatment of a film made of $\underline{3}$ and $\underline{4a}$ containing equivalent amounts of epoxy groups and amino groups showed reasonable changes in the IR spectrum, i.e. the characteristic bands of the amino group (3350 and 3440 cm^{-1}) disappeared and a broad band of the hydroxy group appeared around 3480 cm^{-1}. Bands of the methylene (3050 cm^{-1}) and methyn (2995 cm^{-1}) of the epoxy group decreased considerably. The film did not dissolve in solvents such as DMF, THF etc., which indicate that besides $\underline{5}$, some crosslinking to $\underline{6}$ took place simultaneously. The thermal treatment was done at relatively high temperature since the reaction takes place in a solid state. It may be reacted at lower temperature, however, we did not try at other temperatures.

In Figs. 5 and 6, are shown the adsorption isotherms of the alkali metal picrates in methanol. RC-12 contains 4.3 % of benzo-15-crown-5 and RC-22 6.4 % of benzo-18-crown-6 by weight. It should be noted that the polymeric crown ethers thus immobilized showed the expected selectivities, i.e. RC-12 for K and Rb, and RC-22 for Cs. It also should be noted that the side chain is much more flexible and hydrophilic in this case.

In Figs. 7 and 8 are shown the results of the column elution of a mixture of KCl and NaCl in methanol, containing 10 ppm of each metal. The solution was passed at a space velocity of 4.2 hr^{-1} and the content of each metal in each fraction of 25-30 ml was determined by the atomic absorption. The separation is remarkable especially in the case of RC-22. Complete separation and quantitative analysis are to be expected if a sufficiently small amount of sample is applied.

The advantage of this method is that the immobilization takes place simultaneously with the insolubilization, allowing its application to any surface such as a film, a fiber, porous beads, etc.

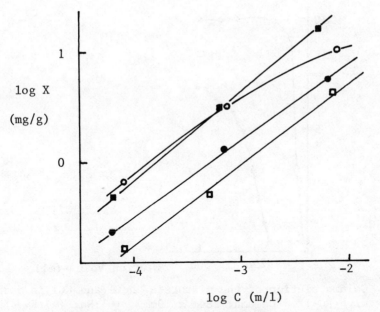

Fig.5. Adsorption isotherm of picrates with RC-12.

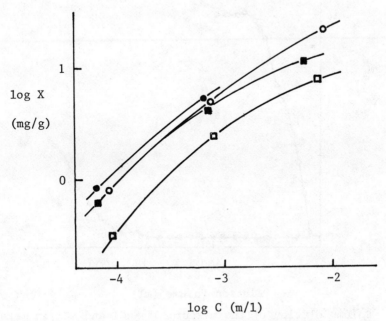

Fig. 6. Adsorption isotherm of picrates with RC-22.

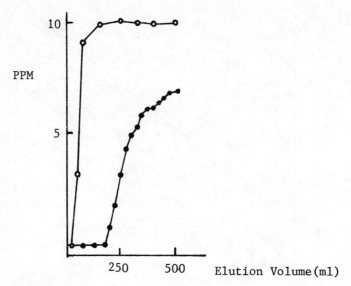

Fig. 7.   Column elution of the mixture of NaCl and KCl in methanol
with RC-12. (Column: 28cm x 10mm), (o):Na; (•):K

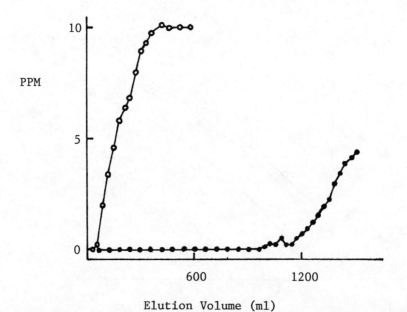

Elution Volume (ml)

Fig. 8.   Column elution of the mixture of NaCl and KCl in methanol
with RC-22.   (Column: 32cm x 10mm), (o):Na; (•):K

Another advantage is that packing materials of different selectivities can be prepared by using a particular kind of crown ether depending on the distance between the crown ligands.

We are also interested in the application of these materials in the analysis of salts in aqueous solution, and also as phase transfer catalysts.

REFERENCES

1. S. Kopolow, T.E. Hogen Esch and J. Smid, Macromolecules, $\underline{4}$, 359(1971).
2. S. Kopolow, T.E. Hogen Esch and J. Smid, Macromolecules, $\underline{6}$, 133(1973).
3. K. Kimura, T. Maeda and T. Shono, Polym. Bull., $\underline{1}$, 403(1979).
4. A.J. Varma, T. Majewicz and J. Smid, J. Polym. Sci., Polym. Chem. Ed., $\underline{17}$, 1573(1979).
5. K. Kimura, T. Maeda and T. Shono, Anal. Lett., A$\underline{11}$, 821 (1978).
6. K. Kimura, T. Maeda and T. Shono, Talanta, $\underline{26}$, 945(1979).
7. K. Yagi, J.A. Ruiz and M.C. Sanchez, Makromol. Chem., Rapid Commun., $\underline{1}$, 263(1980).
8. E. Blasius, P.-G. Maurer, Makromol. Chem., $\underline{178}$, 649(1977).
9. G. Dotsevi, Y. Sogah, D.J. Cram, J. Am. Chem. Soc., $\underline{98}$, 3038(1976).
10. M. Cinquini, S. Colona, H. Molinari, F. Montanari, P. Tundo, J. Chem. Soc., Chem. Commun., 394(1976).
11. K. Kimura, M. Nakajima and T. Schono, Anal. Lett. A$\underline{13}$, 741 (1980).
12. M. Hiraoka, "Crown Compounds", Kodansha, p299 (in Japanese).
13. K. Yagi, J.A. Ruiz, M.C. Sanchez and C. Guerrerro, Rev. Soc. Quim. Mex., to be published.
14. M. Bourgoin, K.H. Wong, J.Y. Hui and J. Smid, J. Am. Chem. Soc., $\underline{97}$, 3462(1975).
15. S.C. Shah, S.L. Kopolow and J. Smid, Polymer, $\underline{21}$, 188 (1980).

SULFUR-SUBSTITUTED POLYOXYETHYLENES

SEQUENTIAL ETHER-THIOETHER COPOLYMERS

L.J. Mathias and J.B. Canterberry
Department of Polymer Science
University of Southern Mississippi
Hattiesburg, Mississippi 39406-0076

Polyoxyethylene (POE) and polythioethylene (PTE) display very different physical and chemical properties despite a close similarity in structure. POE, for example, melts at 68°C and PTE at 216°C.[1] POE is soluble in many organic solvents and in water while PTE is generally insoluble. The crystalline conformations are also very different with POE adopting a (7/2) helix with oxygen lone-pair electrons oriented inward around the interior and PTE assuming a (2/0) glide-plane structure in which adjacent sulfur lone-pairs are oriented at 180° to each other.[3-4]

The dominant factors effecting these differences in properties are the different atomic and molecular dimensions centered around the heteroatoms.[1] The values of the van der Waals radii used most frequently are in the range of 1.8-1.9Å for sulfur and 1.4-1.6 for oxygen. The average C-S bond length is ca. 30% greater than the C-O value and the C-S-C bond angle is 12° smaller than the C-O-C angle. The covalent bond length is 1.82Å for thioethers and 1.43Å for ethers. Thus, sulfur forms longer but more sharply bent thioether units compared to oxygen ether moieties.

Investigations of the parent homopolymers, homologous analogs and some sequential ether-thioether copolymers have been reported.[5-10] These have led to a clearer understanding of the fundamental differences in behavior resulting from basic molecular dimensions and inter- and intramolecular interactions. We previously reported an efficient general synthesis of sequential ether-thioether copolymers which made available a wide variety of

359

structural repeat units.[11] It seemed worthwhile to examine the
behavior of these copolymers empirically by evaluating trends in
physical and chemical properties within and between families consisting of homologous repeat units. We report here on the
synthesis and characterization of several such families containing
sequences of alkyl ether and thioether groups, and include some
initial observations on their ability to complex and solublize
transition metal salts.

EXPERIMENTAL

All solvents and chemicals were reagent grade materials.
Tetrahydrofuran (THF) was dried over 4A sieves. Oligooxyethylenes were purchased from Polysciences, Inc. and Aldrich Chemical
Company, vacuum distilled and stored over 4A sieves. The various
dithiols were purchased from Aldrich Chemical Company and Fair-
field Chemical Company and distilled before use.

^{1}H NMR spectra were obtained on a Varian EM390 and ^{13}C spectra
on a JOEL FX 90Q FT-NMR. Thermal behavior was evaluated with an
electrothermal melting apparatus, a Perkin-Elmer DSC-1B and a
Leitz polarizing microscope equipped with a Mettler hot stage.
Viscosities were determined at $30^\circ \pm 0.2^\circ$C in chloroform with a
Cannon-Ubbelohde #50 semimicro viscometer. Microanalyses were
carried out by M-H-W Laboratories of Phoenix, Arizona.

Bisisourea Synthesis

A typical procedure involved combining distilled oligooxyeth-
ylene (0.1 mole) with N,N'-diisopropylcarbodiimide (0.21 mole) in
dry THF and adding a catalytic amount of cupric chloride. The
yellow or light green mixture was stirred overnight at room
temperature. The reaction mixture was then diluted with ten
volume-equivalents of hexane and vacuum filtered through a three-
inch pad of basic alumina solvent-packed into a 150 ml fritted-
glass funnel. The colorless filtrate was evaporated under vacuum
to the viscous product and kept in vacuo until use.

Polymer Synthesis

Two methods of thermal polymerization were employed. Method
A was used for dithiols with low boiling points and involved
equimolar mixtures of the bisisourea, the dithiol and dry potassium fluoride in the absence of solvent. The reaction mixture
was heated at 140°C under a nitrogen atmosphere for three days.
Polymer was isolated by extracting the reaction mixture with
chloroform and precipitating into methanol. Filtration and
vacuum drying gave white to off-white products in fair to good
yields.

   Method B involved combining equimolar mixtures of comonomers
in a round-bottom flask fitted with a septum for evacuation.
After vacuum introduction the flask was immersed in an oil bath
preheated to 110°C. The oil bath temperature was gradually raised
to 180° over a 3 hour period. The vessel was then cooled and
polymer isolated as in Method A except hot chloroform was needed
in some cases to solublize the polymeric products.

   For several polymers obtained by these methods the N,N'-
diisopropylurea side-product was not completely removed by the
reprecipitation sequence. The urea crystals were clearly present
as an entrapped separate phase under the microscope and displayed
a characteristic transition on the DSC. These polymers were
purified by Soxhlet extraction with methanol for three days.

RESULTS AND DISCUSSION

Polymer Synthesis

   The general polymerization method employed[11] was based on
isourea intermediates which have proven useful in a number of
synthetic applications.[12] As indicated below, the two-step procedure involves a mild, high-yield addition of oligooxyethylenes
to N,N'-diisopropylcarbodiimide to give a bisisourea followed by
step-growth thermal polymerization with various dithiols. The
two major drawbacks of this method are, first, that any water
present in the oligooxyethylenes will react with either the
carbodiimide during synthesis or later with the isourea to destroy
active functionality; second, the bisisoureas cannot be purified
by distillation. They can, however, be chromatographed under
mild conditions to remove the copper catalyst and excess carbodiimide. The bisisoureas described here were generally obtained
quantitatively and were either used immediately or carefully
stored under vacuum.

   The three major advantages of this two-step procedure are,
first, that the bisisoureas are obtained quantitatively from the
oligooxyethylenes; second, the polymerizations require no strong
acid or base catalyst and give a neutral, easily-separable by-
product; and finally, the variety and combinations of bisisoureas
and dithoils capable of polymerization with these methods allow
synthesis of a large number of polymers of known repeat unit
structure.

   This procedure compares favorably with previously reported
syntheses of polythioethers and copolymers in terms of yeild
and polymer properties. These include a number of reactions
between alkyl bishalides and inorganic or organic thiol salts[13],

phase transfer catalyzed systems,[14],[15] and several acid-catalyzed ring-opening or dehydration polymerizations of monomers containing ether and thioether groups.[6-10]

$$HO \{ CH_2CH_2O \}_n H \qquad n = 2,3,4$$

$$\begin{array}{c} CH_3 \\ \diagdown \\ CH_3 \end{array} CH - N = C = N - CH \begin{array}{c} CH_3 \\ \diagup \\ CH_3 \end{array}$$

in
THF | CuCl

$$\begin{array}{c} i\text{-Pr}-HN \\ \diagdown \\ i\text{-Pr}-N \end{array} C - OCH_2CH_2 \{ OCH_2CH_2 \}_n O - C \begin{array}{c} NH - i\text{-Pr} \\ \diagup \\ N - i\text{-Pr} \end{array}$$

$$n = 1,2,3$$

HS—R—SH

$$\longrightarrow \quad 2 \; i\text{-PrNHCNH}i\text{-Pr} \\ \qquad\qquad\qquad \underset{O}{\overset{\|}{}}$$

$$\{ S - CH_2CH_2 \{ OCH_2CH_2 \}_n S - R \}$$

Final polymer yields were generally less than quantitative. Repeated reprecipitation and Soxhlet extraction removed residual urea by-product and low-molecular weight polymer fractions. Tables 1 and 2 list the various comonomers employed along with melting points, viscosities and yields. The polymers were obtained as white-to-yellow solids, soft waxes or viscous oils depending on the sulfur and oxygen contents and ratios. Microanalysis data were obtained for several polymers (Table 3) and were consistant with expected structures. Spectroscopic characterization included infrared, ^{1}H and ^{13}C NMR. The IR and ^{1}H NMR spectra for all polymers were very similar and provided little unique information. Figure 1 is a typical IR spectrum and displays the expected bands for C-H, C-O-C and C-S-C groups. ^{1}H NMR displayed the expected integration ratios for hydrogens of the various methylene groups. Those adjacent to ether and thioether groups were clearly separated (Table 4) although coupling patterns overlapped and were often broad.

The most useful characterization method was ^{13}C NMR which generally allowed separation and identification of all unique carbons in the repeat units. Figures 2-4 illustrate this for

Table 1. Copolymer melting points, intrinsic viscosities (in dl/g) and yields for 2-, 3-, 4- and 6-carbon dithiols.

1   $HSCH_2CH_2SH$
2   $HSCH_2CH_2CH_2SH$
3   $HSCH_2CH_2CH_2S$
4   $HSCH_2CH_2CH_2CH_2CH_2S$

A $HOCH_2CH_2OCH_2CH_2OH$
B $HOCH_2CH_2OCH_2CH_2OCH_2CH_2OH$
C $HOCH_2CH_2OCH_2CH_2OCH_2CH_2OCH_2CH_2OH$

	MP^1	$[\eta]^1$	YIELD %
1 + A	87-91	.23	
1 + B	54-57	.13	81.4
1 + C	38-39	.55	34.5
2 + A	OIL	.12	
2 + B	R.T.	.96	
2 + C	OIL	.10	
3 + A	38-40	.26	40.8
3 + B	OIL	.18	71.9
3 + C	OIL	.27	95.4
4 + A	30	.22	61.1
4 + B	OIL	.12	
4 + C	R.T.	.12	28.8

(1) °C.
(2) IN $CHCL_3$ AT 30°C.

Table 2. Copolymer yields, melting points and intrinsic viscosities (in dl/g) for 4-thia- and 4-oxa-1,7-dithiaheptanes.

5   $HSCH_2CH_2SCH_2CH_2SH$
6   $HSCH_2CH_2OCH_2CH_2SH$

A $HOCH_2CH_2OCH_2CH_2OH$
B $HOCH_2CH_2OCH_2CH_2OCH_2CH_2OH$
C $HOCH_2CH_2OCH_2CH_2OCH_2CH_2OCH_2CH_2OH$

	yield %	mp°	$(\eta)^\bullet$
5 + A	85	122-5	i
6 + A	52	44-5	.21
5 + B	67	97-9	.11
6 + B	76	32-4	.20
5 + C	82	68-9	.18

°·C°

•·in $CHCl_3$ at 30°C

i· insoluable

Table 3. Microanalysis data for selected copolymers.

	ANAL (%)		CAL (%)	
	C	H	C	H
1A	44.60	7.72	43.87	7.36
5A	40.25	6.67	42.83	7.13
5B	45.71	7.71	44.74	7.51
5C	45.87	7.61	46.14	7.68
6A	45.98	7.78	46.15	7.68

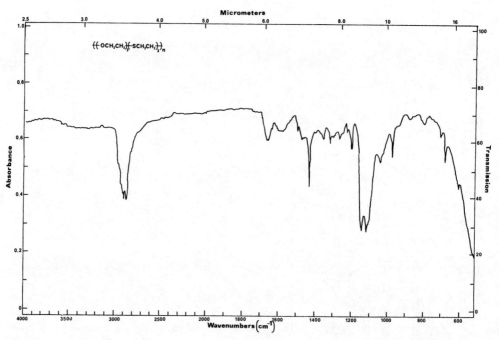

Figure 1. Typical IR spectrum of an ether-thioether copolymer.

Table 4. ^{1}H NMR chemical shifts and coupling patterns (in parantheses); T = triplet, Q = quintet and M = multiplet.

	CH-O	CH-S	CH-C
1A	2.6-2.8 (M)	3.5-3.7 (T)	
1B	2.6-2.8 (T)	3.5-3.7 (M)	
1C	2.6-2.8 (T)	3.5-3.7 (M)	
2A	2.5-2.8 (M)	3.4-3.7 (M)	1.6-2.0 (M)
2B	2.5-2.8 (T)	3.5-3.7 (M)	1.6-2.0 (Q)
2C	2.5-2.8 (T)	3.5-3.7 (M)	1.6-2.0 (Q)
3A	2.5-2.8 (M)	3.5-3.7 (T)	1.6-2.0 (M)
3B	2.5-2.8 (M)	3.5-3.8 (M)	1.6-1.8 (M)
3C	2.5-2.8 (M)	3.5-3.8 (M)	1.6-1.8 (Q)
4A	2.4-2.8 (M)	3.5-3.8 (M)	1.2-1.8 (M)
4B	2.5-2.8 (M)	3.5-3.8 (T)	1.2-1.8 (M)
4C	2.5-2.8 (M)	3.5-3.8 (M)	1.2-1.8 (M)

simple and complex structures. The ether and thioether carbons are clearly separated, of course, but even up to three individual peaks within each type of group are distinguishable. The small differences in chemical shift for similar carbons result from short- and long-range heteroatom effects through 2-5 bonds. Table 5 summarizes the unique carbon peaks observed for most of the copolymers synthesized. Individual peak assignments indicated in the Figures are tentative; the Table values are listed simply from highest to lowest chemical shift.

One of the more interesting properties of the various families of copolymer sequences are the melting temperatures. These polymers are highly crystalline according to x-ray diffraction and polarizing microscope studies. Thus, the melting points observed visually and by differential scanning calorimetry are true crystalline melting transitions. Figures 5 and 6 illustrate trends observed when either the number of thioether or ether repeat units is held constant, and the number of comonomer units increased. In general, increasing the number of adjacent thioethylene units increases the melting point while increasing the number of adjacent oxyethylene groups decreases it.

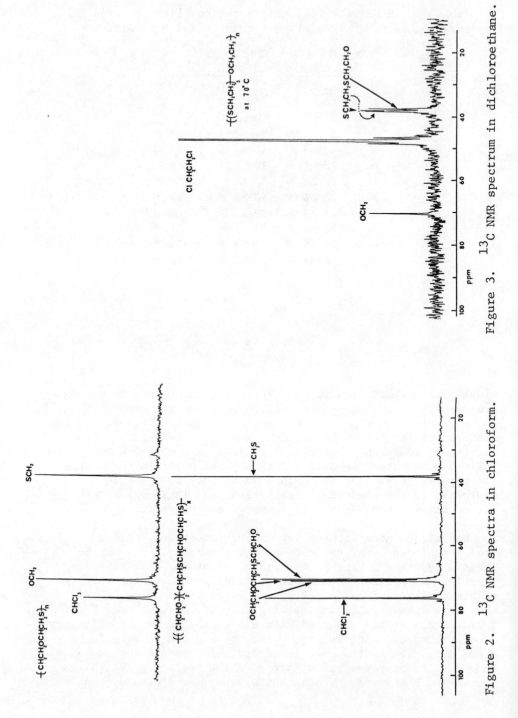

Figure 3. ^{13}C NMR spectrum in dichloroethane.

Figure 2. ^{13}C NMR spectra in chloroform.

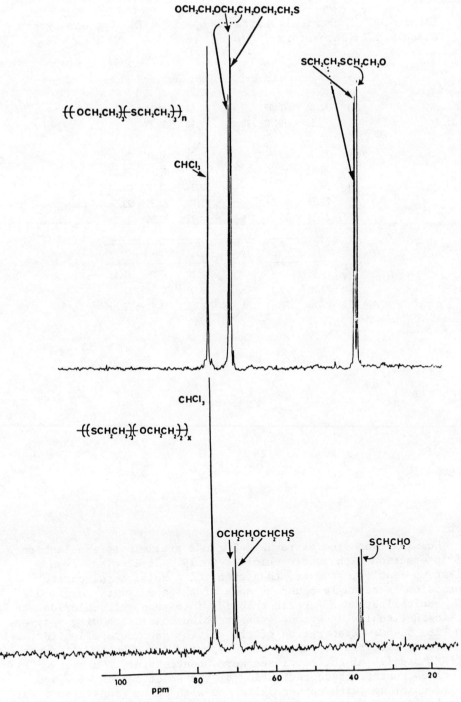

Figure 4. ^{13}C NMR of trithiacopolymers in chloroform.

Table 5. ^{13}C chemical shifts in ppm from TMS for copolymers dissolved in chloroform.

	C-O			C-S		OTHER
1A	70.94			32.82	31.78	
1B	71.43	70.49		32.80	31.54	
1C	71.18	70.70	70.46	32.80	31.73	
2A						
2B	71.44	70.74		31.91	31.76	30.01
2C	70.41	69.96	69.68	30.78	30.61	29.01
3A	70.74			32.14	31.54	28.71
3B	70.76	69.98		31.89	30.30	28.56
3C	71.11	70.71	70.41	32.21	31.56	28.84

	C-O			C-S		
5A[a]	70.21			32.23	31.85	31.10
5B	70.97	70.16		32.50	32.18	31.47
5C	70.76	70.21	70.00	32.23	31.91	31.20
6A	70.43			31.69		
6B	70.70	70.38	69.64	31.58		

[a] IN CH_2CLCH_2CL AT 70°C

Complexation studies to date include preliminary examination of interaction with cupric chloride $CuCl_2$. It is clear from clean-up results, however, that even the alkalai metal cation potassium is strongly bound by some of the copolymers; e.g., 5A-5C. Solublization of cupric chloride in carbon tetrachloride was followed visually by strong green-to-blue coloration with polymers 1A, 2B, 5A-5C. Evaporation of the solvent gave colored solid complexes with the copper evenly dispersed throughout. Attempts at NMR studies of the complexes were unsuccessful. Current work involves further characterization of the solution behavior and solid complexes with copper salts and with other transition metal complexes.

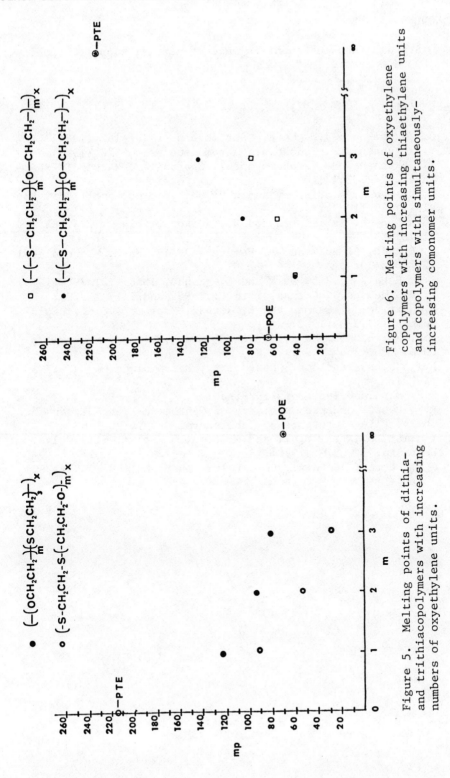

Figure 6. Melting points of oxyethylene copolymers with increasing thiaethylene units and copolymers with simultaneously-increasing comonomer units.

Figure 5. Melting points of dithia- and trithiacopolymers with increasing numbers of oxyethylene units.

ACKNOWLEDGEMENTS

Grateful acknowledgement is made of partial support of this work by 3M Company.

REFERENCES

1. D. Bhaumik and J.E. Mark, Macromolecules $\underline{14}$, 162 (1981).
2. Y. Takahashi, H. Tadokoro, Macromolecules $\underline{6}$, 672 (1973).
3. Y. Takahashi, H. Tadokoro and Y. Chatani, J. Macromol. Sci.-Phys. $\underline{B2(2)}$ 361 (1968).
4. H. Hasegawa, W. Claffey and P.H. Geil, J. Macromol. Sci.-Phys. $\underline{B13(1)}$ 89 (1977).
5. Y. Gotoh, H. Sakakihara and H. Tadokoro, Polym. J. $\underline{4}$, 68 (1973).
6. E. Riande, S. Boileau, P. Hemery and J.E. Mark, Macromolecules $\underline{12}$, 702 (1979).
7. R.R. Rahalkar, J.E. Mark and E. Riande, Ibid. $\underline{12}$, 795 (1979).
8. E. Riande and J. Guzman, Ibid. $\underline{12}$, 952 and 1117 (1979).
9. L. Garrido, J. Guzman and E. Riande, Macromol. Chem. Rapid Commun. $\underline{2}$, 379 (1981).
10. W.J. Welsh, J.E. Mark, J. Guzman and E. Riande, Macromol. Chem. $\underline{183}$, 2565 (1982).
11. L.J. Mathias and J.B. Canterberry, Macromolecules $\underline{13}$, 1723 (1980).
12. L.J. Mathias, Synthesis $\underline{1979}$, 561.
13. R.W. Lenz, "Organic Chemistry of Synthetic High Polymers," Ch. 6, Wiley-Interscience Publishers, New York, 1967.
14. Y. Imai, A. Kato, M. Ii and M. Udea, J. Polym. Sci. Polym. Lett. Ed. $\underline{17}$, 579 (1979).
15. G.R. Pettit, J. Polym. Sci. Polym. Chem. Ed. $\underline{18}$, 345 (1980).

POLY(ETHYLENE GLYCOL)S AS PHASE-TRANSFER CATALYSTS

SYNTHESIS OF DERIVATIVES

J. Milton Harris, Nedra H. Hundley, Thomas G. Shannon,
and Evelyn C. Struck

Department of Chemistry
The University of Alabama in Huntsville
Huntsville, AL 35899

Polyethylene glycols (PEG's) and their derivatives are finding a rapidly expanding range of chemical, biomedical, and industrial applications resulting from their low cost and useful properties such as solubility in aqueous and organic solvents, metal complexing ability, biological compatibility, and ease of chemical modification.[1] The linear polymers have been employed as soluble matrices for liquid-phase peptide synthesis,[2] ligands for water-soluble transition metal complexes,[3] drug carriers,[4] and water-soluble cell fusion agents.[5] Other reasons for the current interest in PEG's include their reduction of enzyme immunogenicity,[6] and their utility in aqueous two-phase polymer cell separations (phase partitioning).[7]

In addition to these uses, PEG's, their derivatives, and the oligomeric glymes are being increasingly used as phase transfer catalysts (PTC's). In the present paper we describe the chemistry we have developed for synthesis of PEG derivatives, briefly review the subject of PTC by PEG's and related molecules, and describe our work on the use of PEG's and derivatives as soluble, recoverable PTC's.

SYNTHESIS OF PEG DERIVATIVES

We have used five general routes, utilizing four key intermediates, for the synthesis of PEG derivatives, Scheme 1.[8-10] Some of the compounds prepared are given in Scheme 2. Note that the symbol PEG-X represents $XCH_2O(CH_2CH_2O)_nCH_2X$ so that PEG-CH_2OH, for example, symbolizes the diol $HOCH_2-CH_2O(CH_2CH_2O)_nCH_2-CH_2OH$.

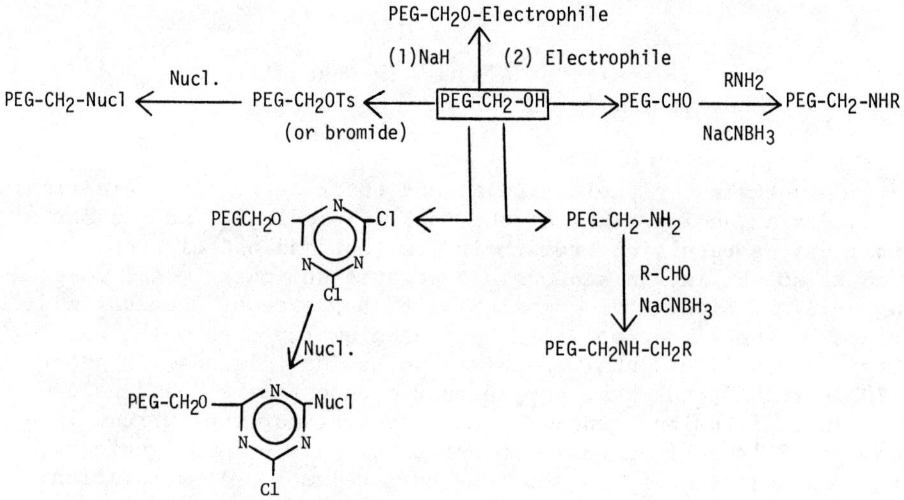

SCHEME 1.   Routes Used for Synthesis of PEG Derivatives

PEG-CH$_2$OH   PEG-CH$_2$OTs   PEG-CHO   PEG-CH$_2$Br

<u>1</u>   <u>2</u>   <u>3</u>   <u>4</u>

PEGCH$_2$NH(CH$_2$)$_6$NH$_2$

<u>5</u>

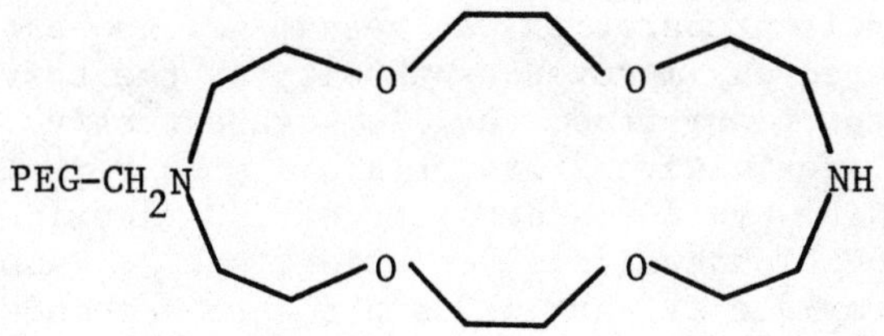

<u>6</u>

PEG-CH$_2$NH-Glucose   PEG-CH$_2$NH-chitosan   glass-Si(CH$_2$)$_3$NH-PEG

<u>7</u>   <u>8</u>   <u>9</u>

PEG-CH$_2$NH(CH$_2$)$_6$NH-CH$_2$CD   PEG-NH$_2$ PEG-CH$_2$OR   PEG-CH$_2$-NHR

<u>10</u>, CD = cyclodextrin   <u>12</u>   <u>13</u> R = octadecyl   <u>14</u>

PEG-CH$_2$O$_2$CR   PEG-CH$_2$PPh$_2$ (PEGCH$_2$PPh$_2$)$_x$RhH(CO)(PPh$_3$)$_{3-x}$

<u>15</u> R = fatty acid   <u>16</u>   <u>17</u>

(PEGCH$_2$O)$_2$PPh   [(PEGCH$_2$O)$_2$PPh]$_x$RhH(CO)(PPh$_3$)$_{3-x}$

<u>18</u>   <u>19</u>

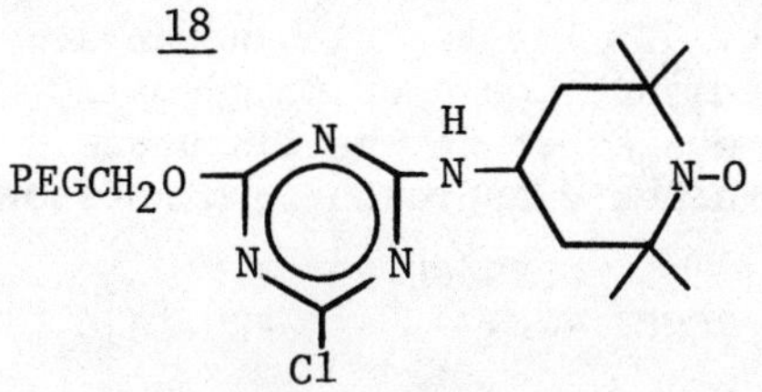

<u>20</u>

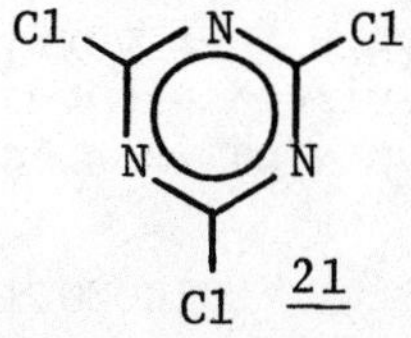

<u>21</u>

<u>Some PEG</u>

<u>Derivatives</u>

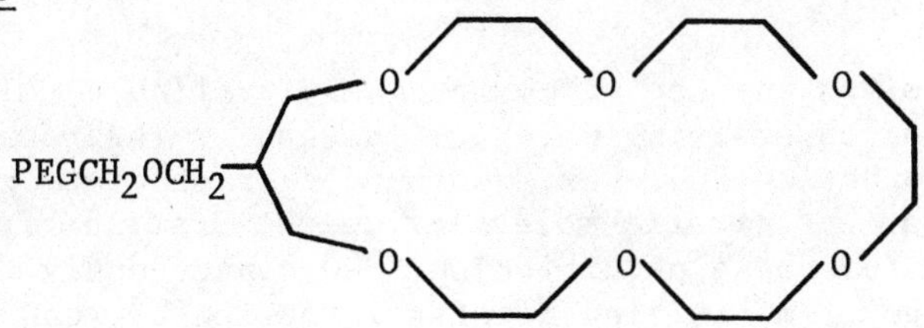

<u>12</u>

SCHEME 2.

All of our work discussed here has been published in references 8-10.

Of the derivatives prepared, the tosylate $\underline{2}$,[2] bromide $\underline{4}$,[11] amine $\underline{5}$,[11] and fatty-acid ester $\underline{12}$[12] have previously been reported, but their preparation was reexamined as part of our work. We developed an improved synthesis for the tosylate since, in our hands, the reported method[2] (eq. 1) gave extensive chain cleavage. In our route, derivative $\underline{2}$ was obtained with very little chain degradation by sodium-hydride generation of PEG alkoxide and subsequent reaction with p-toluenesulfonyl chloride, eq. 2. Another problem with use of the tosylate is that it is prone to decomposition upon storage for extended periods. The mesylate (eq. 3) and bromide (eq. 4) on the

$$PEG-CH_2OH \;+\; TsCl \;\xrightarrow[\;CH_2Cl_2\;]{\text{pyridine}}\; PEG-CH_2OTs \qquad\qquad (1)$$

$$PEG-CH_2OH \;\xrightarrow{\;NaH\;}\; PEG-CH_2O^-Na^+ \;\xrightarrow{\;TsCl\;}\; PEG-CH_2OTs \qquad (2)$$

$$PEG-CH_2OH \;+\; CH_3SO_2Cl \;\xrightarrow[\;CH_2Cl_2\;]{\;Et_3N\;}\; PEG-CH_2OMs \qquad (3)[13]$$

$$PEG-CH_2OH \;+\; SOBr_2 \;\xrightarrow{\text{toluene}}\; PEG-CH_2Br \qquad\qquad (4)[11]$$

other hand, were prepared without chain cleavage, and were found to be stable upon storage.

Our preparation of the PEG ester derivatives involved generation of the fatty acid chlorides using oxalyl chloride followed by $\underline{\text{in situ}}$ condensation with PEG, eq. 5. The reaction proceeded smoothly and was of particular convenience for preparing small quantities of radioisotope-labelled monoester for use in phase partitioning studies. An obvious synthetic approach to monoesters

$$RCOOH \;+\; ClCOCOCl \;\xrightarrow[\;Et_3N\;]{\text{toluene}}\; RCOCl \;\xrightarrow{PEG-CH_2OH}\; PEG-CH_2O_2R \quad (5)$$

involves the use of commercially available (Aldrich) PEG monomethyl ether as starting material instead of the common dihydroxy analogues. In practice, however, we found this ether to have an extremely broad and erratic-molecular weight distribution which precludes its use for phase partitioning. We consequently adopted the alternative route of minimizing diester formation by running the reaction with excess PEG and separating the PEG monoester derivative from the reaction mixture. Residual amounts of free fatty acid also constitute a serious impediment to phase-partitioning studies and must also be removed from the monoester. The free fatty acid was removed

by gel filtration on Sephadex LH20, and the monoester was separated
from the unsubstituted PEG by hydrophobic chromatography on octyl-
sepharose.

PEG amine was prepared from the bromide by Johansson's method,
eq. 6,[11] and by reductive amination of PEG aldehyde, eq. 7. Size
exclusion chromatography (HPLC-SEC) revealed no noticable alteration in molecular weight distribution.

$$PEG-CH_2Br \quad + \quad H_2N(CH_2)_6NH_2 \xrightarrow{\text{ethanol}} PEGCH_2NH(CH_2)_6NH_2 \quad (6)^{11}$$

$$PEG-CHO \quad + \quad NH_4OAc \xrightarrow[CH_3OH]{NaCNBH_3} PEG-CH_2NH_2 \quad (7)$$

Preparation of PEG aldehyde proved problematic primarily because
of the pronounced ability of PEG to form complexes with metallic
oxidizing agents. An initial preparation with pyridinium chloro-
chromate gave aldehyde product, which could be purified with
variable success, by chromatography on alumina. Oxidation with
DMSO-acetic anhydride and substitution on the diethyl acetal of
bromoacetaldehyde (eq. 8) provided two excellent methods for preparation of the PEG aldehyde.

$$PEG-CH_2OH \quad + \quad BrCH_2CH(OEt)_2 \xrightarrow[\text{(2) HCl(aq)}]{\text{(1) } KO\underline{t}Bu/C_6H_6} PEGCH_2CH_2CHO \quad (8)$$

The final key intermediate synthesized was the s-triazine
derivative, Scheme 1, prepared from reaction of s-triazine with
PEG-CH$_2$OH. Additional nucleophiles could be substituted for the
second and third halogens by increasing reaction temperature.[14]

UTILIZATION OF INTERMEDIATES

The four intermediates shown in Scheme 1 can be used to prepare
many derivatives by the five routes shown. In this section we
will briefly discuss some of the derivatives we have prepared with
these methods.

Reductive amination of the aldehyde 3 with various amines
provides a ready route to the preparation of derivatives.[15] For
example, reductive amination of 3 in methanol using 1,10-diaza-
18-crown afforded the water soluble crown derivative 6. Similarly,
the PEG-carbohydrate conjugates 7 and 8 were obtained by reductive
amination with glucosamine and chitosan, respectively. PEG-coated
glass beads 9 were obtained from aldehyde 3, sodium cyanoborohydride,
and aminopropyl-derivatized, controlled-pore glass. We are pre-
sently examining this same reaction for coupling PEG and proteins.
Also, as mentioned above, PEG amine was prepared by reductive
amination with ammonium acetate.

Conversely, PEG amine and the diamine 5 can be used to reductively aminate various aldehydes. As one example of this approach the C6-monoaldehyde of β-cyclodextrin, available from another study,[16] was reductively aminated with both of these amines to afford the cyclodextrin conjugates 10 and 22.

Compounds 6, 7, 10 and 22 are of interest as polymeric affinity ligands which we are currently examining for their ability to bind to cell membranes. The PEG conjugate 8 was prepared as part of our efforts to design branched-chain, water-soluble chitosan derivatives[17] for rheological studies.[18] Polymer-coated glass surfaces, such as 9, may be useful as solid cell separation media, "anti-adhesive" glass,[19] and for eliminating electroosmosis adjacent to glass surfaces.

Compounds 12-14 were prepared by direct displacement reactions of the PEG tosylate or mesylate. In the ether preparations, the 19-crown alcohol and octadecanol were first treated with sodium hydride. In these reactions the sulfonate esters were substantially more reactive than the bromide, and offered the additional advantage that their reactions could be monitored by disappearance of the sulfonate UV absorption using HPLC.

The crown-PEG derivative is of interest in our phase-transfer studies discussed below. The octadecyl ether and amine were designed to replace the corresponding ester in phase partitioning studies. The ester is limited in that it is highly susceptible to bacterial degradation.

The phosphorous derivative 16 was prepared by reaction of lithium diphenyl phosphide with PEG bromide, and the related derivative 18 was prepared by reaction of PEG itself with dichlorophenyl phosphine. Derivatives 16 and 18 were used to prepare the polymer-bound rhodium compounds 17 and 19, respectively. These rhodium complexes were of interest as water-soluble, recoverable catalysts. Whitesides and coworkers have prepared a rhodium derivative similar to 17 using PEG with molecular weight of approximately 1000 g/mol.[3] We found the products derived from such low molecular weight PEG's to be amorphous and less readily recoverable (by precipitation) in comparison to our higher molecular weight (6800 g/mol) derivatives. Catalytic studies of 17 and 19 are in progress.

An alternative procedure of coupling to PEG is exemplified by the nitroxide spin-labelled derivative 21 which was obtained from the PEG-s-triazine (Scheme 1) by reaction with 22. By simple choice of reaction temperature, this method offers the advantage of providing both mono- and di-substitution of the triazine moieties. Derivative 21 is being employed for esr studies of interaction of PEG affinity ligands with cell membranes.

$$H_2N - \underset{\underline{22}}{\text{(ring structure)}} - N-O$$

22

PHASE TRANSFER CATALYSIS BY PEG'S AND RELATED MOLECULES

The dimethyl ethers of oligoethylene glycols (glymes) have long been known to enhance the reactivities of _soluble_ metal salts. For example, in 1960 Zaugg and coworkers described the rate enhancement of sodiomalonate ester alkylation in benzene solution by glymes and a large number of other aprotic, polar additives.[20] In 1965 Ugelstad and coworkers conducted a detailed study of alkylation of potassium and sodium phenoxides in tetrahydrofuran and glymes as solvents, and documented impressive rate enhancements in the larger glymes.[21] Similarly, Brown long ago noted the utility of glymes as solvents for borohydride reductions.[22] There has also been long term interest, of a physical-chemical rather than a synthetic nature, in the metal-complexing ability of glymes and PEG's.[23]

More recent research has focused on the use of PEG's and glymes under phase transfer conditions as replacements for expensive (and sometimes toxic) crowns and crypts. Generally, it appears that the noncyclic polyethers are the catalysts of choice. Onium salts, of course, are also relatively inexpensive phase transfer catalysts (PTC's), but tend to be less effective than PEG's in nonpolar solvents.

A careful, direct comparison of the effectiveness of PEG's, glymes, crowns, crypts, and ammonium salts as PTC's has been done by Stott, Bradshaw, and Parish.[24] These workers examined the nucleophilic substitution reaction of p-nitrobenzyl bromide in chloroform with aqueous sodium thiocyanate, and found the following relative rates: 18-crown-6 (1.44), dibenzo-18-crown-6 (1.00), crypt 222 (5.10), tetrabutylammonium perchlorate (1.29), decaglyme (1.41), and PEG 20,000 (0.83). The author's conclusion that the onium salts and PEG are the catalysts of choice is made obvious by their observation that the crypts cost \$30-150/g, 18-crown-6 \$1.00/g, Adogen 464 (an ammonium salt) 2.8¢/g, and Carbowax 20M 3.2¢/g.

PEG catalysis of nucleophilic substitution reactions has been examined by several other groups.[13,25-27] In all cases accelerations by PEG were significant and comparable to those obtained with alkyl ammonium salts and crown ethers.

$$C_6H_5CH_2Cl \;+\; KSCN(aq) \xrightarrow[\text{PEG 1500}]{\text{CHCl}_3} C_6H_5CH_2Cl \qquad (9)^{25}$$

$$C_6H_5CH_2Br \;+\; KO_2CCH_3(s) \xrightarrow[\text{PEG 33,000}]{\text{benzene}} C_6H_5CH_2OAc \qquad (10)^{26}$$

$$C_6H_5O^-Na^+ \;+\; n\text{-}C_4H_9Br \xrightarrow[\text{PEG 33,000}]{\text{dioxane}} C_6H_5OC_4H_9 \qquad (11)^{26}$$

$$C_6H_5CH_2Br \;+\; KX \xrightarrow[\text{glyme 400}]{\text{benzene or CH}_3CN} C_6H_5CH_2X \qquad (12)^{27}$$

$$X = SH,\; SCN,\; N_3,\; OAc,\; CN,\; F$$

$$(13)^{13}$$

There has also been interest in using PEG in place of crown ethers in potassium permanganate oxidations of alkenes in benzene (the Sam-Simmons procedure).[25,26,28,29] In the most extensive study, Lee and Chang[28] examined the oxidation of 1-decene, trans-5-decene, and cyclododecene. All three alkenes gave the expected carboxylic acids, and the trans-5-decene and cyclododecene gave appreciable amounts of the diones. As catalysts, Adogen 464 (a methyltrialkyl-ammonium chloride), dicyclohexano-18-crown-6, and a polydisperse glyme whose chief component was the dimethylhexaethylene glycol. Methylene chloride or benzene were used as solvents. The glyme and Adogen do indeed solubilize permanganate in benzene or methylene chloride just as the crown was known to do. Further, these inex-pensive PTC's proved to be as active as the expensive crown, and again were shown to be the catalysts of choice. PEG is much more soluble in benzene than in water so it was necessary to use powdered permanganate when benzene was the organic layer. However, the reverse solubility order holds for PEG in methylene chloride and water, so aqueous permanganate could be used for oxidation in methylene chloride. It is advantageous to use aqueous permanganate since powdering is not necessary and since with solid permanganate oxidant is lost as a brown precipitate (probably MnO_4^{2-}). One advantage benzene does have over methylene chloride is that the glyme catalyst can be removed from the reaction mixture by extrac-

tion with water; with the crown or ammonium salt products are initially contaminated by the PTC. However, as we show later in this
review, removal of larger PEG's from methylene chloride can be
achieved easily by precipitation with ethyl ether; thus it would
be interesting to examine the use of larger PEG's in the methylene
chloride-water permanganate oxidations.

Similarly, Balsubramanian and coworkers investigated the oxidation of benzyl alcohol,[25] and Yamazaki and coworkers investigated
the oxidation of <u>trans</u>-stilbene.[26] The first group observed a yield
of 85%. The second group obtained a yield of 67% that compares well

$$C_6H_5CH_2OH \ + \ KMnO_4(s) \ \xrightarrow[\text{PEG MW ?}]{\text{benzene}} \ C_6H_5COOH \qquad\qquad (14)^{25}$$

$$C_6H_5CH \!=\! CHC_6H_5 \ + \ KMnO_4(s) \ \xrightarrow[\text{PEG 33,000}]{\text{benzene}} \ C_6H_5COOH \qquad (15)^{26}$$

with dibenzo-18-crown-6 (48%) but not so well with the Sam and
Simmons 100% yield using dicyclohexyl-18-crown-6.[29]

Hogan and Gandour have examined the catalytic effect of glymes
and crowns on ester aminolysis in chlorobenzene.[30] Although the
origin of the catalysis is not clear, it appears to involve ether
complexation of a tetrahedral, ammonium intermediate and related
transition state. Crowns were known to bind much more strongly
than glymes with tert-butylammonium ion, yet the glymes were un-
expectedly found to be more effective catalysts. The reaction rate
was found to be linearly related to the number of oxygens in the
glyme but not in the crowns. Glymes having up to twenty two ethyleneoxide units were studied.

It is also interesting that glymes can retard certain reactions.
For example Bartsch and Juri[31] have found that glymes slow the thermal
decomposition of p-tert-butylbenzenediazonium tetrafluoroborate in
1,2-dichloroethane. The dimethylether of PEG 1000 (approximately
22 ethyleneoxide units) was found to be the most effective noncyclic
ether, being a factor of five less effective than 18-crown-6. The
rate retardation is said to result from formation of a stable complex
with the diazonium ion.

Reductions in reaction yields by PEG and 18-crown-6 were observed by Yamazaki and coworkers[26] for borohydride reductions of
aldehydes and ketones in tetrahydrofuran and for the reaction of
Grignard reagents with ketones and nitriles in benzene. The dele-
terious effect on the borohydride reductions was especially interesting because of previous observations of enhancement by crown ethers.[32]
Yamazaki and coworkers found 18-crown-6 and PEG to have comparable

retarding effects. The authors postulated that the polyethers acted by complexing the alkali metal, reducing its effective electronegativity, thus lowering reducing power in accord with known relationships. Reduction in yields of the Grignard reactions by PEG and 18-crown-6 were dramatic; for example, the yield from reaction of PhMgCl and PhCN was decreased from 70% to 18% by addition of PEG. The effect in this case was proposed to result from formation of insoluble complexes between the Grignard reagent and the polyether.

In a recent work, Kimura and Regen have examined the utility of PEG's for phase transfer catalysis of 2-bromooctane dehydrobromination in benzene-aqueous potassium hydroxide.[33] Tetraethylene glycol exhibited activity comparable to that of tetrabutylammonium hydrogen sulfate and 18-crown-6. Smaller glycols were inactive. Pentaethylene glycol and PEG's of average molecular weights of 600 and 3400 were very effective, being superior to the ammonium salt and the crown. The following product yields are illustrative: triglyme, 0%; tetraglyme, 41%; pentaglyme, 70%; PEG 600, 82%; PEG 3400, 78%; ammonium salt, 37%; and 18-crown-6, 22%. These results, when coupled with the known thermal instability of ammonium salts[33] and the expense of crowns clearly point out the utility of PEG's as PTC's.

It is interesting to note that PEG is effective in the Kimura and Regen work at transferring potassium ion into benzene solution despite the known[13,28] great preference of PEG for water. Recall that Lee and Chang found that PEG was ineffective at transferring $KMnO_4$ from aqueous solution into benzene, and had to use solid-liquid PTC or switch to methylene chloride as solvent. A further observation of Kimura and Regen is pertinent. They note that PEG monomethyl ether is only half as effective as PEG, and further that the dimethyl ether is essentially inactive. Thus it appears that the hydroxyl group of the PEG is actively involved as the alkoxide basic site. The permanganate oxidations obviously differ in that the anion in this case must also be transferred into the benzene phase.

Thus there are many examples of PEG acting as a PTC (refer also to other contributions to the Symposium). It is of interest to inquire as to the effectiveness of other polyethers. Yamazaki and coworkers have examined the effect of various polymeric additives on the reaction between sodium phenoxide and n-butyl bromide in dioxane. Poly(oxetane), poly(tetrahydrofuran), and poly(ethylvinylether) were found to be ineffective, while PEG gave a rate enhancement of approximately 100. Interestingly, polyvinylpyrrolidone of several molecular weights was even more effective than PEG.[26]

POLY(ETHYLENE GLYCOL)S AS SOLUBLE, RECOVERABLE PHASE-TRANSFER CATALYS'

PEG's of molecular weights from 1500 g/mol to 6800 g/mol can be quantitatively precipitated from benzene, acetone, acetonitrile, methanol, or methylene chloride by addition of ethyl ether. This property offers the possibility of designing various soluble but recoverable materials such as catalysts and synthetic intermediates. A soluble, recoverable PTC provides an interesting alternative to the popular practice[34,35] of immobilizing PTC's on insoluble polymer backbones. In the present section we describe our work on PEG's and PEG-crowns (see above) as soluble, recoverable PTC's.[8]

We have examined the partitioning of PEG-6800 between water and benzene and methylene chloride at 25°C. For methylene chloride about 75% (depending on concentration) of the PEG favors the organic phase, while for benzene less than one percent of the PEG partitions to the organic phase. Thus it appears that methylene chloride and PEG will be useful for liquid-liquid PTC, while benzene and PEG will generally be limited to solid-liquid PTC (see previous section).

In order to further assess the ability of the crown-polymers 6 and 12 and of PEG to act under liquid-liquid conditions as PTC's, we determined the extent of transfer of sodium and potassium picrate from water into methylene chloride in the presence of these phase transfer agents (PTA's). The polymers were essentially identically effective, transferring approximately 5% of sodium picrate and 20% of potassium picrate into the methylene chloride (0.181 g/100 mL of solvent and 0.0954 mM and 0.085 mM in potassium and sodium picrate, respectively). Interestingly, PEG 1000 was comparable to PEG 6800 for sodium complexation (3.9% vs. 3.6% transferred) but not for potassium complexation (14.5% vs. 19.2% transferred) when compared on a equal-mass basis. The preference for potassium over sodium is surprising, and has been noted by Yamazaki and coworkers.[26]

The ability of PEG-6800, PEG-3400, 18-crown-6, 6, and 12 to catalyze the reaction of potassium acetate with benzyl bromide was determined, eq. 16; the accelerations observed were as follows: 18-crown-6, 150; PEG-6800, 19; PEG-3400, 14; 6, 26, and 12, 20 (benzyl bromide, 1.7M; KOAc, 3.4M, and 0.2g of PTA/10 ml solution).

$$C_6H_5CH_2Br \ + \ CH_3COOK(s) \xrightarrow[PTC]{CH_3CN} C_6H_5CH_2O_2CCH_3 + KBr \qquad (16)$$

Several of these results are noteworthy. First, the PTC's are compared on an equal-mass basis. If compared on a molar basis the PEG's appear much more effective relative to 18-crown-6. Secondly, PEG-6800 is a more effective catalyst than PEG-3400. This result relates to that mentioned earlier for the metal-picrate partitioning. Finally, the PEG-crowns 6 and 12 are only marginally more effective than PEG itself, and it is doubtful that their synthesis is justified for this purpose.

As a final point regarding the utility of the PEG materials as

recoverable PTC's it remains to demonstrate their recoverability.
We have carefully measured the extent of recovery of the PEG's and
the crown derivatives from methylene chloride and acetonitrile by
ether precipitation. In every case recovery was essentially quantitative, with only minor loss of material on glassware surfaces.

We are continuing our work on the use of PEG's as soluble,
recoverable catalysts and carriers for synthetic intermediates.

REFERENCES

1. (a) "Handbook of Water-Soluble Gums and Resins," McGraw-Hill,
    New York, 1980, Chapter 18; (b) F. E. Bailey and J. V. Koleske,
    "Poly(ethylene oxide)," Academic Press, New York, 1976.
2. V. N. Pillai, M. Mutter, E. Bayer, and I. Gatfield, J. Org. Chem.,
    45, 5364 (1980).
3. R. G. Nuzzo, S. L. Haynie, M. E. Wilson, and G. M. Whitesides,
    J. Org. Chem., 46, 2861 (1981).
4. P. Ferruti, M. C. Tanzi, L. Fusconi, and R. Cecchi, Makromol.
    Chem., 182, 2183 (1981).
5. K. Honda, Y. Maeda, S. Sasakawa, H. Ohno, and E. Tsuchida, Bioch.
    Bioph. Res. Comm., 101, 165 (1981).
6. (a) K. V. Savoca, A. Abuchowski, T. van Es, F. F. Davis, and
    N. C. Palczuk, Biochim. Biophys. Acta, 578, 47 (1979); (b) Y.
    Ashihara, T. Kono, S. Yamazaki, Y. Inada, Biochem. Biophys.
    Res. Commun. 83, 385 (1978).
7. P. A. Albertsson, "Partition of Cell Particles and Macromolecules,'
    2nd. Ed., Wiley, New York, 1971.
8. J. M. Harris, N. H. Hundley, T. G. Shannon, and E. C. Struck,
    J. Org. Chem., 47, 0000 (1982).
9. J. M. Harris, M. Yalpani, E. C. Struck, J. M. Van Alstine, T. G.
    Shannon, M. S. Paley, and D. E. Brooks, J. Org. Chem., submitted
    for publication.
10. J. M. Harris, N. H. Hundley, T. G. Shannon, and E. C. Struck,
    Polymer Preprints, 23, 193 (1982).
11. A. F. Buckmann, M. Morr, and G. Johansson, Makromol. Chem., 182,
    1379 (1981).
12. V. P. Shanbhag and G. Johansson, Bioch. Bioph. Res. Comm., 61,
    1141 (1974).
13. D. J. Brunelle and D. A. Singleton, personal communication.
14. M. J. Adams and L. D. Hall, Carb. Res., 68, C17 (1979).
15. R. F. Borch, Org. Syn., 52, 124 (1972).
16. M. Yalpani, D. E. Brooks, and J. M. Harris, unpublished results.
17. L. D. Hall and M. Yalpani, J. Chem. Soc., Chem. Commun., 1153
    (1980).
18. M. Yalpani, D. E. Brooks, M. Tong, and L. C. Hall, unpublished
    results.
19. C. Persiani, P. Cukor, and K. French, J. Chr. Sci., 14, 417
    (1976).

20. H. E. Zaugg, B. W. Horrom, and S. Borgwardt, J. Am. Chem. Soc.,
     82, 2895 (1960).
21. J. Ugelstad, A. Berge, and H. Listou, Acta Chem. Scand., 19,
     208 (1965).
22. H. C. Brown, "Boron in Organic Chemistry," Cornell Univ. Press,
     Ithaca, N. Y., 1972.
23. For references to this work see: (a) S. Yanagida, K. Takahashi,
     and M. Okahara, Bull. Chem. Soc., Japan, 50, 1386 (1977); (b)
     I. M. Panayotov, D. K. Dimov, C. B. Tsvetanov, V. V. Stepanov,
     S. S. Skorokhodov, J. Poly. Sci., Poly. Chem., 18, 3059 (1980);
     (c) S. Kopolow, T. E. Hogen Esch, and J. Smid, Macromol., 6,
     133 (1973).
24. P. E. Stott, J. S. Bradshaw, and W. W. Parish, J. Am. Chem. Soc.,
     102, 4810 (1980).
25. D. Balasubramanian, P. Sukumar, and B. Chandani, Tetrahedron
     Lett., 3543 (1979).
26. N. Yamazaki, A. Hirao, and S. Nakahama, J. Macromol. Sci.-Chem.,
     A13, 321 (1979).
27. H. Lehmkuhl, R. Rabet, K. Hauschild, Synthesis, 184 (1977).
28. D. G. Lee and V. S. Chang, J. Org. Chem., 43, 1532 (1978).
29. D. J. Sam and H. F. Simmons, J. Am. Chem. Soc., 94, 4024 (1972).
30. J. C. Hogan and R. D. Gandour, J. Am. Chem. Soc., 102, 2865
     (1980).
31. R. A. Bartsch and P. N. Juri, Tetrahedron Lett., 407 (1979).
32. (a) J. W. Zubrick, B. I. Dunbar, and H. D. Durst, Tetrahedron
     Lett., 71 (1975); (b) T. Matsuda and K. Koida, Bull. Chem.
     Soc. Japan, 46, 2259 (1973).
33. Y. Kimura and S. L. Regen, J. Org. Chem., 47, 2493 (1982).
34. S. L. Regen, Angew. Chem., Int. Ed. Engl. 18, 421 (1979). D. C.
     Sherrington in "Polymer-Supported Reactions in Organic Synthesis,"
     P. Hodge, D. C. Sherrington, Eds., Wiley, New York, 1980, p 180.
35. M. Tomoi, W. T. Ford, J. Am. Chem. Soc., 102, 7140 (1980); 103,
     3821 (1981).

PHOTO-INDUCED NUCLEOPHILIC SUBSTITUTION OF ANISOLES IN THE PRESENCE OF POLYETHYLENE GLYCOL (PEG): USEFULNESS OF PEG ON PHOTOCHEMICAL REACTIONS[1]

Nobutaka Suzuki, Yasuo Ayaguchi, Kazuo Shimazu,
Toshikuni Ito, and Yasuji Izawa

Department of Industrial Chemistry
MIE University, Tsu Japan

INTRODUCTION

Chemistry in cyclic polyethers (crown ethers) attracts keen interests of many investigators[2] because of the rather high selectivity of the reactions by the activated anion species in many circumstances. Many interesting photochemical reactions were performed in crown ether. [2]

Polyethylene glycol (PEG) is an excellent substitute for crown ether because of its low-price, low-toxicity, non-volatility, and solubility both in organic solvents and water (easily removable). It has been used as a co-solvent as described in our previous reports[1] and by others:[3]. Santaniello et al.[3b] had investigated various reactions in the presence of PEG #400 as a solvent for reactions of organic halides with inorganic reagents, such as $K_2Cr_2O_7$, KOAc, KCN, KI, and PhOK and a reaction of a ketone with $NaBH_4$ in excellent yields; Pillai et al.[3c] had applied PEG (mw: 6000) to a peptide synthesis; Sukata[3e-f] had applied PEG #400-4000 to the Williamson synthesis and alkylation of phenylacetonitrile in the two phase reactions.

We want to describe here that much higher selectivity could be obtained in PEG employed instead of crown ether on the photochemical nucleophilic substitution reaction of anisole (1) to give o- and p-cyanoanisoles (3) and the higher p-CN/o-CN ratios than in crown ether and also to describe the similar reactions of o- and p-dimethoxybenzenes (5) to give the corresponding o- and p-cyanoanisoles (3) in quantitative yields. Few photochemical reactions have been conducted in PEG, or acyclic polyethers.[4]

Photochemical nucleophilic substitutions of aromatic ring systems in protic solvents have been well documented[5] (Scheme 1). When a crown ether is present the photocyanation proceeds better in aprotic solvents than in protic solvents[6] (Scheme 2), while addition of an electron acceptor, such as terephthalonitrile (2), improves both the yields of the photocyanation products and the specificity of substitution[7] (Scheme 3). This report concerns the photochemical cyanation of anisole (1) and dimethoxybenzenes (5) with KCN in an aprotic solvent (CH_2Cl_2), in the presence and absence of 2 when polyethylene glycol (PEG) is present instead of crown ether (CE) (Schemes 4 and 7).

Scheme 1. Photochemical Substitution in Protic Solvents.[8-11]

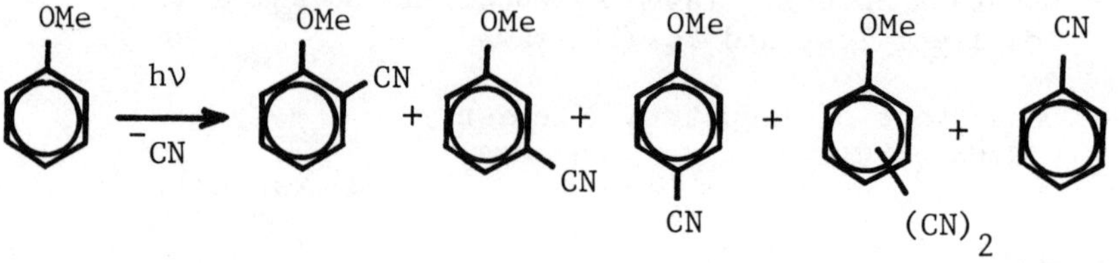

						ref.
MeOH (N$_2$)	---	---	---	---	3%	8
MeOH (Air)	53	0.2	47	---	---	9
t-BuOH/H$_2$O						
(1:2)	13.8	---	28.6	10.6	---	10
(1:3)	24.7	---	24.7	(2,4-) 10.5	---	11
				(2,6-) 3		

Scheme 2. Photochemical Substitution in Aprotic Solvents.[6]

		with 18-CE-6 in MeCN or MeOH		without CE in protic solv.	Bu$_4$NCN in MeCN	ref.
biphenyl (I)	KCN → hν	(4-CN)	50%	28%	7%	6
		(2-CN)	---	14		
		((CN)$_2$)	10	---		
naphthalene (II)		(1-CN)	15	6	30	
anthracene (III)		(9-CN)	10	---		
		(9,10-(CN)$_2$)	10	---		
phenanthrene (IV)		(9-CN)	25	---	20	

Scheme 3. Photochemical Substitution in the
Presence of an Electron Acceptor.[7]

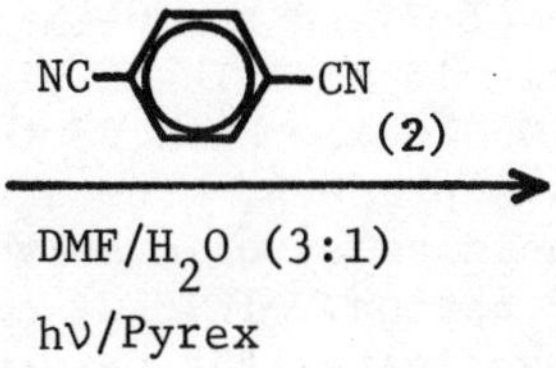

II $\xrightarrow[\substack{\text{DMF/H}_2\text{O (3:1)} \\ \text{h}\nu/\text{Pyrex}}]{\text{NC}-\text{C}_6\text{H}_4-\text{CN (2)}}$ (monoCN)   30%

                                    $((\text{CN})_2)$ 15%

III   (9-CN)   6%

IV   (9-CN)
                                     $(9\text{-CN-}9,10\text{-H}_2)$ 49%

in the absence of 2;   yields $\leq$ 1/15.

RESULTS AND DISCUSSIONS

1. Photochemical Substitution of Anisole (1) with KCN

    A solution of anisole (1) in CH_2Cl_2 with 3 equiv. of KCN and
PEG (#200 to #1000 were tested) or 15-CE-5 was irradiated in the
presence/absence of 15 equiv. of terephthalonitrile (2) with a high-
pressure mercury lamp (300 W) through a quartz or Pyrex cell (Scheme
4). The irradiated solutions were analysed by GLC after being washed
with water to remove the PEG and the other water-soluble components.
The results are shown in Table 1.

Scheme 4.

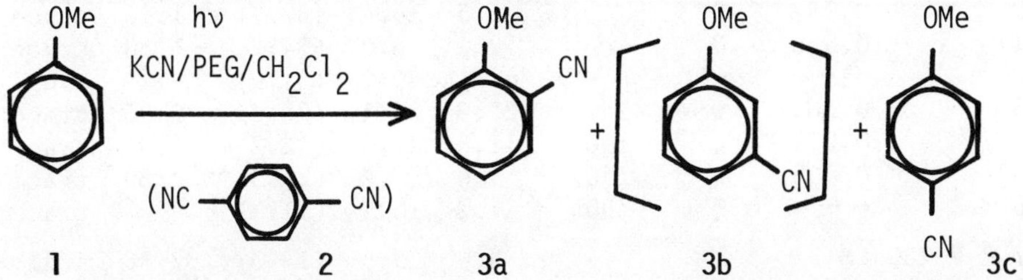

OMe $\xrightarrow[\substack{(\text{NC}-\text{C}_6\text{H}_4-\text{CN})}]{\substack{\text{h}\nu \\ \text{KCN/PEG/CH}_2\text{Cl}_2}}$ OMe-CN + [OMe-CN] + OMe-CN

1   2   3a   3b   3c

    In a quartz cell, when 2 was absent, consumption of 1 with PEG
present was comparable to that with CE. However, the conversion
yields of 3 decreased in the order CE>PEG#1000>PEG#200 present. The
ratio of yields of [3c]/[3a] (p-CN/o-CN) decreased in the order
PEG#200>PEG#1000>CE.

    In the absence of 2 with PEG present, 1 gave o-, p-, and a trace
of m-cyanoanisoles [3a, 3c, and 3b] as substitution products but in
the presence of 2, 1 gave only the o- and p-cyanoanisoles 3a and 3c
in yields that increased with irradiation time. However, the

consumption of **1** was lower than in the absence of **2**; in CE the consumption of **1** in the presence of **2** was higher than that in the absence of **2**. Conversion yields of **3** and the p-CN/o-CN ratio with PEG present were higher than those in CE. These yields were excellent with PEG#200 (PEG#200>PEG#1000>CE). With CE the meta-isomer **3b** was also formed even when **2** was present. These results reveal that PEG activates the ⁻CN anion in an aprotic solvent, as does CE, by forming a complex with the K⁺ cation. In PEG, the acceptor **2** forms a complex with **1** and then activates the para-position of **1** rather than the ortho-position, and totally deactivates the meta-position. These are useful properties of PEG.

Table 1. Photocyanation of Anisole (**1**) in PEG (hν/3h)

Run	PEG	Cell Q or P	2 (mM)	Yield[a] & C.Y.[b] of 3 (%)			[3c]/[3a]	3b
1	# 200 (1M)	Q	–	12.2	39.0	(31.3)	1.71	trace
2	# 400	Q	–	10.8	35.4	(30.2)	1.44	trace
3	# 600	Q	–	8.4	33.9	(24.4)	1.56	trace
4	#1000	Q	–	3.3	15.3	(21.6)	1.58	trace
5	# 200	Q	150	17.3	30.9	(56.0)	2.29	trace
6	# 400	Q	150	9.2	23.3	(39.3)	2.70	trace
7	# 600	Q	150	8.0	35.7	(22.4)	3.33	trace
8	#1000	Q	150	8.5	34.4	(24.7)	2.64	trace
9	# 200 0.1M	Q	–	4.5	18.4	(24.4)	1.57	trace
10	0.05	Q	–	5.5	16.4	(32.9)	1.70	trace
11	0.025	Q	–	2.1	6.0	(34.2)	1.80	trace
12	——	Q	–	n.d.	0	(31.9)	–	n.d.
13	0.1M	Q	150	20.3	28.1	(72.6)	2.06	trace
14	0.025	Q	150	11.6	27.4	(42.2)	2.05	trace
15	0.005	Q	150	1.6	3.2	(27.5)	2.80	trace
16	——	Q	150	1.2	10.7	(11.0)	–	trace
17	# 200 1M	P	–	3.6	25.5	(14.3)	2.86	n.d.[c]
18		P	150	9.6	34.9	(27.6)	6.12	n.d.
19[d]		Q	–	8.3	12.3	(68.5)	1.69	trace
20[e]		Q	–	9.6	15.3	(63.3)	1.63	trace
21[d]		Q	150	8.1	13.2	(61.5)	2.63	n.d.
22[e]		Q	150	7.5	12.0	(62.3)	2.53	n.d.
23	CE	Q	–	26.5	58.0	(46.0)	0.38	trace
24	CE	Q	150	10.9	20.0	(54.0)	0.75	trace

[a]Yield of **3** based on initial amt. of **1**. [b]Conv. yield based on **1** consumed. Consumption of **1** is given in parentheses.
[c]Not detected. [d]In air. [e]In argon.

The lower yields obtained with PEG than with CE may be explained as follows. The CE incorporates the K^+ cation tightly and therefore liberates a naked anion ($^-$CN) which may react efficiently with electrophiles, while PEG, which is assumed to have a complex spiral structure,[2,12] incorporates the cation less effectively. While a mixture of 1 and 2 gave no extra absorption in its UV spectrum, a new fluorescence emission band was present. This suggests that a charge transfer complex (exciplex, 4) between 1 and 2 is formed, which may explain both the enhancement of the yields of 3 and the higher preference for para-substitution.[5,7]

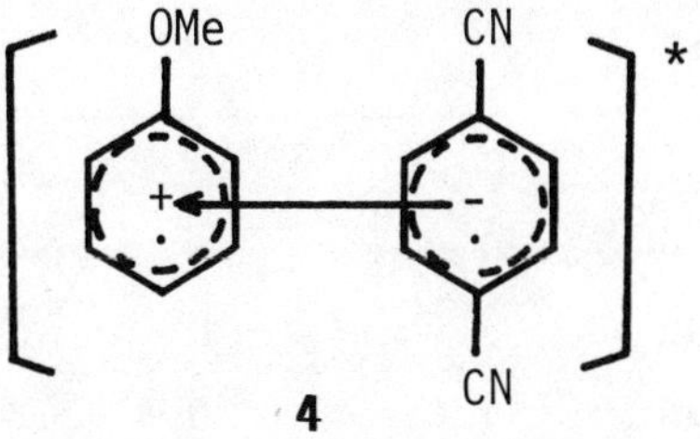

4

The total conversion yields of 3 decreased with the irradiation times. This implies that further reactions [e.g., cyanation of 3][5,13] which may be faster than cyanation of 1 may be occurring.[11] In addition, cyanoanisoles (3) decompose photochemically under the experimental conditions (Table 2). The photochemical decay rate of 3a was much faster than that of 3b and 3c in quartz cells and this tendency is more marked in Pyrex cells. The fact that the p-CN/o-CN ratio was much larger in Pyrex cells than in quartz cells may be explained by the facts above. Since the decay rate of 3b was comparable with that of 3c, the absence of 3b in the reaction reflects the fact that the rate of its formation was very low.

Scheme 5. Further Reactions of 3.

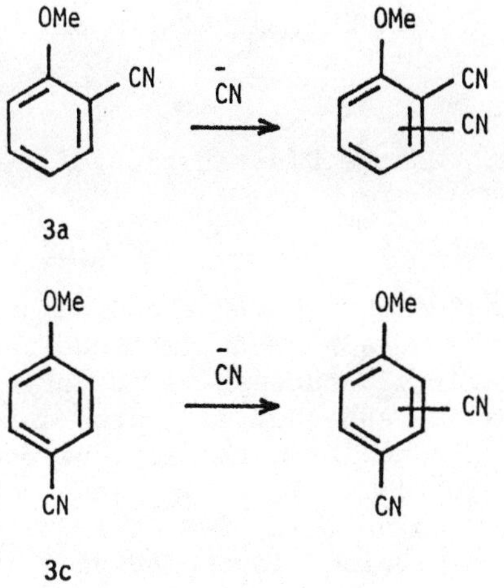

Quenching and fluorescence experiments showed that the present reaction proceeds through the T_1 excited state of **1**, in agreement with Nilsson's work,[9] when **2** is absent, and the S_1 state of the charge-transfer complex (**4**) when **2** is present. The charge distribution described in the literatures[5,11,13] of **1** in the S_1 and T_1

Table 2.   Photodegradation of the Cyanoanisoles (3)

			Conv. of **3** (%)		
Run	**3**	**2** (mM)	2h	4h	6h
1	**3a**	–	10.7	12.3	9.9
2	**3a**	150	41.0	59.7	69.2
3	**3c**	–	0	0.3	0
4	**3c**	150	10.4	13.0	13.3

excited states and the radical cation (Chart 1) cannot, however, adequately explain the different reactions in the presence and absence of 2 which suggests at least that 4 has a different electron distribution to that described.

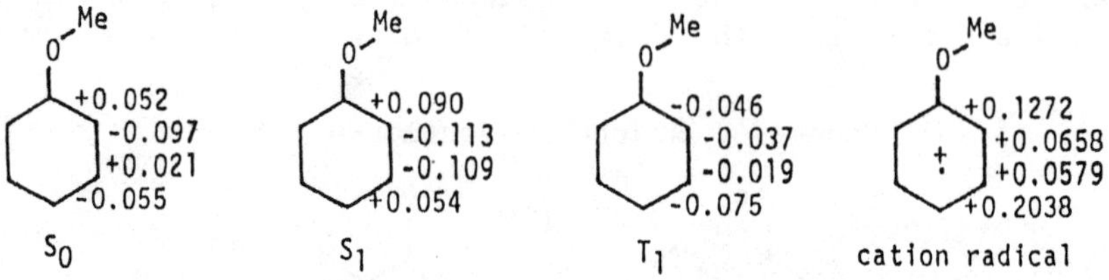

Chart 1.   Charge Distribution of 1[5,11,13]

The mechanism in the absence of **2** can be explained by Havinga's theory[11] (Scheme 6) that the $\pi-\pi*$ triplet excited state of **1**, being produced photochemically _via_ successive intersystem crossings, is attacked by the $^-$CN anion and then liberates an electron to give a cation radical. This radical reacts with one more $^-$CN anion to afford monocyanoanisoles (**3**). In the presence of **2**, **1** may react through **4**. The formation of **4** may result in the higher conversion yields of **3** than those obtained in the absence of **2**.

Scheme 6. Havinga's Mechanism[11]

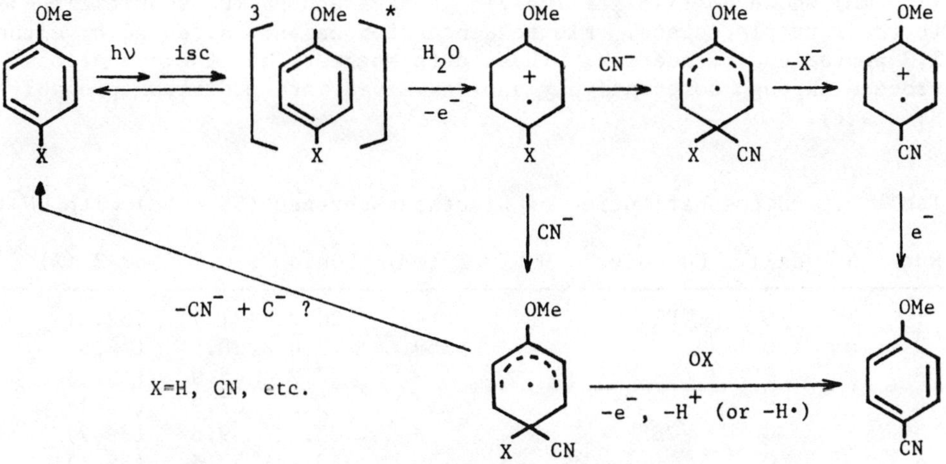

X=H, CN, etc.

2. Photochemical Substitution of Dimethoxybenzenes (5) with KCN

Similar photochemical cyanation reactions were investigated for o- and p-dimethoxybenzenes (5a and 5b) with KCN in an aprotic solvent, in the presence/absence of 2 when PEG is used as the co-solvent (Scheme 7). The results are shown in Table 3.

Scheme 7.

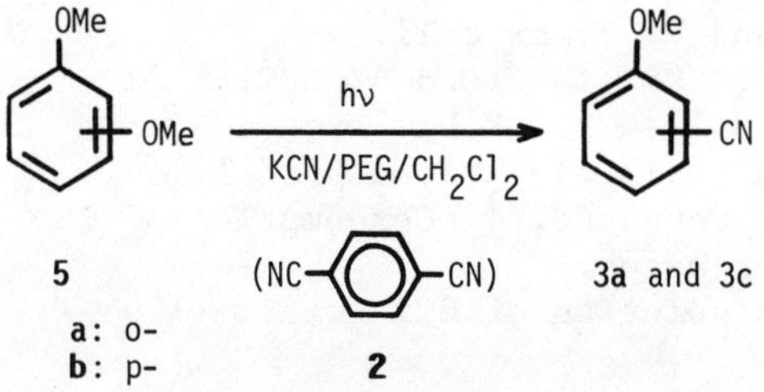

In a quartz cell, when 2 was absent. Consumption of 1 in the presence of PEG was slightly slower than that in the presence of crown ether. However, the conversion yields of 3 with PEG were comparable to that for 3a and superior to those for 3c with crown ether (see runs #1, 5, 7, and 12 in Table 3).

In comparisons with the results obtained by Havinga and his co-workers[11] in $\underline{t}$-BuOH-H_2O, both the consumption and the conversion yield were worse for 5a with PEG, but those for 5c were much better.

Both the conversion of 5 and formation of 3 were neither
quenched with 1,3-pentadiene (∿50 mM) nor sensitized with acetone
(∿50 mM) whose E_T (∿78 kcal/mol)[14] is high enough for sensitizing 5
to their triplet state.   Fluorescence of 5 was not affected by either
1,3-pentadiene or acetone.   These data suggest that the cyanations
proceed through comparatively shorter-lived intermediates (probably
S_1 state).[14]

Table 3.   Photosubstitution of Dimethoxybenzenes (5a & 5c) with CN^-.

Run	5	cell[a]	Co-solv.[b]	MCN[c]	2 (mM)	Yield[d]	& C.Y.[e]	of 3 (%)	
1	5a	Q	PEG	K	–	20.2	31.7	(63.8)	
2		Q			150	7.6	28.5	(26.6)	
3		P			–	7.2	45.9	(15.7)	
4		P			150	5.5	67.9	(8.1)	
5		Q	CE		–	35.5	39.6	(89.7)	
6		Q			150	33.4	57.5	(58.1)	
7		Q	PEG	Na	–	22.8	37.2	(61.3)	
		Q[f]	BH		–		73.2	(30.2)	g)
8	5c	Q		K	–	29.1	32.5	(89.7)	
9		Q			150	63.5	72.6	(87.5)	
10		P			–	27.1	41.6	(65.1)	
11		P			150	71.5	96.6	(74.0)	
12		Q	CE		–	26.5	28.4	(93.3)	
13		Q			150	57.9	73.9	(78.3)	
14		Q	PEG	Na	–	23.6	27.7	(85.0)	
		Q[f]	BH		–		22.2	(61.6)	g)

[a]Q: quartz cell and P: Pyrex cell.
[b]PEG: #200/CH_2Cl_2;   CE: 18-Crown-6/CH_2Cl_2;
BH: t-BuOH/H_2O (3:1).   [c]K: KCN;   Na; NaCN.
[d]Yield of 3 based on initial amounts of 1.
[e]Yield based on 1 consumed.[11]   Consumption of 1
is given in parentheses.
[f]Filter: 1 cm of a solution of 6 in water (100
mg/1).
[g]See reference 11.

When 2 was present.   Both consumption of 5a and conversion
yield of 3a with PEG were a half of those with crown ether.
However, those with PEG for the reaction of 5a were very similar to
crown ether.   Photolyses of 5c with PEG were fairly superior to those
with crown ether and could be used for the preparation of 3c.   Both
the conversion of 5 and formation of 3 were quenched by 1,3-
pentadiene[14] ($k_q\tau_0$: 11 $l \cdot mol^{-1}$ for 5a and 6 for 5c; 6 for 3a and
16 for 3c) and sensitized with acetone slightly.[14]   These data

suggest that the cyanations in the presence of **2** proceed through comparatively longer-lived intermediates[14] (probably T_1 state of **5** and/or of a charge transfer complex of **5** and **2** like **7**).[11]

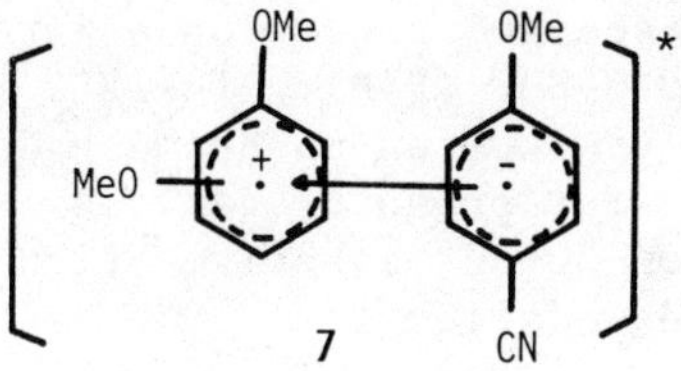

In the both cases. Substitution of KCN with NaCN did not affect the consumption of **5**, but did the conversion yields of **3a** (increased) and **3c** (decreased). This reflects the tolerance of PEG for the size of the cationic guests incorporated in it compared with crown ether.

Polonged irradiation caused a slight decrease of the conversion yields of **3**. This implies that further reactions [e.g., cyanation of **3**][5] may be faster than cyanations of **5**[11] as shown in Scheme 5.

Although in Pyrex cells with PEG consumption of **5** was much slower than that in quartz cells, conversion yields of **3** increased remarkably, especially for **5c** (see runs #1 through 4 and 8 through 11 in Table 3). This phenomenon may result from the further photocyanation of **3**[11] in a quartz cell (see Table 2).

The photochemical decay rate of **3a** was much faster than that of **3c** in Pyrex cells and this tendency is more marked in quartz cells, especially in the presence of **2** (Scheme 5).

Suspected transformation of **5** and **2** to yield corresponding **3** was excluded for these photocyanation reactions (Scheme 8).

Scheme 8. Assumed Photochemical Disproportionation Reaction of **5**.

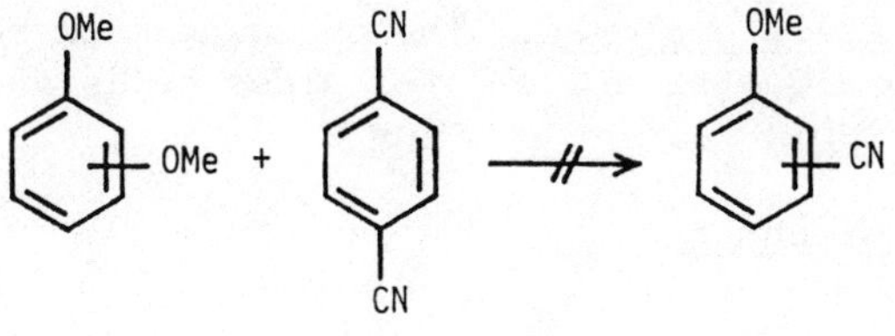

EXPERIMENTAL

Instrumentation

All mps were taken with a microscope hot stage and are uncor-
rected. UV-visible spectra were recorded on a Shimadzu UV-200,
fluorescence/phosphorescence spectra on a Hitachi MPF-2A, and GLC
on a Yanagimoto Yanaco G-80F (FID). The GLC conditions used: Column
A (3.0 mm x 2.0 m, 5% PEG-20M on 60-80 mesh Gasport in a stainless
steel column) was used for product analyses. Column B (3.8 mm x
3.0 m, 25% Apiezon L on 30-60 mesh fire brick in a glass column) was
used to analyse the degradation of **1**.

Materials. Anisole (**1**), terephthalonitrile (**2**) m-cyanoanisole (**3b**),
o- and p-dimethylbenzenes (**5a and c**), 15-crown-5, 18-crown-6,
benzonitrile, and KCN, were purchased. CH_2Cl_2 was distilled from
CaH_2. PEG #200-#1000 (Nakarai Chem. Co) were dried azeotropically
by benzene. Ortho- and para-cyanoanisoles (**3a**[15] and **3c**[15]) were
prepared according to the methods of the literatures. 1,3-Pentadiene
and acetone were distilled directly before use.

Photosubstitution of Anisole (**1**)

General experimental procedure was as follows. A solution of
1 (10 mM) in CH_2Cl_2 (10 ml) with KCN (30 mM) and PEG (1 M) [#200 (2
g) to #1000 (3 g)] or 15-crown-5 was irradiated externally with a
high-pressure mercury lamp (Eikosha: PlH-300; 300W) through a water-
cooled quartz or Pyrex cell. Experiments were carried out in the
presence or absence of **2** (15 mol. equiv.). The irradiated solution
was analysed by GLC after washed with a certain amount of water and
the results are given in the Table 1.

Effects of Molecular Weight of PEG. Several PEG with various
mws. (#200-#1000, which show the average mws.) were employed for the
photochemical reactions under the similar conditions other than the
PEG. The results are shown in Table 1.

Concentration Effects of PEG #200. Effects of the concentration
of PEG #200 on the photochemical reaction were studied under the
similar conditions other than the concentration of PEG #200. (See
Table 1).

Effects of Cells employed. The photochemical reactions were
compared through a quartz and a Pyrex glass cells under the similar
conditions (See Table 1).

Effects of Circumstances. The photochemical reactions were
compared under argon or air in Pyrex tubes. (See Table 1).

Quenching with 1,3-Pentadiene. Solutions sampled as above with a solution of certain amounts of 1,3-pentadiene (distilled directly before use: bp 42-44 °C) were irradiated and the resulted mixtures were analysed by GLC. Formation of **3** was quenched when **2** was absent, but was not quenched when **2** was present.

Quenching of Fluorescence. Relative quantum yields of fluorescence of solutions sampled as above with a solution of certain amounts of 1,3-pentadiene were measured and the Stern-Volmer plotts were drawn. 1,3-Pentadiene did not quenched fluorescence from **1**.

Sensitization with Acetone. No sensitization was observed when the certain amount of acetone was added to the solutions sampled as above.

Degradation of the Cyanoanisoles. Photodegradation of **3a** and **3c** (10 mM) was observed under the similar conditions used for **1** (See Table 2).

Photosubstitution of Dimethoxybenzenes (5a and 5c)

General Experimental Procedures. A solution of **5** (10 mM), KCN (30 mM), and PEG (1 M) or 18-crown-6 in CH_2Cl_2 (10 ml) in the presence or absence of **2** (150 mM) was irradiated externally with a high pressure mercury lamp (Eikosha: P1H-300; 300W) through a water-cooled quartz or Pyrex cell. The irradiated solutions were analysed by GLC after being washed with water to remove PEG and KCN. The results are given in the Table 3.

An Attempt of Photochemical Transformation of 5 and 2 to 3.
Solution of **5** (10 mM) and **2** (10 mM) were irradiated and the mixtures were analysed by GLC. No formation of cyanoanisoles (**3**) was observed.

SUMMARY

Polyethylene glycol (PEG) can replace crown ether as a co-solvent in the photochemical substitution reactions of anisole (**1**) and dimethoxybenzenes (**5**) with KCN in CH_2Cl_2 in the presence or absence of terephthalonitrile (**2**).

ACKNOWLEDGEMENTS

We are grateful to the Ministry of Education, Science, and Culture of Japan for the financial support (Grant-in-Aid).

REFERENCES

1. Preliminary communication: N. Suzuki, K. Shimazu, T. Ito, and
    Y. Izawa, Chem. Commun., 1253 (1980).
2. "Chemistry of Crown Ethers," Kagaku Zokan, Vol. 74, R. Oda,
    T. Shono, and I. Tabushi eds., Kagaku Dojin, Kyoto, 1978.
3. For dark reactions using PEG: (a) W. P. Weber and G. W. Gokel,
    in "Phase Transfer Catalysis in Org. Synth.," Springer-
    Verlag, Berlin, 1978, p. 3; (b) E. Santaniello, A. Manzocci,
    and P. Sozzani, Tetrahedron Lett., 4581 (1979); (c) V. N. R.
    Pillai, M. Mutter, E. Baeyer, and I. Gatfield, J. Org. Chem.,
    45, 5364 (1980); (d) K. Sukata, J. Synth. Org. Chem. Japan,
    39, 86 (1981); (e) ibid., 39, 443 (1981); (f) ibid., 39, 1131
    (1981).
4. S. Yanagida and M. Okahara, in ref. 2, p. 149.
5. J. Cornelisse and E. Havinga, Chem. Rev., 75, 353 (1975);
    J. Cornelisse, Pure Appl. Chem., 41, 433 (1975).
6. R. Beugelmans, H. Ginsburg, A. Lecas, M.-T. LeGoff, J. Pusset,
    and G. Roussi, Chem. Commun., 885 (1977); R. Beugelmans,
    M.-T. LeGoff, J. Pusset, and G. Roussi, Chem. Commun., 377
    (1976).
7. K. Mizuno, C. Pac, and H. Sakurai, Chem. Commun., 553 (1975).
8. J. A. Barltrop, N. J. Bunce, and A. Thomson, J. Chem. Soc., (c),
    1142 (1967).
9. S. Nilsson, Acta Chem. Scand., 27, 329 (1973).
10. J. A. J. Vink, C. M. Lok, J. Cornelisse, and E. Havinga, Chem.
    Commun., 710 (1972); J. A. J. Vink, P. L. Verheijdt,
    J. Cornelisse, and E. Havinga, Tetrahedron, 28, 5081 (1972).
11. J. D. Heijer, O. B. Shadid, J. Cornelisse, and E. Havinga,
    Tetrahedron, 33, 779 (1977).
12. F. E. Bailey, Jr. and J. V. Koleske, "Poly(ethylene Oxide),"
    Academic Press, New York, 1976; F. Vögtle and E. Weber, Angew.
    Chem. Int. Ed. Engl., 18, 753 (1979); for CE: J. S. Bradshaw
    in "Synth. Multidentate Macrocyl. Compd.," eds. R. M. Izatt
    and J. J. Christenesen, Academic Press, New York, 1975.
13. G. Lodder and E. Havinga, Tetrahedron, 28, 5583 (1972).
14. N. J. Turro, "Modern Mol. Photochem.," Benjamin/Cummings, Menlo
    Park, C. A., 1978, p. 351-354.
15. H. H. Hodgson and F. Heyworth, J. Chem. Soc., 1131 (1949).

POLYETHYLENE GLYCOLS AS OLIGOMERIC HOST SOLVENTS: APPLICATIONS TO OXIDATION AND REDUCTION REACTIONS

Enzo Santaniello

Istituto di Chimica, Facoltà di Medicina
Università di Milano, Via Saldini, 50
I-20133 Milano, Italy

INTRODUCTION

Polyethylene glycols (PEG) are oligomeric diols of the general formula $HO-(CH_2-CH_2-O)_n-H$, which are commercially available also under the trade name of Carbowax. They can be regarded as protic solvents with aprotic sites of binding constituted by the oxyethylene units $-(CH_2-CH_2-O)-$. A few inorganic salts as well as many organic substrates are soluble in low molecular weight PEG ($\bar{M}$ ranging from 200 to 600), which have been recently proposed as solvents for organic reactions[1]. PEG were named "host" solvents, since they have the capability of forming complexes with some cations. This fact has been sustained by [1]H-N.M.R. studies[2] and isolation of a few solid complexes[3]. Further, PEG have a viscosity which is compatible with usual laboratory operations and offer a few advantages such as ready availability at low price, low vapour pressure and biodegradability.
From the reactions products can be recovered by traditional work-up such as addition of a volume of water almost equal to the one of PEG and extraction with a suitable solvent. Also, if the product is soluble in alkanes, extraction with n-pentane is possible from the bulk of reaction and for some product direct distillation is possible from the low volatile PEG.

397

Scheme 1

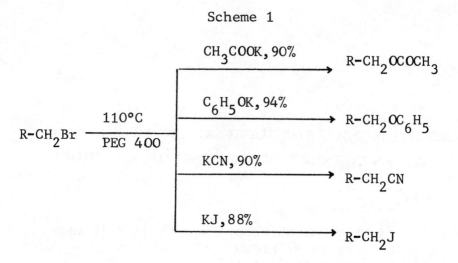

PEG 400 was the solvent of choice, since it apparently reconciled its role of solvent with good experimental conditions. A few substitution reactions were examined, therefore, in PEG 400 (Scheme 1) and in all the cases yields of isolated products were good and times short if the appropriate temperature was chosen.

OXIDATIONS

Among the few cases of oxidants examined, $KMnO_4$ was soluble in PEG 400, but the solutions were unstable and not further investigated. Potassium dichromate, $K_2Cr_2O_7$, is relatively soluble in PEG 400 and solutions about 0.15 molar could be prepared. These stable solutions oxidized benzyl bromide[1] as well as allylic and benzylic alcohols[4] with good yields of isolated products (Scheme 2). It should be mentioned that the aldehydes obtained by oxidation of allylic alcohols showed a 30% isomerization of the double bond.
In any event, the oxidations by means of solutions of potassium dichromate in PEG 400 can be compared to other similar systems which have been reported to bring into organic solutions the dichromate anion.

Scheme 2

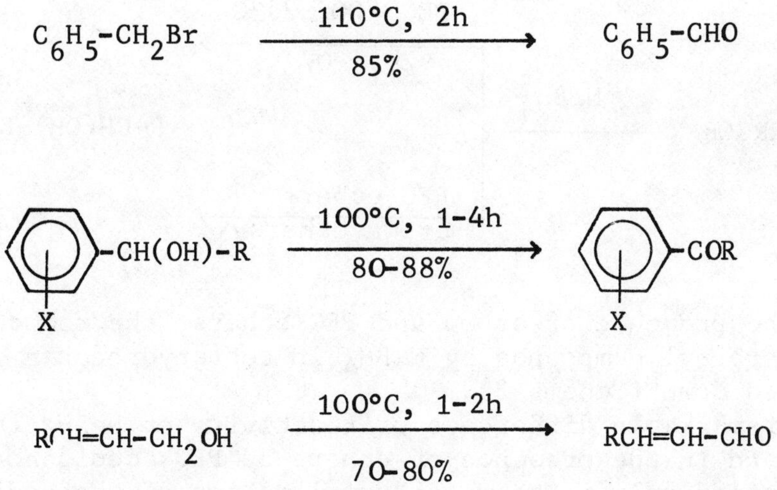

$$C_6H_5-CH_2Br \xrightarrow[\text{85\%}]{110°C, \ 2h} C_6H_5-CHO$$

$$X-C_6H_4-CH(OH)-R \xrightarrow[\text{80-88\%}]{100°C, \ 1-4h} X-C_6H_4-COR$$

$$RCH=CH-CH_2OH \xrightarrow[\text{70-80\%}]{100°C, \ 1-2h} RCH=CH-CHO$$

In fact, oxidations of allylic and benzylic halides can be also performed by sodium chromate, Na_2CrO_4, in HMPT in the presence of a crown ether[5] or by bis(tetrabutylammonium) dichromate (TBADC) in organic solvent[6]. Benzylic and allylic alcohols can be selectively oxidized also by potassium dichromate in DMSO[4] or TBADC in dichloromethane[7]. The main feature of all the above oxidations is that they do not require the presence of acids or bases and represent a unique case of "neutral" chromic oxidation. Should an intermediate chromate ester be formed in these reactions, this should happen after the relatively easy cleavage of the C-O bond of benzylic and allylic alcohols.

REDUCTIONS

Preliminary studies on the use of PEG 400 as solvent contained also the valuable information that carbonyl compounds could be reduced by sodium borohydride in the above solvent[1]. In fact, it was reported that 2-octanone was reduced at room temperature in excellent yields, the result being in contrast with the observation that

Scheme 3

PEG 400, 25°C
2h, 90%
NaBH4
RCOR'
R-CH(OH)-R'
PEG ethers
crown ethers

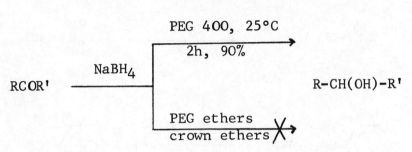

in the presence of crown and PEG ethers, the reduction of carbonyl compounds by $NaBH_4$ in tetrahydrofuran was slowed down[8] (Scheme 3).
The remarkable difference of reactivity of $NaBH_4$ in PEG 400 and in the presence of ethers of PEG, could not be explained only in terms of complexation of the sodium cation and further investigations were undertaken in order to try define scope and limitations of the reducing system. As first and important result, it was found that sodium borohydride could reduce alkyl and aryl esters to the corresponding alcohols in PEG 400 at 65°C[9] (Scheme 4).

Scheme 4

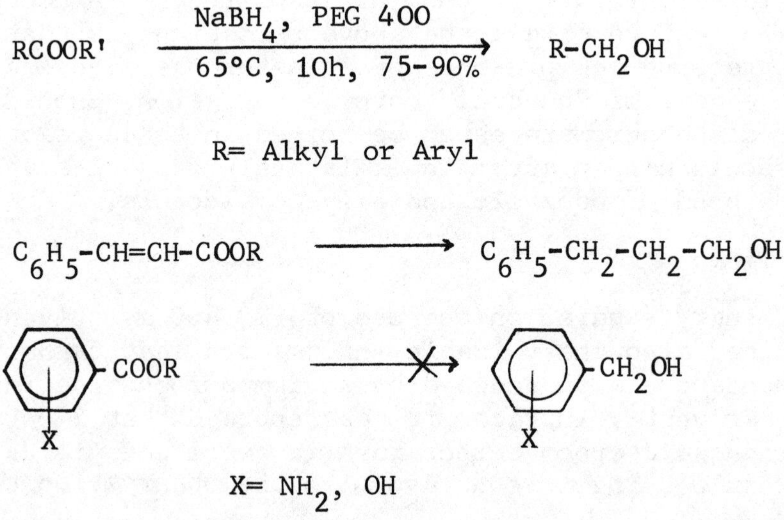

NaBH4, PEG 400
RCOOR'
65°C, 10h, 75-90%
R-CH2OH
R= Alkyl or Aryl
C6H5-CH=CH-COOR
C6H5-CH2-CH2-CH2OH
COOR
CH2OH
X
X
X= NH2, OH

In the case of aromatic esters, a few groups such as nitro and bromo survived at the reduction conditions, whereas hydroxy and amino groups completely suppressed the reaction. Considering the case of ethyl cinnamate, the fully saturated alcohol was the sole product of the reduction. In all the examined cases a molar ratio of $NaBH_4$/ester of 3:1 was always used in order to achieve the best yields.

Additional examples of reduction were sought and it was found that also primary and secondary alkyl and benzyl halides were reduced to the corresponding hydrocarbons[10] (Scheme 5). The average temperature was 70°C and alkyl iodides reacted faster than bromides, whereas alkyl chlorides showed the slowest reactivity. Also benzylic halides were easily reduced and secondary bromides afforded only the corresponding alkane with no evidence of any product of elimination. In all the cases examined an excess of sodium borohydride was needed (2-10 moles of borohydride per mole of halide, depending on the type of substrate).

Scheme 5

$$R-CH_2X \xrightarrow[\text{70°C, 3-10h, 75-90\%}]{NaBH_4, \text{ PEG 400}} R-CH_3$$

R= Alkyl or Aryl

X= Cl, Br, J

$$R-CHBr-R' \xrightarrow[\text{70°C, 12h, 80\%}]{NaBH_4, \text{ PEG 400}} R-CH_2-R'$$

Scheme 6

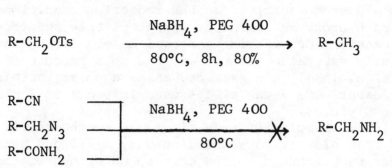

Furthermore, preliminary results indicate that also primary tosylates can be reduced to the corresponding alkanes, whereas nitrogen containing functions such as nitriles, azides and amides were not reduced at any extent[10] (Scheme 6).

MECHANISM OF THE REDUCTION

During the reduction of esters by $NaBH_4$ a vigorous evolution of hydrogen was observed and it has been established that two moles of the gas were generated from one mole of sodium borohydride in PEG 400 at 65-80°C. At higher temperatures, a viscous liquid is formed which is more resistant to the final hydrolysis by acids. Only from these preliminary observations, it could be reasonably assumed that at 65-80°C a polymeric specie such as a dialkoxyborohydride $[BH_2(OR)_2]^-$ is formed. This stable complex borohydride, which could be named $Na(PEGBH_2)$, is soluble in PEG itself and can exhibit an enhanced reactivity, if compared with $NaBH_4$ in hydroxylic solvents. Further, $Na(PEGBH_2)$ can be formed

$$-O-\overline{B}H_2-O-CH_2CH_2-O-(CH_2-CH_2-O)_{n-2}-CH_2-CH_2-O-\overline{B}H_2-O-$$

$$Na(PEGBH_2)$$

also by reaction of one mole of PEG 400 and one of sodium borohydride at 80°C for 1-2 hours. A viscous liquid is produced, which can reduce alkyl halides in THF very rapidly and in excellent yields.[10]

In order to gain more informations on the nature of the complex borohydride Na(PEGBH$_2$), the reduction of esters was examined using NaBH$_4$ in PEG 200, 300 and monomethoxy PEG. From these results it appears that methyl benzoate is completely reduced in PEG 200, whereas the same reduction in PEG 300 is complicated by a partial hydrolysis of the ester to benzoic acid. Finally, if one of the two hydroxy groups of PEG is blocked as ether no reduction at all takes place (Scheme 7).

From these data it is evident that the chain length of PEG does not play a decisive role for the formation of Na(PEGBH$_2$), whereas the presence of the two hydroxy groups is a necessary prerequisite for the accomplishment of the reduction. As a consequence, it could be perhaps speculated that the polymeric Na(PEGBH$_2$) either is formed more easily or is more stable than the monomer which could be generated by reaction of sodium borohydride with $H_3CO-(CH_2-CH_2-O)_n H$.

Scheme 7

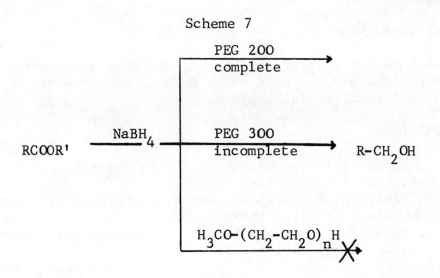

CONCLUDING REMARKS

Use of PEG has been limited until now to classical phase-
transfer catalysis and also in this case only a few
applications have been reported.[11,12] Also for their
catalytic effect PEG have been compared to crown ethers
and the examples are restricted to substitution reactions.
In fact, PEG can be regarded as non cyclic crown ethers
and opening of the ring is accompanied by decrease of
the complexation constant, due to conformational entropy
factors as well as to solvation effects on the ligand.[11]
However, from the results described in this report PEG
have also proven to be valuable solvents for classical
substitution reactions,[1] oxidations[4] and reductions.[1,9,10]
These inexpensive and readily available solvents can
offer the already cited advantages and in a few cases
even peculiar reactions can be observed.
Still much work remains to be accomplished in order to
clarify the exact role of PEG with respect to the effect
of complexation of the cations in the reported reactions.
If reductions by sodium borohydride are concerned, the
nature of the reducing specie involved awaits detailed
investigations and a complete picture of the reductions
obtainable by this new reducing system seems desirable.
We are actively investigating reductions as well as other
aspects of use of PEG.

ACKNOWLEDGEMENTS

I am indebted to prof. A. Fiecchi for many stimulating
discussions and to the Italian Council for Research
(C.N.R.) for finacial support.

REFERENCES

1. E. Santaniello, A. Manzocchi and P. Sozzani,
   Tetrahedron Lett., 4581 (1979).
2. S. Yanagida, K. Takahashi and M. Okahara, Bull. Chem.
   Soc. Jpn., 51, 1294 (1978).
3. S. Yanagida, K. Takahashi and M. Okahara, Bull. Chem.
   Soc. Jpn., 51, 3111 (1978).

4. E. Santaniello, P. Ferraboschi and P. Sozzani, Synthesis, 646 (1980).
5. G. Cardillo, M. Orena and S. Sandri, J. Chem. Soc. Chem. Comm., 190 (1976).
6. D. Landini and F. Rolla, Chem. Ind., 213 (1979).
7. E. Santaniello and P. Ferraboschi, Synth. Commun., 10, 75 (1980).
8. A. Hirao, S. Nakahama, M. Takahashi and N. Yamasaki, Makromol. Chem., 179, 2343 (1978).
9. E. Santaniello, P. Ferraboschi and P. Sozzani, J. Org. Chem., 46, 4584 (1981).
10. E. Santaniello, A. Manzocchi and P. Ferraboschi, to be published.
11. F. Montanari, D. Landini and F. Rolla "Topics in Current Chemistry" (F.L. Boschke, Ed.), Springer-Verlag, Berlin Heidelberg, 1982, p. 170.
12. S. Slaoni, R. LeGoaller, J.L. Pierre and J.L. Luche, Tetrahedron Lett., 23, 1681 (1982).

KOENIGS KNORR REACTION CATALYZED BY SUPPORTED CROWN ETHERS:

AN EXAMPLE OF THE POLYMER INFLUENCE

Alain Ricard, Joël Capillon and Claude Quivoron

Laboratoire de Physico-Chimie Macromoléculaire
E.S.P.C.I. - 10, rue Vauquelin
75231 Paris Cedex 05 France

I. INTRODUCTION

The complexing ability of crown ethers has been extensively investigated and their remarkable complexation behavior has found various applications in analytical and synthetic chemistry[1]. For instance, many chemical reactions can be influenced by the use of these compounds. By complexing the cation of a metal salt, an enhancement of the reactivity of the anion may result.

The perpective of attractive applications in polymer chemistry prompted the synthesis of soluble or gel type polymeric crown ethers.

I.1. Soluble polymers

Physico-chemical studies and extraction techniques were carried out on soluble polymeric crown ethers. The pioneering work in this field was made by Smid[2] who prepared several polymers or copolymers based on vinylbenzocrown ether and reported extensive data on the complexation of alkali picrate salts in various solvents including water. These polymers exhibit a high complexation of alkali cation, a selectivity related to the dimension of the cavity size of the macrocycle, a cooperative binding of the macrocyclic moieties of the polymer and a polyelectrolyte behavior.

Gramain[3] prepared polyesters and polyamides by poly-condensation reaction of diazamacrocycle derivatives with various acid dichlorides. The binding properties are strongly dependent on the polymeric structure. Physico-chemical studies have also been carried out on water soluble polydiazacrown ethers and the results have been compared to those of the monomeric analogues[4].

Polyureas[5], with a diazamacrocycle in the backbone, has been synthesized and the film forming material has a marked interaction with alkali ions.

Physico-chemical studies have essentially been made on soluble polymers and various papers have pointed out a greater ability of the polymer to complex ions than monomer when a cooperative effect is observed. This is encountered when the cation diameter is larger than the crown size[6-8].

In synthetic chemistry Smid has shown that polyvinylcrown ethers are effective catalysts in decarboxylation reaction[9].

I.2. Resins

Application of gel type polymeric crown ethers have been focused in 2 areas.
- in analytical chemistry for the separation of cations.
- in organic synthesis as supported reagents.

Blasius reported a good Na^+/K^+ separation with various exchangers[10]. The structure of the exchanger was investigated thoroughly and the macrocycle was either incorporated in a matrix obtained by polycondensation reaction or grafted to a polystyrene backbone. An hydrophilic gel with pendent diazamacrocycle grafted to a polyacrylamide backbone has also been used as a packing for the separation of various cations by HPLC[11].

In preparative chemistry the use of a macrocycle or a cryptand immobilized on an insoluble support presents many advantages : easy work up, easy product purification and recycling of an expensive reagent. However the ion binding and the catalytic activity of the supported macrocycle depend on many variables : spacer length, fonctionnality, polarity and structure of the resin[12-14].

To study these parameters, resins of different structures have been synthesized depending on the chemical constitution, on the degree of crosslinking and on the experimental conditions. The role of the diffusion in the complexation and in the catalysis has been investigated on gel type or macroporous resins. Steric and polar factors can change significantly both the direction and the magnitude of the product distribution.

Macrocycle and cryptand anchored to a resin are more effective catalysts than open chain polyether in solid liquid phase transfer reaction. But compared to small analogue molecules, it has been pointed out that they display:

- a weaker interaction of alcali ions with the resin ;
- a modification of the selectivity ;
- a diminution of the activity of the supported macrocycle when
  used as a catalyst in a substitution reaction.

The influence of the length of the spacer on the activity of supported crown ethers and cryptands used as catalysts in anion promoted phase transfer reactions has been studied by Montanari[14]. There is an increase in activity approaching that of the small analogue catalysts when they are bonded with a long chain to the polymeric support.

II. THE KOENIGS-KNORR REACTION

The role of the support in heterogeneous reaction is still not clear and research in this field is growing. We have been interested in studying some of the parameters by testing the anionic activation ability of a diaza-18-crown-6 network which forms complexes with a large number of cations,and by the comparison with small analogue molecules.

In a previous study of chromatographic separation of various cations we have shown that the diaza-18-crown-6 grafted to an acrylamide gel had a great affinity for cations like Ag^+, Hg^{2+}, Cd^{2+} [11]. Therefore we applied this complexation ability to study a model reaction in which one of these salts and a macrocycle or a cryptand were involved.

In recent years Knoechel and coworkers[15] reported on the Koenigs-Knorr reaction and studied the influence of a macrocycle on the reaction. This reaction is important because of its synthetic utility for glycoside formation. The authors showed that the macrocycle promotes a fast reaction with better yield and that the product distribution is sensitive to the nature of the macrocycle and to the bulkiness of the alcohol.

The bromide is converted to a mixture of β-nitrate and
β-glycoside derivatives. The reaction was carried out with silver
nitrate and a macrocycle or a cryptand as a catalyst. The reaction
was studied with several alcohols. The nitrate derivative formation
depends on both the alcohol hindrance and the affinity of the crown
ether or the cryptand for Ag^+. The more bulky the alcohol and the
more complexed the silver ion, the greater is the formation of the
nitrate.

Therefore, the interest in studying this reaction lies
in its sensitivity to the microenvironment and ion complexation. As
Ag^+ cation forms strong complex with the diaza 18 crown 6 we expected a selectivity modification by carrying out the Koenigs-Knorr
reaction in the presence of the immobilized ligand.

We reexamined this reaction using free crown ethers,
polymeric crown ethers and crown ethers covalently attached to a
resin. We aimed at probing the catalytic activity of the polymeric
ligands and studying the influence of these polymers on the yield
and the product distribution by comparison with small ligands. The
crown ethers used were either poor (macrocyclic polyether) or rather
good complexing agent (diaza macrocycle) for silver ion and are shown
in Figure 1.

The diaza-18-crown-6 was attached either on a polyacrylamide[11] or a polystyrene gel[16] to probe the influence of
the support polarity. The gels are non porous materials with nearly
the same capacity (~ 0.4meq/g). In addition we used a polyamide
with the macrocycle incorporated in the backbone. The polyamide,
insoluble in alcohol, was obtained from diaminodibenzo 18 crown 6
and isophtaloyl dichloride.

- With methanol, a small and very reactive molecule, the formation
  of the glycoside is quantitative when small molecules or polymers are added.

$$\text{Sugar + MeOH} \quad \xrightarrow[\text{AgNO}_3]{\text{crown ether}} \quad \beta\text{-glycoside}$$

The reaction is fast and achieved within 2 min at 0°C.

- With t-butyl alcohol a mixture of β nitrate and β glycoside is
  obtained. The product distribution is independent of the ligand
  concentration and of the time for both the small and polymeric
  molecules. Except when stated, alcohol is the solvent.

$$\text{Sugar + t-BuOH} \quad \xrightarrow[\text{AgNO}_3]{\text{crown ether}} \quad \beta\text{ONO}_2 + \beta\text{glycoside}$$

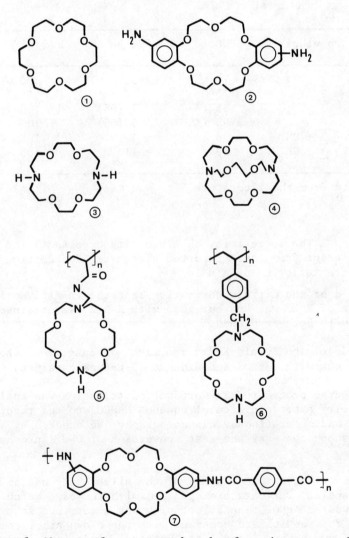

Fig. 1. Macrocycle, cryptand and polymeric macrocycle.

The reaction is fast and achieved within 5mn at 25°C. The reaction with a bulky alcohol is competitive and leads to selectivity differences depending on the macrocycle.

The results are shown in table 1.

Table 1. Reaction selectivity in the presence of crown ethers.

crown ethers	βONO_2	$\beta Ot-Bu$	yield
1	70	30	100
2	70	30	44
3	93 (4)*	7 (96)*	100
4	95 (37)*	5 (63)*	100
3 + DMF	96	4	100
3 + CH_2Cl_2	89	11	100

*data from the literature[15].

The selectivity obtained with cryptands 3 and 4 are quite different from those reported elsewhere. In addition we disagree on the following points :

- the yield of the nitrate derivative reported in diglyme in the presence of K 22 was bad but good with K 222. We obtained good yields with both cryptands.

- the yield of the Koenig-Knorr reaction was only given when it was carried out with t-BuOH and dibenzo-18-C-6 as catalyst. (50%).

- experimental procedure. The product distribution was analyzed by polarimetry after addition of aqueous NAHCO$_3$ to the reaction mixture and separation by chromatography. We checked by IR spectroscopy the easy and fast conversion of ONO_2 into hydroxyl group.

The physical state of the silver salt has an influence on the reaction. Our data are obtained with the same batch of finely powdered $AgNO_3$. The selectivity was determined from NMR analysis. The results are coherent with the crown ether complexing ability. Crown ethers 3 and 4 are better ligands for silver ion than crown ethers 1 and 2. They form cryptand separated ion pairs which favor nitrate derivative formation.

The result with compound 3 plus DMF shows that there is no selectivity modification when the reaction is carried out in a solvent of similar constitution to the amide gel.

        The product distribution obtained with the polymeric
crown ethers are shown in Table 2.

        Table 2. Selectivity with polymeric crown ethers.

crown ethers	βONO_2	$\beta Ot-Bu$	yield %
$\underline{5}$	57	43	100
$\underline{6}$	59	41	100
$\underline{7}$	59	41	90
$\underline{5}$ + DMF	95	5	100
$\underline{6}$ + CH_2Cl_2	75	25	100
$\underline{6}$ + $\phi-CH_3$	73	27	100

        As depicted in Table 1 large differences in selectivi-
ty were observed when the reaction was carried out in the presence
of a macrocycle depending on the nature of the macrocycle.
Likewise polymers lead to the same selectivity. Glycoside and nitrate derivatives are formed in nearly equal quantity, the same experimental conditions be compared to small analogues the reaction leads
to a lower proportion of the nitrate derivative when the polymers
are used as catalysts. The product distribution is not time dependent
and the same results are obtained after a day time period.

        The selectivity is nevertheless greatly affected when
a swelling solvent of the gel is added. The solvents used are good
solvents of the crown ethers but with no specific interaction with
the polymer backbone. The formation of the βONO_2 increases. This
result points out that there is no loss of the complexing ability
of the grafted crown ether and that the complexes formed are similar to the complexes with the small macrocycle.

        Our interpretation of this reaction is based on steric
effect of the polymer.

        There is a competition between anion and t-butyl
alcohol. The sugar and nitrate ion are separated by the plane of
the macrocycle.

        When the cryptand is immobilized, we assume that the
backbone or the framework of the polymer hinders the anion to
react at the brominated carbon and a greater proportion of the
glycoside is obtained. When the reaction is carried out in CH_2Cl_2

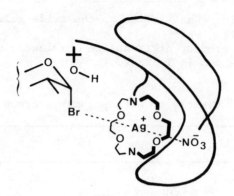

or toluene the amount of nitrate derivative increases. In these
solvents, the polystyrene resin is swollen and the reagents
become more accessible. The amount of nitrate derivative increases
also when the polyacrylamide resin is swollen in DMF.

III. CONCLUSION

        The reaction studied has shown that the polymer frame-
work of the catalyst is not inactive :

- with an alcohol of small dimension the formation of the glycoside
  is good whether the crown ether is supported or not.

- bulky alcohols are sensitive to the environment of the macrocycle
  induced by the polymer structure.

        Free crown ethers are not good catalytic agents to
give in good yields hindered glycosides. But the use of supported
crown ethers gives the opportunity to synthesize these compounds.
More information on the role of the network is needed and investigation with polymers of different texture will be useful.

REFERENCES

1. Synthetic Multidentate Macrocycle compounds, Academic Press,
   New York (1978).
2. a) S. Kopolow, T.E. Hogen-Esch and J. Smid, Macromolecules,
      6, 1 : 133 (1973).

b) L.H. Wong , J. Smid, J. Amer. Chem. Soc., 99, 17 : 5637 (1977).

c) R. Sinta, J. Smid, Macromolecules, 13, 2 : 339 (1980).

d) L.H. Wong, J. Smid, Polymer, 21, 2 : 188 (1980).

e) S. Shah, S. Kopolow, J. Smid, Polymer, 21, 2 : 195 (1980).

3. Ph. Gramain, M. Kleiber, Y. Frere, Polymer, 21 : 915 (1980).

4. Ph. Gramain and Y. Frere, Polymer, 21 : 921 (1980).

5. L.J. Mathias and K.B. Al-Jumah, J. Polym. Sci., Polym. Chem. Ed., 18 : 2911 (1980).

6. M. Bourgoin, K.H. Wong, J.Y. Hui and J. Smid, J. Amer. Chem. Soc., 98 : 5198 (1976).

7. K. Kimura, T. Maeda and T. Shono, Talanta, 26 : 945 (1979).

8. J. Smid, Makromol. Chem., Suppl. 5 : 203 (1981).

9. M. Shirai, J. Smid, J. Polym. Sci., Polym. Lett. Ed., 18, 10 : 659 (1980).

10. a) E. Blasius and P. Maurer, J. Chromatogr., 125 : 511 (1976).

b) E. Blasius, K.P. Janzen, W. Adrian, G. Klautke, R. Lorscheider, P. Maurer, V. NGuyen, T. NGuyen-Tien, G. Scholten and J. Stockemer, Z. Anal. Chem., 284 : 337 (1977).

c) E. Blasius, K.P. Janzen, M. Keller, H. Lander, T. NGuyen-Tien, and G. Scholten, Talanta, 27 : 107 (1980).

d) E. Blasius, K.P. Janzen, W. Adrian, W. Klein, H. Klotz, H. Luxenburger , E. Mernke, V.B. NGuyen, T. NGuyen-Tien, R. Rausch, J. Stockemer and A. Toussaint, Talanta, 27 : 127 (1980).

e) E. Blasius, K.P. Janzen, H. Luxenburger, V.B. NGuyen, H. Klotz and J. Stockemer, J. Chromatogr., 167 : 307 (1981).

11. P. Kutchukov, A. Ricard and C. Quivoron, Europ. Polym. Journal, 16 : 753 (1980).

12. a) G. Manecke, P. Reuter, J. Mol. Catal., 13, 3 : 355 (1981).

b) G. Manecke, A. Kraemer, Makromol. Chem., 182, 11 : 3017 (1981).

c) G. Manecke, P. Reuter, Makromol. Chem., 182, 7 : 1973 (1981).

13. R. Sinta, J. Smid, J. Amer. Chem. Soc., 103, 23 : 6962 (1981).

14. a) M. Cinquini, S. Colonna, H. Molinari, F. Montanari and P. Tundo, J. Chem. Sco. Chem. Comm., 394 (1976).

b) H. Molinari, F. Montanari and P. Tundo, J. Chem. Soc. Chem. Comm., 639 (1977).

c) H. Molinari, F. Montanari, S. Quici and P. Tundo, J. Amer. Chem. Soc., 100 : 3920 (1978).

d) F. Montanari and P. Tundo, J. Org. Chem., 46, 2125 (1981).

15. a) A. Knoechel, G. Rudolph and J. Thiem, Tetrahedron Lett., 551 (1974).

b) A. Knoechel, G. Rudolph, Tetrahedron Lett., 3739 (1974).

16. M. Ricard, D. Villemin, A. Ricard, Tetrahedron Lett., 21 : 47 (1980).

CONTRIBUTORS

Yasuo Ayaguchi
Dept. of Industrial Chemistry
MIE University
Tsu, Japan

R. Bacskai
Chevron Research Company
Richmond, California

Samuel T. Bajah
Department of Chemistry
University of Ibadan
Ibadan, Nigeria

Howard S. Blaxall
Department of Chemistry
Wright State University
Dayton, Ohio

S. Boileau
Laboratorie de Chimie Macromolec-
    ulaire associe au CNRS
College de France, Paris, France

R. W. Brooker
School of Textile Engineering
Georgia Institute of Technology
Atlanta, Georgia

J. D. Caldwell
Department of Chemistry
Louisiana State University
Baton Rouge, Louisiana

J. B. Canterberry
Department of Polymer Science
Uni. of Southern Mississippi
Southern Station, Mississippi

Joel Capillon
Laboratoire de Physico-Chimie
    Macromoleculaire
Paris, France

Charles E. Carraher, Jr.
Department of Chemistry
Wright State University
Dayton, Ohio

P. Caubere
Universite de Chimie Organique I
Nancy, France

Jin-Hae Chang
Department of Chemistry
Korea University
Seoul, Korea

Tai Chun Cheng
Corporate R&D
Raychem Corporation
Menlo Park, California
    and
Central Research Laboratory
Firestone and Rubber Company
Akron, Ohio

E. Chiellini
Centro CNR Macromolecole Stercor-
    dinate Otticamente Attive,
Istituti di Chimica Organica e
    Chimica Organica Industriale
Universita di Pisa, Pisa, Italy

F. L. Cook
School of Textile Engineering
Georgia Institute of Technology
Atlanta, Georgia

417

W. H. Daly
Department of Chemistry
Louisiana State University
Baton Rouge, Louisiana

Warren T. Ford
Department of Chemistry
Oklahoma State University
Stillwater, Oklahoma

S. D'Antone
Centro CNR Macromolecole Stercor-
    dinate Otticamente Attive,
Istituti di Chimica Organica e
    Chimica Organica Industriale
Universita di Pisa, Pisa, Italy

Jean M. J. Frechet
Department of Chemistry
University of Ottawa
Ottawa, Ontario, Canada

J. C. Gatier
Centre de Recherche du Bouchet,
SNPE
Vert le Petit, France

J. Milton Harris
Department of Chemistry
The University of Alabama in
    Huntsville
Huntsville, Alabama

Charles R. Harrison
Chemistry Department
University of Lancaster
Lancaster, Great Britain

Philip Hodge
Chemistry Department
University of Lancaster
Lancaster, Great Britain

Nedra H. Jundley
Department of Chemistry
The University of Alabama in
    Huntsville
Huntsville, Alabama

Barry J. Hunt
Chemistry Department
University of Lancaster
Lancaster, Great Britain

Yoshio Imai
Department of Textile and
    Polymeric Materials
Tokyo Institute of Technology
Meguro-ku, Tokyo, Japan

Toshikuni Ito
Dept. of Industrial Chemistry
MIE University
Tsu, Japan

Yasuji Izawa
Dept. of Industrial Chemistry
MIE University
Tsu, Japan

Leonard M. Jambaya
Department of Chemistry
Wright State University
Dayton, Ohio

Jung-Il Jin
Department of Chemistry
Korea University
Seoul, Korea

J. Kelly
Dept. of Pure and Applied Chemistry
University of Strathclyde
Glasgow, Scotland, U.K.

Ezzatollah Khoshdel
Chemistry Department
University of Lancaster
Lancaster, Great Britain

Yuhsuke Kawakami
Dept. of Synthetic Chemistry
Nagoya University
Chikusa, Nagoya 464 Japan

Raymond J. Linville
Department of Chemistry
Wright State University
Dayton, Ohio

L. J. Mathias
Department of Polymer Science
Uni. of Southern Mississippi
Southern Station, Mississippi

J. A. Moore
Department of Chemistry
Rensselaer Polytechnic Institute
Troy, New York

Melissa D. Naas
Department of Chemistry
Wright State University
Dayton, Ohio

T. G. N'Guyen
Laboratorie de Chimie Macromolec-
    ulaire associe au CNRS
College de France, Paris, France

E. M. Partain III
Department of Chemistry
Rensselaer Polytechnic Institute
Troy, New York

K. V. Phung
Department of Chemistry
Louisiana State University
Baton Rouge, Louisiana

Claude Quivoron
Laboratoire de Physico-Chimie
    Macromoleculaire
Paris, France

Jerald K. Rasmussen
Central Research Laboratories
3M, 3M Center
St. Paul, Minnesota

Alain Ricard
Laboratoire de Physico-Chimie
    Macromoleculaire
Paris, France

Maria Cristina Sanchez
Departamento de Quimica, ICET,
Uni. Autonoma de Guadalajara
Lomas del Valle
Guadalajara, Jalisco, Mexico

Enzo Santaniello
Instituto di Chimica, Facolta di
    Medicina
Universita di Milano, Via Saldini
Milano, Italy

Thomas G. Shannon
Department of Chemistry
The Uni. of Alabama in Huntsville
Huntsville, Alabama

D. C. Sherrington
Dept. of Pure and Applied Chemistry
University of Strathclyde
Glasgow, Scotland, U.K.

Kazuo Shimazu
Dept. of Industrial Chemistry
MIE University
Tsu, Japan

Roger Sinta
Polymer Research Institute
Chemistry Department
College of Environmental Science
    and Forestry
State University of New York
Syracuse, New York

Johannes Smid
Polymer Research Institute
Chemistry Department
College of Environmental Science
    and Forestry
State University of New York
Syracuse, New York

Howell K. Smith II
Central Research Laboratories
3M, 3M Center
St. Paul, Minnesota

R. Solaro
Centro CNR Macromolecole Stercor-
    dinate Otticamente Attive,
Istituti di Chimica Organica e
    Chimica Organica Industriale
Universita di Pisa,
Pisa, Italy

Evelyn C. Struck
Department of Chemistry
The Uni. of Alabama in Huntsville
Huntsville, Alabama

Nobutaka Suzuki
Dept. of Industrial Chemistry
MIE University
Tsu, Japan

T. Tang
Department of Chemistry
Louisiana State University
Baton Rouge, Louisiana

Pietro Tundo
Instituto di Chimica Organica
   dell'Universita
Torino, Italy

Mitsuru Ueda
Department of Polymer Chemistry
Yamagata University
Yonezawa, Yamagata, Japan

Paolo Venturello
Instituto di Chemica Organica
   dell'Universita
Torino, Italy

Janette Waterhouse
Chemistry Department
University of Lancaster
Lancaster, Great Britain

Koji Yagi
Departamento de Quimica, ICET,
Universidad Autonoma de
   Guadalajara
Lomas del Valle
Guadalajara, Jalisco, Mexico

Yuya Yamashita
Dept. of Synthetic Chemistry
Nagoya University
Chikusa, Nagoya 464 Japan